Saman K. Halgamuge, Lipo Wang (Eds.)

Classification and Clustering for Knowledge Discovery

T0137298

Studies in Computational Intelligence, Volume 4

Editor-in-chief
Prof. Janusz Kacprzyk
Systems Research Institute
Polish Academy of Sciences
ul. Newelska 6
01-447 Warsaw
Poland
E-mail: kacprzyk@ibspan.waw.pl

Further volumes of this series
can be found on our homepage:
springeronline.com

Vol. 1. Tetsuya Hoya
*Artificial Mind System – Kernel Memory
Approach,* 2005
ISBN 3-540-26072-2

Vol. 2. Saman K. Halgamuge, Lipo Wang
(Eds.)
*Computational Intelligence for Modelling
and Prediction,* 2005
ISBN 3-540-26071-4

Vol. 3. Bożena Kostek
*Perception-Based Data Processing in
Acoustics,* 2005
ISBN 3-540-25729-2

Vol. 4. Saman K. Halgamuge, Lipo Wang
(Eds.)
*Classification and Clustering for Knowledge
Discovery,* 2005
ISBN 3-540-26073-0

Saman K. Halgamuge
Lipo Wang
(Eds.)

Classification and Clustering for Knowledge Discovery

 Springer

Dr. Saman K. Halgamuge

Associate Professor & Reader
Mechatronics & Manufacturing
Research Group
Department of Mechanical and
Manufacturing Engineering
The University of Melbourne
Victoria 3010
Australia
E-mail: saman@unimelb.edu.au

Dr. Lipo Wang

Associate Professor
School of Electrical and Electronic
Engineering
Nanyang Technological University
Block S1, 50 Nanyang Avenue
Singapore 639798
Singapore
E-mail: elpwang@ntu.edu.sg

ISBN 978-3-642-06542-2   e-ISBN 978-3-540-32404-1

ISSN print edition: 1860-949X
ISSN electronic edition: 1860-9503

Springer is a part of Springer Science+Business Media
springeronline.com
© Springer-Verlag Berlin Heidelberg 2010
Printed in The Netherlands

Preface

Knowledge Discovery today is a significant study and research area. As we try to find answers to many research questions in this area we hope that knowledge can be extracted from various forms of data around us. The International Conference on Fuzzy Systems and Knowledge Discovery (FSKD) held in Singapore in 2002, in conjunction with two other conferences, has led to the publication of these two edited volumes. This volume contains methods and applications in classification and clustering for Knowledge Discovery. The other volume entitled "Computational Intelligence for Modelling and Prediction", from the same publisher, includes the papers on Modelling and Prediction.

Unsupervised data analysis via clustering plays a vital role in knowledge discovery, particularly if we do not know what relationships in data are to be expected. The Self Organizing Feature Map (SOM) is among the most well known methods used for clustering data. The first 5 chapters of this volume provide a collection of recent research in distributed clustering, the SOM and their recent extensions:

Chapter 1 proposes a new SOM-based method for exploratory data analysis, dimensionality reduction and variable selection, and demonstrates its application in a real data set;

Chapter 2 compares the SOM and Fuzzy C means algorithms on generating application clusters from network traffic data;

Chapter 3 investigates the use of a Growing SOM, a variant of the SOM in which the user does not have to define the map size a priori, in monitoring shift and movement in data;

Chapter 4 explores the use of a Hierarchical Growing SOM in providing content-based organization of information repositories and in facilitating intuitive browsing;

Chapter 5 proposes a new algorithm for clustering large distributed spatial databases.

If labeled data or data with known associations are available, we may be able to use supervised data analysis methods, such as classifying neural networks, fuzzy rule-based classifiers, and decision trees. The next 6 chapters are about methods of supervised data analysis:

Chapter 6 proposes a probabilistic approach for finding frequent combinations of fuzzy items in a database and shows that the number of iterations needed is less than that of the a priori method;

Chapter 7 suggests a new data mining method capable of identifying commonality as well as individuality in patterns;

Chapter 8 compares five classifiers including neural networks and Bayesian classifiers for scoring features of human sleep recordings extracted from EEG and EMG;

Chapter 9 proposes an information fusion algorithm that can handle multicriteria fuzzy decision-making problems in a flexible manner;

Chapter 10 proposes the simultaneous use of several fuzzy rule-based pattern classifiers on classification data sets;

Chapter 11 presents a hybrid expert system incorporating fuzzy logic, relational databases and multimedia systems, with applications addressing control and management issues of the parthenium weed problem.

A book on knowledge discovery must have the essence of various applications. We capture the range of applications varying from health to telecommunications in the last 11 chapters:

Chapter 12 describes the development of a software tool for the analysis of website data generated from E-Commerce systems;

Chapter 13 investigates fuzzy rule-based classifiers, decision trees, Support Vector Machines, Genetic Programming and an ensemble method to model intrusion detection systems;

Chapter 14 describes a method based on fuzzy rules to recognize handwritten alphanumeric characters. This is distinguished from other methods by its simplicity and adaptivity, as handwriting can vary significantly from person to person;

Chapter 15 proposes methods based on Neural Networks, including Self Organising Maps, Fuzzy Systems, and Evolutionary methods, to discover web access or usage patterns obtained from web server log files;

Chapter 16 proposes a web usage mining algorithm to discover multiple-level browsing patterns from linguistic data;

Chapter 17 describes a fuzzy decision agent based on personal ontology for meeting scheduling support;

Chapter 18 presents a longitudinal comparison of the effect of supervised and unsupervised learning on the grouping of hospital patients based on a four year study conducted in Australia;

Chapter 19 analyses missing persons data collected from police officers using a rule-based system that derives argumentations to supplement officer intuition;

Chapter 20 compares a number of clustering strategies for cluster formation in Sensor Networks;

Chapter 21 presents a routing protocol, based on fuzzy logic, for wireless ad hoc networks, extending the Zone Routing Protocol;

Chapter 22 suggests a new approach to classify sociological and market research data using fuzzy classification.

Our thanks go to the Department of Mechanical and Manufacturing Engineering, University of Melbourne, Australia, and the School of Electrical and Electronic Engineering, Nanyang Technological University, Singapore, which supported our project in various ways. We thank the many people who supported the International Conference on Fuzzy Systems and Knowledge Discovery (FSKD) held in Singapore in 2002. We especially thank its honorary chair Prof Lotfi Zadeh, who suggested the conference's name and motivated us through out the process. Most of the papers in this book reflect the extended work from this conference.

Our special thanks go to Mr. Chon Kit (Kenneth) Chan and Mr. Siddeswara Mayura Guru for managing the paper submission and formatting process, and Mr. Sean Hegarty for proofreading. We also acknowledge the partial support of the Australian Research Council under grants entitled "Pattern Recognition and Interpretation in Sequence Data" and "Sequence Data Analysis". We are grateful to all the authors for enthusiastically submitting high quality work to this publication, and Prof Janusz Kacprzyk and Springer Verlag for realizing this book project.

March 8, 2005   *Saman K. Halgamuge*
                                                                         Lipo Wang

Contents

The Many Faces of a Kohonen Map

A Case Study: SOM-based Clustering for On-Line Fraud Behavior Classification

V. Lemaire and F. Clérot

Telecommunication and Neural Techniques Group, France Telecom Research and Development, FTR&D/TECH/SUSI, 2 Avenue Pierre Marzin, 22307 Lannion cedex France
{vincent.lemaire,fabrice.clerot}@francetelecom.com

Abstract. The Self-Organizing Map (SOM) is an excellent tool for exploratory data analysis. It projects the input space on prototypes of a low-dimensional regular grid which can be efficiently used to visualize and explore the properties of the data.

In this article we present a novel methodology using SOM for exploratory analysis, dimensionality reduction and/or variable selection for a classification problem. The methodology is applied to a real case study and the results are compared with other techniques.

Key words: Self-Organizing Map, Exploratory Analysis, Dimensionality Reduction, Variable Selection

1.1 Introduction

The Self-Organizing Map (SOM) [5] is an excellent tool for data survey because it has prominent visualization properties. It creates a set of prototype vectors representing the data set and carries out a topology preserving projection of the prototypes from the d-dimensional input space onto a low-dimensional grid (two dimensions in this article). This ordered grid can be used as a convenient visualization surface for showing different features of the data.

When the number of SOM units is large, similar units have to be grouped together (clustered) so as to ease the quantitative analysis of the map. Different approaches to clustering of a SOM have been proposed [6, 9] such as hierarchical agglomeration clustering or partitive clustering using k-means. This SOM-based exploratory analysis is therefore a two-stage procedure:

1. a large set of prototypes (much larger than the expected number of clusters) is formed using a large SOM;
2. these prototypes are combined to form the final clusters.

Such an analysis deals with the cases and constitutes only a first step. In this article we follow the pioneering work of Juha Vesanto [10] on the use of Kohonen

V. Lemaire and F. Clérot: *The Many Faces of a Kohonen Map*, Studies in Computational Intelligence (SCI) **4**, 1–13 (2005)
www.springerlink.com

maps for data mining and we propose a second step, which involves a very similar techniques, but deals with the analysis of the variables: each input variable can be described by its projection upon the map of the cases. A visual inspection can be performed to see where (i.e. for which prototype(s) of the SOM) each variable is strong (compared to the other prototypes). It is also possible to compare the projections of different variables. This manual examination of the variables becomes impossible when the number of input variables is large and we propose an automatic examination: the projections of each variable on the map of the cases is taken as a representative vector of the variable and we train a second SOM with these vectors; this second map (map of the variables) can then be clustered, allowing to automatically group together variables which have similar behaviors.

The organization of the article is as follows: the next section deals with the real case studies and in Sect. 1.3 we present our methodology for exploratory analysis with SOM. In Sect. 1.4 we present our methodology for dimensionality reduction and variable selection with SOM. Section 1.5 describes experimental conditions and comparative results between our methodology and others machine learning techniques. A short conclusion follows.

1.2 Case Study

The case study is the on-line detection of the fraudulent use of a post-paid phone card. Post-paid cards are characterized by:

- card number (12 digits written on the card)
- card identifier (4 digits only known by the owner of the card)
- used in public phones (only need to enter the identifier)
- used in any fixed phone (enter the 16 digits for identification)

Here the "fraud" term includes all cases which may lead to a fraudulent non-payment by the caller. The purpose is to prevent non-payments by warning the owners of phone card that the current use of their card is unusual.

The original database contains 15330 individuals described with 226 input variables of various kinds:

- sociological variables
- a series of indicators of traffics;
- variables of descriptive statistics.

Using a large number of these variables in the modeling phase achieves good fraudulent/non-fraudulent classification performances but such models cannot be applied on-line because of computing and data extraction time constraints. It is thus necessary to reduce the number of variables while maintaining good performance.

We split the data into 3 sets: a training set, a validation set and a test set which contain respectively 70%, 15% and 15% of the cases. Whatever the method evaluated below, the test set is never used to build the classifier. 92% of the examples in the database belong to the class "not fraudulent" and 8% belong to the class "fraudulent".

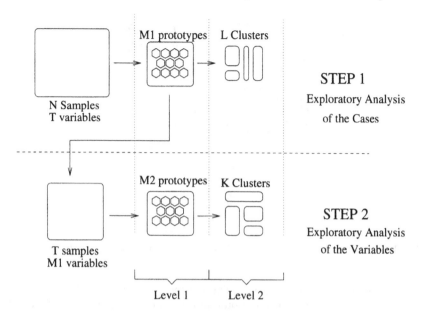

Fig. 1.1. The Two-Step Two-Level Approach

1.3 Methodology

1.3.1 A Two-Step Two-Level Approach

The methodology used in this article is depicted in the Fig. 1.1. The primary benefits of each two-level approach are the reduction of the computational cost [6, 9] and an easier interpretation of the map structure. The second benefit of this methodology is the simultaneous visualization of clusters of cases and clusters of variables for exploratory analysis purposes. Finally, dimensionality reduction and/or variable selection can be implemented.

All the SOM in this article are square maps with hexagonal neighborhoods and are initialized with Principal Component Analysis (PCA). We use a validation set or a cross-validation to measure the error reconstruction and select the map size above which the reconstruction error does not decrease significantly (the reconstruction error for a given size is estimated as an average on 10 attempts).

1.3.2 Top View: Exploratory Analysis of the Cases

The first step of the method is to build a SOM of the cases[1]. The best map size, for the case study was determined to be 12×12. We used the training set and the validation set to achieve a final training of the SOM of the cases.

This map allows to track down the characteristic behaviors of the cases: a standard clustering algorithm can be run on top of the map, revealing groups of cases

---
[1] All the experimentations on SOM have been done with the SOM Toolbox package for Matlab © [11].

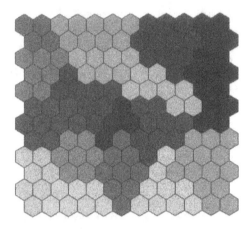

Fig. 1.2. Groups of cases with similar behaviors found using a hierarchical clustering

with similar behaviors (see Fig. 1.2). This clustering is done onto the prototypes of the SOM themselves, not on the prototypes weighted by the number of cases belonging to each prototype. Extensive tests have not shown any significant difference between k-means and hierarchical agglomerative clustering with the Ward criterium for this clustering of the map.

Projecting the class information (see Fig. 1.3, fraudulent use or not in our case study; this information is not used for the construction of the map) on the map allows to investigate the distinctive pro?les of the classes in terms of all the input variables.

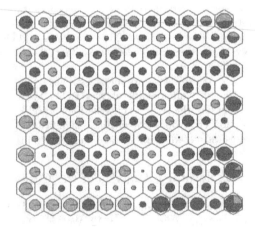

Fig. 1.3. The two populations for each prototype. The size of the pie indicates the number of cases belonging to each prototype. The lighter the color, the less fraudulent the behavior. For each pie the light grey proportion indicates the proportion of fraudulent behavior. We can project other auxiliary data in a similar manner

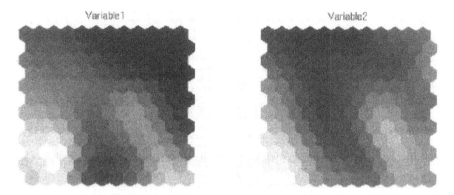

Fig. 1.4. The projections of the first two variables on the map of the cases: the darker the color, the stronger the value for the corresponding prototype

This constitutes the first step. We then proceed to the second step: each input variable is described by its projection upon the map of the cases. Upon visual inspection (see Fig. 1.4), one can determine how a variable behaves on the map and relate this behavior to the clusters of cases.

It is also possible to visually analyze the relationships between different variables. This visual method consists in the visualization of each projection on the map of the cases and to group together variables with similar projections (see Fig. 1.5). However this visual inspection of the variables becomes impossible when the number of input variables grows.

Nevertheless, an automatic clustering of the variables according to their projection can be achieved: the projections of each variable on the map of the cases are taken as a representative vector of the variable and we train a second SOM with these vectors; this second map (map of the variables) can then be clustered, allowing to automatically group together variables which have similar behaviors.

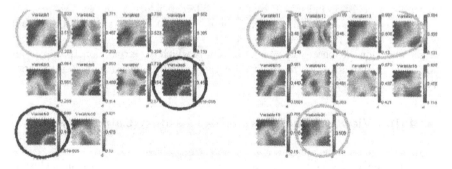

Fig. 1.5. Each subfigure above shows the projection of 10 variables. Visual inspection allows to find some strongly correlated variables (two "obvious" groups in this example) but is of limited efficiency when the number of variables is large

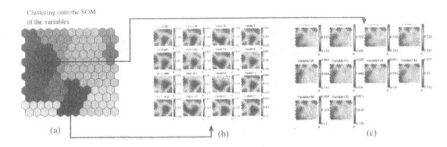

Fig. 1.6. Groups of variables with similar characteristics using K-means clustering

1.3.3 Side View: Exploratory Analysis of the Variables

In the second step, we build a second SOM to explore the variables as follows: each input variable is described by its projection upon the map of the cases, hence by a vector having as many dimensions as the number of neurons of the map of the cases. These variables descriptors are used to train the second map, the map of the variables.

For this SOM we cannot use a validation set since the database is the codebook of the SOM of the cases and is therefore quite small. We use a 5-fold cross-validation [12] method to find the best size of the SOM of the variables. The selected size is 12×12. Knowing the best size of SOM of the variables, we used all the codebooks of the SOM of the cases to perform a final training of the SOM of the variables.

This map allows to explore the relationships between variables and to study the correlation between variables; we also run a standard clustering algorithm on top on this map to create groups of variables with similar behaviors. Again, this clustering is done onto the prototypes of the SOM themselves, not on the prototypes weighted by the number of variables belonging to each prototypes. The clusters found on the map of the variables can be visualized as for the map of cases (see Fig. 1.6).

Figure 1.6 summarizes the results of this analysis of the variables: subfigure (a) shows the map of the variables and its clustering; subfigures (b) and (c) show the projections of the variables for two clusters of variables. The similarity of the projection for variables belonging to the same cluster is obvious and it can be seen that different clusters indeed correspond to different behaviors of the variables.

We used K-means for the clustering onto the SOM of the variables. Here again we cannot use a validation set to determine the optimal K value and we used a 5-fold cross-validation. We chose the value of K^* above which the error reconstruction does not decrease significantly (the result for a given size is an average on 20 attempts). The selected value is $K^* = 11$.

1.3.4 Top View vs. Side View and Exploratory Data Analysis

Figure 1.7 sums up the complete process described above.

At this point, we end up with two clusterings, a clustering of cases and a clustering of variables, which are consistent together: groups of cases have similar behaviors relative to groups of variables and reciprocally, a situation reminiscent of the duality property of PCA. This allows a much easier exploratory analysis and can also be

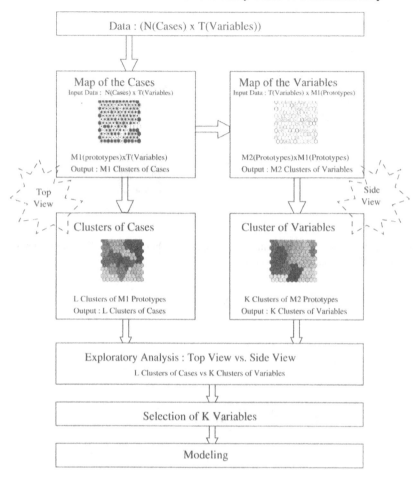

Fig. 1.7. Top View vs. Side View

used for dimensionality reduction and/or variable selection since variables of the same group contribute in the same way to the description of the cases.

The map of the variables allows an easier interpretation of the clusters of cases by re-ordering the variables according to their cluster. We see in Fig. 1.8(a) the clusters on the map of the cases; in Fig. 1.8(b) the mean value of the variables for the cases belonging to the cluster (A) without re-ordering; in Fig. 1.8(c) the mean value of the variables for the cases belonging to the cluster (A) re-ordered according to their cluster. Figure 1.8(c) immediately shows how the different clusters of variables contribute to the formation of the cluster of cases A. Such accurate visual analysis is impossible with the raw ordering of the variables (Fig. 1.8(b)).

Figure 1.9 illustrates the complete exploratory analysis process which can be done using the method described above. The clustering of the SOM of the cases identifies 12 clusters of cases (a) The projection of the class information allows to

Clustering onto the SOM
   of the cases

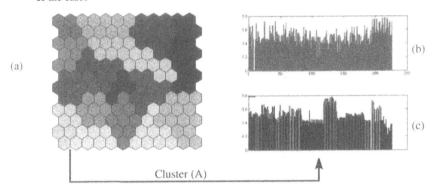

Fig. 1.8. Re-ordering the variables according to their cluster allow an easier interpretation of the cluster of the cases

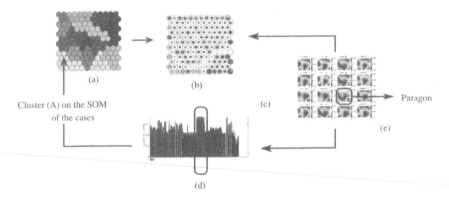

Fig. 1.9. Example of exploratory analysis

visualize the fraudulent behaviors (b). The cluster (A) of the SOM of the cases (in the south-west corner of the map) exhibits fraudulent behavior (b).

The clustering obtained on the SOM of the variables allows the re-ordering of the mean value of the variables for the cases belonging to each cluster (c). For the cluster (A) of the SOM of the cases we see a cluster of stronger variables (d). This group of variable is presented in (e): all the variables are strongly correlated. The grouping of these variables in this cluster is naturally interpreted: these variables represent information about card phone (the amount of communication via a card phone under five temporal observation windows) and indicate that a specific card phone usage pattern is strongly correlated with a fraudulent behavior.

1.4 Dimensionality Reduction vs. Variable Selection

In this article, "dimensionality reduction" refers to techniques which aim at finding a sub-manifold spanned by combinations of the original variables ("features"), while "variable selection" refers to techniques which aim at excluding variables irrelevant to the modeling problem. In both cases, this is a combinatorial optimization problem.

The direct approach (the "wrapper" method) re-trains and re-evaluates a given model for many different feature/variable sets. The "filter" method approximation optimises simple criteria which tend to improve performance. The two simplest optimization methods are forward selection (keep adding the best feature/variable) and backward elimination (keep removing the worst feature/variable) [2, 7].

As we have seen that each cluster of variables gathers variables with very close profiles, we can exploit this clustering for variable selection in a very natural way: we choose one representative variable per cluster, as the "paragon" of the cluster, i.e. the variable which minimizes the sum of the distances to the other variables of the cluster.

We choose to implement dimensionality reduction by building one feature per cluster as a sum of the variables of the cluster (variables are mean-centered and reduced to unit variance before summing). Both techniques are illustrated in Fig. 1.10.

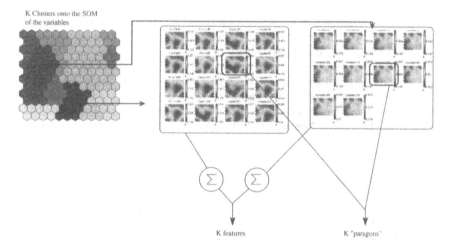

Fig. 1.10. SOM-based dimensionality reduction and variable selection

Both methods reduce the number of input variables to the number K^* of clusters found on the map of the variables. Modeling after variable selection relies on fewer input variables, therefore relying on less information, while modeling after dimensionality reduction relies on fewer features which may gather all the relevant information in the input variables but are often impossible to interpret in an intelligible way.

1.5 Methodology: Comparison and Results

Other machine learning techniques also allow to realize variable selection such as decision trees, Bayesian networks or dimensionality reduction methods such as PCA. We shall compare the methodology described above to such techniques and this section details the experimental conditions for this comparison.

We shall report a comparison of the results obtained on our case-study:

- on the one hand we shall compare the performance of models which use dimensionality reduction: a neural network trained with all the input variables, a neural network which uses the K^* variables found with dimensionality reduction method described below, and a PCA where we kept the first K^* eigenvectors.
- on the other hand we shall compare the performance of models which use a variable selection: a neural network which uses the K^* variables found with the variable selection method proposed below, a Bayesian network, and a decision tree.

1.5.1 Experimental Conditions

Principal Component Analysis

The principal components are random variables of maximal variance constructed from linear combinations of the input features. Equivalently, they are the projections onto the principal component axes, which are lines that minimize the average squared distance to each point in the data set [1]. The Principal Component Analysis (PCA) has been constructed on the training set and projected using the first $K^* = 11$ eigenvectors on the validation set and the test set. This may not be the optimal number of eigenvectors but, for comparison purposes, the number of eigenvectors kept has to correspond to the number of clusters of variables found in Sect. 1.3.3.

Multi-layer Perceptrons

Each neural network, in this article, is a multilayer perceptron, with standard sigmoidal functions, $K^* = 11$ input neurons, one hidden layer with P neurons and one output. We used the stochastic version on the squared error cost function. The training process is achieved when the cost does not decrease significantly as compared to the previous iteration on the validation set. The learning rate is $\beta = 0.001$.

The optimal number P^* of hidden units was selected for the final cost, between 2 and 50 for each case: the neural network trained with all the input variables, the neural network which uses the $K^* = 11$ variables found with the dimensionality reduction method described above, the neural network where we kept the first $K^* = 11$ eigenvectors found with the PCA and the neural network which uses the $K^* = 11$ variables found with the variable selection method proposed above (the result for a given size of neural network is the average estimated on 20 attempts). The P^* values are respectively 12, 10, 6 and 10.

Decision Tree

We used a commercial version of the algorithm C4.5 [8]. The principal training parameters and the pruning conditions are:

- the splitting on the predictor variables continues until all terminal nodes in the classification tree are "pure" (i.e., have no misclassification) or have no more than the minimum of cases computed from the specified fraction of cases (here 100) for the predicted class for the node;
- the Gini measure that reaches a value of zero when only one class is present at a node (with priors estimated from class sizes and equal misclassification costs).

With these parameters the number of variables used by the decision tree is 17, that is more than $K^* = 11$.

Bayesian Network

The Bayesian network (BN) found is similar to the "Naïve Bayes" which assumes that the components of the measurement vector, i.e. the features, are conditionally independent given the class. Like additive regression, this assumption allows each component to be modeled separately. Such assumption is very restrictive but on this real problem a naïve Bayes classifier gives very good results (see [4]). The BN uses 37 variables[2], which is more than three times more than $K^* = 11$.

1.5.2 Results

The various classification performances are given below in the form of lift curves. The methodology described in the article gives excellent results (see Fig. 1.11).

Regarding the variable selection method, the performances of the neural network trained with the selected variables are better than the performances of the Decision Tree and the Bayesian Network. As compared to the neural network trained with all the input variables (226 variables), the neural network trained with the selected variables only ($K^* = 11$ variables) shows a marginal degradation of the performance for very small segments of the population and even has a slightly better behavior for segments larger than 20% of the population. The SOM-based dimensionality reduction method has a performance similar to the PCA-based method.

These comparisons show that, on this real application, it is possible to obtain excellent performances with the methodology described above and in particular with the variable selection method, hence allowing a much simpler interpretation of the model as it only relies on a few input variables.

---

[2] The BN was built by Prof. Munteanu and coworkers, ESIEA, 38 rue D. Calmette Guérin 53000 Laval France, in the framework of the contract "Bayes-Com" with France Telecom. The research report is not available.

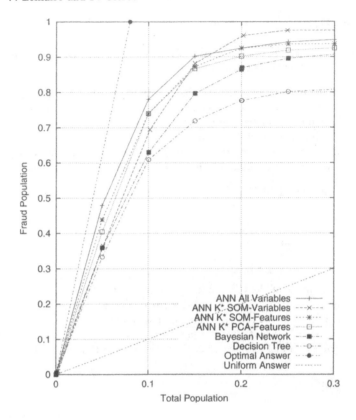

Fig. 1.11. Detection rate (%) of the fraudulent users obtained with different learning methods (ANN: Artificial Neural Network), given as a lift curve

1.6 Conclusion

In this article we have presented a SOM-based methodology for exploratory analysis, dimensionality reduction and/or variable selection. This methodology has been shown to give excellent results on a real case study when compared with other methods both in terms of visualization/interpretation ability and classification performance.

We have successfully applied the methodology described in this article on a variety of problems. It allows:

- to track down characteristic behavior of cases;
- to visualize synthetically various behaviors and their relationships;
- to group together variables which contribute in a similar manner to the constitution of the clustering of cases;
- to analyze the contribution of every group of variables to every group of cases;
- to realize a selection of variables;
- to make all interpretations with the initial variables.

Another example of the application of this methodology can be found in [3]. The authors show how SOM can be used to build successive layers of abstraction starting from low-level traffic data to achieve an interpretable clustering of customers and how the unique visualisation ability of SOM makes the analysis quite natural and easy to interpret.

References

1. Christopher M. Bishop. *Neural Network for Pattern Recognition.* Oxford University Press, 1996.
2. A. Blum and P. Langley. Selection of relevant features and examples in machine learning. *Artificial Intelligence,* pp. 245–271, 1997.
3. F. Clérot and F. Fessant (2003). From IP port numbers to ADSL customer segmentation: Knowledge aggregation and representation using kohonen maps. In *Datamining 2003,* Rio Janeiro, December.
4. David J. Hand and Keming Yu. Idiot's bayes – not so stupid after all. *International Statistical Review,* 69(3):385–398, 2001.
5. Teuvo Kohonen. Self-organizing maps. In *Springer Series in Information Sciences,* volume 30. Springer, Berlin, Heildelberg, 1995.
6. J. Lampinen and E. Oja. Clustering properties of hierarchical self-organizing maps. *Journal of Mathematical Imaging and Vision,* 2(3):261–272, 1992.
7. P. Langley. Selection of relevant features in machine learning. In AAAI Press, editor, *AAAI Fall Symposium on Relevance,* New Orleans, 1994.
8. J.R. Quinlan. *C4.5: Programs for Machine Learning.* Morgan Kaufmann, 1993.
9. Juha Vesanto. Som-based data visualization methods. *Inteligent Data Analysis,* 3(2):111–126, 1999.
10. Juha Vesanto. *Data Exploration Process Based on the Self-Organizing Map.* PhD thesis, Helsinki University of Technology, 2002.
11. Juha Vesanto, Johan Himberg, Esa Alhoniemi, and Juha Parhankangas. SOM toolbox for Matlab 5. Report A57, Helsinki University of Technology, Neural Networks Research Centre, Espoo, Finland, April 2000. http://www.cis.hut.fi/projects/somtoolbox/.
12. S.M. Weiss and C.A. Kulikowski. *Computer Systems That Learn.* Morgan Kaufmann, 1991.

Profiling Network Applications
with Fuzzy C-means and Self-Organizing Maps

Timo Lampinen[1], Mikko Laurikkala[1], Hannu Koivisto[1] and Tapani Honkanen[2]

[1] Tampere University of Technology, Finland
   firstname.lastname@tut.fi
[2] TeliaSonera Corp., Finland
   tapani.honkanen@teliasonera.com

Abstract. This study addresses the problem of generating application clusters from network traffic data. These clusters are used for classification and profiling purposes. The methods are the self-organizing map (SOM) and the fuzzy c-means clustering (FCM) algorithm. Application profiles produce significant information about the network's current state and point out similarities between different applications. This information will be later used to manage the network resources. Methods are compared in the light of the results obtained.

Key words: telecommunication network, fuzzy, neural network, data analysis

2.1 Introduction

The complexity of network infrastructure and the number of network communication applications, protocols and end-users are increasing rapidly. This has given rise to new challenges both from the traffic modelling and the network management point of view.

Traffic modelling and corresponding empirical research have advanced considerably during recent years. Measurement has become common. Several large scale measurement programs are now established, c.f. [5] or [17]. The vast amount of data and its complexity have also caused large scale challenges for data analysis. The current state of art in empirical research is analyzed for example in [7] or in Darryl Veitch's excellent presentation [22]: it seems that traffic is so complex that just to see what is going on we need to measure and study it as if it were an unknown natural phenomenon.

From the network management point of view the situation is fortunately better. Still, due to this development the efficient management of network resources is a complicated task. With traditional network management methods it is difficult to obtain a comprehensive view of the state of the network and simultaneously discover important details from the traffic.

Data mining and intelligent data analysis methods provide promising tools that have been recently used for this type of tasks. With these it is possible to discover the key characteristics and regularities within the network traffic.

Timo Lampinen et al.: *Profiling Network Applications with Fuzzy C-means and Self-Organizing Maps*, Studies in Computational Intelligence (SCI) **4**, 15–27 (2005)
www.springerlink.com

Some examples consider using classical statistical analysis and time series models for backbone traffic classification [18], LVQ for application recognition and CoS (Classes of Service) classification tasks [10] or wavelet based analysis of network traffic anomalies [2]. The SOM map has been studied to find fraudulent user profiles for cellular phones [9]. Traditional neural networks have successfully been used to recognize a limited set of applications [21]. The challenges have also encouraged the development of more efficient data mining methods, c.f. [4].

This study addresses the problem of generating application clusters from network traffic data. These clusters are used for classification and profiling purposes. The methods are the self-organizing map (SOM) and the fuzzy c-means clustering (FCM) algorithm. Preliminary results of this have been presented in [15]. Also some comparative results using rough sets are presented, based on preliminary results [13, 14]. The clustering is based on measured NetFlow data. Application profiles produce significant information about the network's current state and point out similarities between different applications. This information will be later used to manage the network resources. Methods are compared in the light of the results obtained. Each one has its advantages, and the choice of method should be used in accordance with the classification problem.

The next chapter presents the basic concepts and theory behind the analysis. Chapter 3 focuses on the problem at hand and presents the solutions. Chapter 4 reviews the obtained results. Chapter 5 discusses the interpretation of the results and comparison of the methods.

2.2 Theoretical Background

This chapter summarizes the theoretical backgrounds of the fuzzy C-means algorithm as well as the self-organizing map. We also present the nearest prototype classifier that is used for comparing the results.

2.2.1 Fuzzy C-means Clustering

Fuzzy C-means (FCM) algorithm, also known as Fuzzy ISODATA, was introduced by Bezdek [3] as an extension to Dunn's algorithm [8]. The FCM-based algorithms are the most widely used fuzzy clustering algorithms in practice.

Let $\mathbf{X} = \{x_1, x_2, \ldots, x_N\}$, where $x_i \in \mathbb{R}^n$ present a given set of feature data. The objective of FCM-algorithm is to minimize the Fuzzy C-Means cost function formulated as

$$J(\mathbf{U}, \mathbf{V}) = \sum_{j=1}^{C} \sum_{i=1}^{N} (\mu_{ij})^m \|x_i - v_j\|^2 \tag{2.1}$$

$\mathbf{V} = \{v_1, v_2, \ldots, v_C\}$ are the cluster centers. $\mathbf{U} = (\mu_{ij})_{N \times C}$ is a fuzzy partition matrix, in which each member μ_{ij} indicates the degree of membership between the data vector x_i and the cluster j. The values of matrix $\mathbf{U}$ should satisfy the following conditions

$$\mu_{ij} \in [0, 1] \quad \forall i = 1, \ldots, N \quad \forall j = 1, \ldots, C \tag{2.2}$$

$$\sum_{j=1}^{C} \mu_{ij} = 1, \quad \forall i = 1, \dots, N \tag{2.3}$$

The exponent $m \in [1, \infty]$ is a weighting factor which determines the fuzziness of the clusters. The most commonly used distance norm is the Euclidean distance $d_{ij} = \|x_i - v_j\|$, although Babuska suggests that other distance norms could produce better results [1].

Minimization of the cost function $J(\mathbf{U}, \mathbf{V})$ is a nonlinear optimization problem, which can be minimized with the following iterative algorithm:

Step 1: Initialize the membership matrix $\mathbf{U}$ with random values so that the conditions (2.2) and (2.3) are satisfied. Choose the appropriate exponent m and the termination criteria.

Step 2: Calculate the cluster centers $\mathbf{V}$ according to the equation:

$$v_j = \frac{\sum_{i=1}^{N} (\mu_{ij})^m x_i}{\sum_{i=1}^{N} (\mu_{ij})^m}, \quad \forall j = 1, \dots, C \tag{2.4}$$

Step 3: Calculate the new distance norms:

$$d_{ij} = \|x_i - v_j\|, \quad \forall i = 1, \dots, N \quad \forall j = 1, \dots, C \tag{2.5}$$

Step 4: Update the fuzzy partition matrix $\mathbf{U}$:
If $d_{ij} > 0$ (indicating that $x_i \neq v_j$)

$$\mu_{ij} = \frac{1}{\sum_{k=1}^{C} \left(\frac{d_{ij}}{d_{ik}}\right)^{\frac{2}{m-1}}} \tag{2.6}$$

$$\mu_{ij} = 1$$

Step 5: If the termination criteria has been met, stop
Else go to **Step 2**

A suitable termination criterion could be to evaluate the cost function (2.1) and to see whether it is below a certain tolerance value or if its improvement compared to the previous iteration is below a certain threshold [11]. Also the maximum number of iteration cycles can be used as a termination criterion.

2.2.2 Self-Organizing Map

The self-organizing map (SOM) [12], first introduced by Kohonen, is a powerful clustering and data presentation method. A SOM consists of a grid shaped set of nodes. Each node j contains a prototype vector $m_j \in \mathbb{R}^n$, where n is also the dimension of the input data.

SOM is trained iteratively. For each sample x_i of the training data, the best matching unit (BMU) in the SOM is located

$$c = \arg\min_j \{\|x_i - m_j\|\} \tag{2.7}$$

Index c indicates the corresponding prototype (BMU) vector m_c. The Euclidean norm is usually chosen for the distance measure. The BMU m_c and its neighboring prototype vectors in the SOM grid are then updated towards the sample vector in the input space.

$$m_j(t+1) = m_j(t) + h_{cj}(t)\left[x_i - m_j(t)\right] \tag{2.8}$$

Equation (2.8) states that each prototype is turned towards the sample vector x_i according to the neighborhood function $h_{cj}(t)$. The most commonly used neighborhood function is the Gaussian neighborhood

$$h_{cj}(t) = \alpha(t) \cdot \exp\left(-\frac{\|n_c - n_j\|^2}{2\sigma^2(t)}\right) \tag{2.9}$$

The neighborhood is centered on the BMU. Norm $\|n_c - n_j\|$ is the topological distance between prototypes c and j in the SOM grid. The factors $\alpha(t)$ and $\sigma(t)$ are the learning rate factor and the neighborhood width factor, respectively. Both of these factors decrease monotonically as training proceeds.

2.2.3 Nearest Prototype Classifier

The nearest prototype (NP) classifier [3] can be used to classify test data into a set of clusters (or prototypes). The BMU vector (2.7) is calculated with the NP method. For each sample vector we can locate the NP as follows.

$$\text{sample } x_i \text{ belongs to the cluster } j \tag{2.10}$$
$$\Leftrightarrow d(x_i, v_j) = \min_{1 \leq k \leq C}\{d(x_i, v_k)\}$$

The distance $d(x_i, v_j)$ is Euclidian distance. We introduce a modified version of the NP classifier, which has a tolerance value for the distance $d(x_i, v_j)$.

$$\text{sample } x_i \text{ belongs to the cluster } j \tag{2.11}$$
$$\Leftrightarrow d(x_i, v_j) = \min_{1 \leq k \leq C}\{d(x_i, v_k)\} < d_{NP}$$

The tolerance value ensures that each sample vector is within a reasonable distance from the closest cluster center. Separation of outliers becomes more efficient this way. One way to choose the value for d_{NP} is a certain percentage of the absolute vector length of the cluster center v_k.

2.3 Analysis

As mentioned in the introduction, the research topics are twofold: Firstly, generating application clusters from network traffic data and using these clusters for classification and profiling purposes. Another topic is how this information will be used to manage the network resources.

Another topic considers the development and comparison of intelligent data analysis methodologies. The case studies presented in this study are targeted for both purposes. For this a Netflow [6] based data collection was performed in a rather small test network (see Fig. 2.1).

NetFlow network management data contains statistical information of the traffic of the network. A flow is defined as a unidirectional sequence of packets between given source and destination endpoints [6]. Different endpoints are separated with source and destination IP addresses, port numbers, protocol and ToS numbers. A

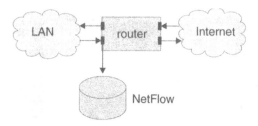

Fig. 2.1. Measurement setup. The Netflow collector is recording traffic directed from the LAN to the public Internet

session is a collection of flows with same endpoints. An example of typical Netflow records is presented in Table 2.1.

The measured NetFlow network data consisted of over 274000 samples of different computer application sessions, gathered from an edge router of a LAN network. The LAN was used for test purposes and contained several types of users and services. The arrangement allowed us only to measure traffic one way, from the LAN network to the Internet. A much larger and bidirectional measurement setup was performed afterwards.

All the data preprocessing and most of the analysis was done with Matlab® software, like Fuzzy Logic Toolbox. The SOM toolbox used in the analysis is available in [20].

The original data was first preprocessed and the essential attributes were selected for the analysis. Some attributes had to be further formed from the original NetFlow features. Preprocessing and analysis is always an iterative process. Good visualization tools helped much in this. A typical example of visualization is presented in Fig. 2.2: colored attribute values within a SOM map.

Table 2.1. Examples of NetFlow records

Time	111	3	3	3
Source IP	193.210.154.x	204.221.165.x	204.221.165.x	204.221.165.x
Destination IP	131.177.155.99	205.263.48.192	204.221.29.42	204.221.29.29
Source port	18184	32863	1130	2146
Destination port	1170	31779	53	53
Protocol	6	6	17	17
Type of service	0	0	0	0
Packets total	4	8	1	1
Bytes	172	616	56	169
Flows	1	4	1	1
First timestamp	851274791	989194127	989194087	989194087
Last timestamp	851274791	989194352	989194087	989194087
Total active time	12	240	0	0

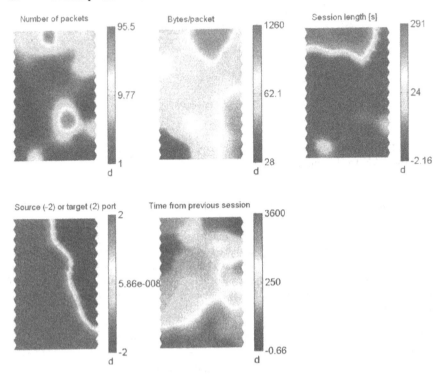

Fig. 2.2. Input attributes on self-organizing map, color coded according to their values

2.4 Results

Selected attributes for each session in SOM/FCM comparison were

- Number of packets
- Bytes/Packet ratio
- Duration of session in seconds
- Feature indicating whether the dominating application was used as a source or destination port, indicating the direction of the first flow.
- Time from previous similar session (same endpoints, max. 3600 sec.)
- Application (TCP/UDP port) number

The application number was not used in the actual clustering process, since it is a categorical attribute and as such can not be used with continuous clustering methods. But it can be used for labeling clusters with both of the used methods.

For the final analysis, 5 weekdays of traffic data were chosen, a total sum of 27801 sessions. The data was divided into training and validation sets, containing 75% and 25% of the total amount, respectively. Most of the applications were only used few times in the 5 day period. Only the 20 most frequently used applications were therefore selected. The data also had to be normalized because of the different ranges in values between attributes.

SOM and FCM algorithms were run several times to make sure that the randomized initializations would not affect the results. According to the Xie-Benin [23] index the number of the clusters for FCM was chosen to be 25. The size of the SOM was made larger, so that even small application clusters would be clearly visible.

2.4.1 FCM Clusters

FCM produced 25 clusters, which were all represented by a cluster center vector. The next step was to label each cluster. This was done with the help of the validation data set. Each sample from the validation set was placed in the most similar cluster according to the NP (2.11) classification. The tolerance value was set to 25% of the Euclidian length of the cluster center. Every cluster was hit with a reasonable amount of validation samples.

The applications that were most common in each cluster gave the cluster two labels. Table 2.2 illustrates the obtained application clusters. Note that only the 2 most common applications are shown for each cluster. Results show that most of

Table 2.2. Application clusters produced by the FCM algorithm. Each cluster is shown together with two of its most common applications and their percentages. The last column is the total number of sample vectors that hit each cluster

Cluster	1st Application		%	2nd Application		%	Total Hits
1	80	HTTP	63.5	53	DNS	32.5	203
2	0	ICMP	100.0				107
3	31779	unknown	71.4	8888	unknown	24.1	399
4	20	FTP	100.0				89
5	53	DNS	99.1	80	HTTP	0.6	797
6	53	DNS	84.5	0	ICMP	7.9	354
7	53	DNS	93.5	434	Mobile-IP	4.5	201
8	0	ICMP	100.0				870
9	53	DNS	65.3	0	ICMP	34.3	236
10	18184	unknown	97.9	20	FTP	1.2	329
11	53	DNS	95.5	1346	unknown	2.8	176
12	53	DNS	71.4	80	HTTP	21.1	133
13	8000	HTTP	90.8	21	FTP	2.8	360
14	137	NetBios	67.7	138	NetBios	12.6	223
15	53	DNS	97.1	80	HTTP	2.7	852
16	0	ICMP	100.0				68
17	8888	unknown	98.7	21	FTP	0.9	224
18	53	DNS	85.8	80	HTTP	7.1	127
19	80	HTTP	90.6	443	HTTP/SSL	3.1	64
20	138	NetBios	95.1	1346	unknown	2.9	102
21	123	NTP	80.2	31779	unknown	12.7	126
22	53	DNS	97.3	20	FTP	0.5	221
23	137	NetBios	50.3	53	DNS	29.4	177
24	135	MS RPC	23.7	1346	unknown	23.7	76
25	53	DNS	80.1	20	FTP	6.1	196

the clusters have only one or two applications in them. The great number of DNS application occurrences in the table can be explained by the fact that the LAN network under analysis was in test use and DNS was the most frequently appearing application among all traffic.

The similarities between DNS and HTTP applications are caused by the uni-directional measurement system, which in most cases allowed us to measure only the request part of the HTTP session. It is important to point out that most of the frequently used applications, such as DNS, NetBios[1], ICMP and NTP only use up a small portion of the network's available resources. These applications are mainly used for network management purposes. The applications normally causing heavy load on the network, such as Napster, Gnutella and FTP, were not commonly used in this LAN network.

2.4.2 SOM Results

The self-organizing map was 40 by 22 nodes in size. Prototype vectors were labeled using the validation data set. Figure 2.3 illustrates the U-matrix of the labeled SOM map. In U-matrix the darker tone implies that the neighboring prototype vectors are more similar with each other than those in the whiter areas of the map. The map clearly has some well separated clusters, for example the top right corner area containing labels 8888 and the area with ports 8000 (HTTP) next to it. U-matrix also reveals how similar the DNS, HTTP and ICMP applications actually are.

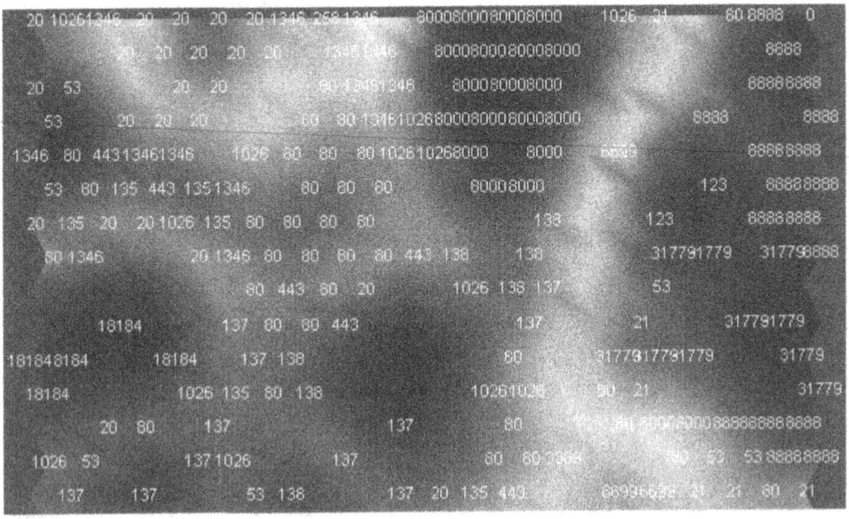

Fig. 2.3. Part of U-matrix with application (port) numbers used as labels

---

[1] Acronym NetBios is in this study used as a synonym for NetBT (NetBios-over-TCP/IP) to denote ports 137–139.

2.4.3 Comparison of SOM and FCM Results

Since the SOM map has 880 prototype vectors and the FCM produced 25 clusters, there is not a straightforward way to compare the two methods. The solution for this is to modify the results into a format where visual comparison is possible. We used the NP classification to attach FCM cluster numbers into the SOM's U-matrix presentation. Each of the prototype vectors is given a label of the most similar cluster. Cluster number is used as a label unless all of the cluster centers are further than the tolerance value (2.11) from the prototype vector. The used tolerance value was set to 15% of the Euclidian length of the prototype vector.

When we look at Figs. 2.3 and 2.4, together with the clustering results from Table 2.2, we notice that most of the major clusters found by FCM are also present in the U-matrix. These clusters cover compact regions in the map. Also in most areas the FCM clusters mostly consist of the same two applications that the original SOM had as labels. Surrounding some of the clusters are fences of unlabeled prototypes indicating that these clusters are distinct and differ considerably from the neighboring prototypes. This presentation provides a way to ensure the quality of both clustering results. In addition to this, the comparison can be used to clarify and find the actual cluster borders and shapes in the SOM.

2.4.4 Rough Set Results

For comparison, results considering rough set based classification are briefly outlined here. Preliminary results have been presented in [13] and [14].

The FCM and SOM methods presented above are targeted for real-valued attributes. Also methods using qualitative attributes are available. Rough sets theory

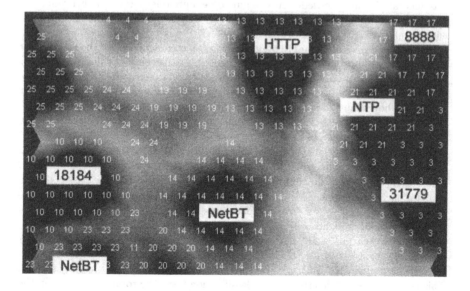

Fig. 2.4. Comparison of SOM and FCM clustering results. SOM prototype vectors are labeled with FCM clusters numbers taken from Table 2.2

introduced by Pawlak [19] in the early 1980's is one of these. This deals with the classificatory analysis of data tables. All variables are expressed as qualitative values. This feature rose interest to study its capabilities for clustering network data. A disadvantage however is that qualitative attributes in turn require discretisation of quantitative attributes. The basic theory is not presented here, for a more detailed analysis, c.f. [24].

The main target was to investigate capabilities of rough sets in network application profiling and to compare its efficiency with FCM clustering. As before, the aim is to classify network traffic sessions (data rows) by applications. The analysis was made using the same data set as others in this study, but using different attributes and slightly different time slices. The results are still comparable. Actual computations were done using ROSETTA software [25].

The previous methods differ slightly in the nature of classification. Rough sets can be used to classify the data straightforwardly by using the application as the decision attribute and approximating the decision classes induced by application. FCM in turn builds clusters of similar applications. One cluster can contain several applications.

There were 14 attributes available in the data. Out of these 14, some were considered to be useless or even harmful to the classification. For example, IP-addresses were rejected because the goal was to recognize different applications from the characteristics of the sessions, not addresses.

The selected condition attributes for rough sets were: time of the day, protocol, packet count, byte count, flow count, duration and total active time. This is quite unlike FCM, where the number of attributes had to be kept smaller for computational reasons and thus the original attributes were processed to yield bytes per packet and bytes per second.

During SOM and FCM studies some a priori knowledge about attribute selection was generated. This knowledge turned out to be quite relevant: no further reduction could be made which means that all attributes were necessary for the classification with rough sets.

Rules were generated simply by interpreting each unique row of the training set as a rule. Some filtering was performed to reject the least relevant rules. Inconsistent rules, i.e. rules with similar condition attributes but different decision attributes, were removed with a simple voting mechanism: each rule was attached to a counter indicating how many data rows supported the rule, and the rule with the largest counter value was selected among the ambiguous ones. After these operations, the size of the rule base was 193 rules.

To see how the generated rule set behaves with new data, the validation set was classified using the rules. Table 2.3 shows results from the validating classification. The fourth column of the table shows a percentage of successfully classified data rows for each application. Most of the applications were classified with a percentage of more than 90. There were some applications showing a poor percentage but this is due to a small amount of data as can be seen from the third column.

The classification of HTTP-traffic (port 80) did not succeed as well as other main applications. It can be noted that 17 rows (7.2%) of HTTP were classified as FTP (port 20), which is quite natural since the two applications are possibly similar. This is the largest single error of the classification test. The total percentage on the bottom line of Table 2.3 is a weighted average of all applications. Taking into account

Table 2.3. Results from rough set based classification of the validation data set

Application	Port	Number of Data Rows	Successfully Classified
ICMP	0	602	89.2%
FTP-default data	20	355	99.7%
FTP-control	21	7	0.0%
DNS	53	2177	98.6%
HTTP	80	235	80.1%
NTP	123	76	93.4%
NetBios-ns	137	66	95.5%
MobileIP-agent	434	150	99.3%
unknown	1321	13	0.0%
Napster	6699	12	16.7%
Napster	6700	5	0.0%
unknown	8888	146	97.9%
unknown	31779	114	94.7%
Total		3958	95.1%

the experimental nature of this study, results from rough sets can be considered fairly good.

2.5 Conclusion

The analyzed data consisted more of network management applications than actual human generated traffic. However the management applications are common to most LAN networks so analyzing them produces valuable information of the networks state. Another point of interest is the efficiency and suitability of the methods. It should be noted that a fair comparison of the results is difficult because of the differences in the classification methods used.

Results from the FCM clustering algorithm were somewhat affected by the fact that the number of samples for some applications was much higher than with other applications. These frequent applications formed separate clusters because the FCM algorithm uses the weighted mean of sample vectors to place the cluster centers. This is why more clusters are on areas where data density is higher. FCM was able to find clusters with just one application which is ultimately the most desirable result. But it is more crucial to be able to find the general characteristics of each application. The clusters with two or more applications are important in finding similarity between applications.

SOM visualized most of the same results as FCM did. The most significant difference was that due to the U-matrix representation it is possible to see the similarity of applications from the map labels. The closer the applications are located on the map grid, the more similar they are in characteristics. Like FCM, SOM concentrates more on the areas where the data density is higher. Applications with fewer sample vectors are easily left with just a few prototype vectors in the map.

However, even the rare applications were usually present in the map which makes it possible to see their relation with other applications.

Comparison between FCM and SOM methods was done by applying the modified NP classification procedure combined with the U-matrix. The produced U-matrix supported the correctness of both of the methods. This procedure proved a prominent way to compare and validate any clustering results to those obtained with SOM.

The experiment reported here shows some important advantages of rough sets theory over many other methods. Rough sets theory showed its power in classifying data into sets inherently induced by a decision attribute. This is desirable when no meaningful metric exists between the classes. This is quite unlike any other classification method studied here.

The data used was not perfectly suitable for any of the methods. Rough sets theory could easily handle qualitative attributes but required discretisation of the quantitative ones, while FCM and SOM had problems with attributes with irregularly distributed values.

The obtained clusters will eventually be used for application profiling. The application profiles combined with the knowledge of the user behavior and the topological information of the network are essential in producing a dynamic model for the network. Application resource requirements will help network administrators and service providers to anticipate, what kind of changes new users and applications will cause in use of network resources.

In the following stage of this research the same type of statistical network was used with data measured from a campus network with up to several thousand users. Also the data chosen for analysis has covered a much longer period of time. Preliminary results are reported in [16].

References

1. Babuska R (1998) Fuzzy Modelling for Control. Kluwer Academic Publishers, The Netherlands.
2. Barford P, Kline J, Plonka D, Ron A (2002) A Signal Analysis of Network Traffic Anomalies. http://citeseer.nj.nec.com/barford02signal.html, Referenced Dec 2003.
3. Bezdek J (1981) Pattern Recognition with Fuzzy Objective Function Algorithms. Plenum Press, USA.
4. Bonabeau E, Dorigo M, Theraulaz G (1999) Swarm Intelligence – From Natural to Artificial Systems. Santa Fe Institute, Studies in the Science of Complexity, Oxford University Press, New York.
5. The Caida Website. http://www.caida.org, Referenced Dec 2003.
6. Cisco Systems (2000) Inside Cisco IOS Software Architecture. Cisco Press, USA.
7. Claffy KC (2000) Measuring the Internet. IEEE Internet Computing, Vol. 4, no. 1, pp. 73–75.
8. Dunn JC (1974) A Fuzzy Relative of the ISODATA Process and its Use in Detecting Compact, Well Separated Clusters. Journal of Cybernetics, Vol. 3, No. 3, pp. 32–57.

9. Hollmén J (2000) User profiling and classification for fraud detection in mobile communication networks. Doctorate thesis, Helsinki University of Technology, Finland.
10. Ilvesmäki M, Luoma M (2001) On the capabilities of application level traffic measurements to differentiate and classify Internet traffic. Internet Performance and Control of Network Systems II, Proceedings of SPIE, Vol. 4523.
11. Jang J-S, Sun C-T, Mizutani E (1997) Neuro-Fuzzy and Soft Computing. Prentice-Hall, USA.
12. Kohonen T (2001) Self-Organizing Maps (Ed.). Springer-Verlag, Germany.
13. Laamanen V, Laurikkala M, Koivisto H (2002) Network Traffic Classification Using Rough Sets. Recent Advances in Simulation, Computational Methods and Soft Computing, WSEAS Press.
14. Laamanen V, Lampinen T, Laurikkala M, Koivisto, H (2002) A Comparison of Fuzzy C-means Clustering and Rough Sets Based Classification in Network Data Analysis. 3rd WSEAS International Conference on Fuzzy Sets and Fuzzy Systems, Interlaken, Switzerland.
15. Lampinen T, Koivisto H, Honkanen T (2002) Profiling Network Applications with Fuzzy C-Means Clustering and Self-Organizing Map. International Conference on Fuzzy Systems and Knowledge Discovery, Singapore, November 2002.
16. Laurikkala M, Honkanen T, Koivisto H (2004) Elephant flows snatch a lion's share of network capacity. International Conference on Next Generation Teletraffic and Wired/Wireless Advanced Networking, St. Petersburg (submitted).
17. The NLANR Measurement and Network Analysis Group Website. http://moat.nlanr.net/, Referenced Dec 2003.
18. Papagiannaki D, Taft N, Zhang Z, Diot C (2003) Long-Term Forecasting of Internet Backbone Traffic: Observations and Initial Models. in Proc. of IEEE INFOCOM 2003.
19. Pavlak Z (1991) Rough Sets – Theoretical Aspects of Reasoning about Data. Kluwer Academic Publishers, The Netherlands.
20. SOM toolbox (2002) http://www.cis.hut.fi/projects/somtoolbox/, Referenced 13th of June 2002.
21. Tan KMC, Collie BS (1997) Detection and classification of TCP/IP Network services. Proceedings of the 13th Annual Computer Security Applications Conference, USA.
22. Veitch D (2001) Traffic Modelling, a New Golden Era. http://www.emulab.ee.mu.oz.au/~darryl, Referenced Dec 2003.
23. Xie XL, Beni G (1991) A validity measure for fuzzy clustering. Transactions on Pattern Analysis and Machine Intelligence, Vol. 13, no. 8, pp. 841–847.
24. Ziarko W, Shan N (1995) Discovering Attribute Relationships, Dependencies and Rules by Using Rough Sets. 8th Hawaii Int. Conf. on System Sciences, Wailea, USA, 3-6 Jan 1995.
25. Öhrn A (2000) ROSETTA Technical Reference Manual. Department of Computer and Information Science. Norwegian University of Science and Technology (NTNU), Trondheim, Norway.

Monitoring Shift and Movement in Data using Dynamic Feature Maps

Damminda Alahakoon

School of Business Systems, Monash University, Australia
damminda.alahakoon@infotech.monash.edu.au

Abstract. Identifying change or *movement* in data can be highly useful to many organisations. Such movement identification will enable an organisation to react quickly to opportunities and also take fast remedial action in case of deviations from goals, etc. The Growing Self Organising Map (GSOM) has been presented as a technique which has a flexible structure suitable for dynamic applications. The GSOM is an unsupervised neural network algorithm which is based on the Self Organising Map (SOM), which provides the user with the ability to specify the *level of spread* of a map. The input dependant structure of the GSOM results in more representative maps compared to the SOM. This paper proposes a method using the GSOM as a base for identifying shift and/or movement in data.

Key words: Self Organising Map, Unsupervised neural Networks, Data Mining, Data exploration

3.1 Introduction

The SOM has been described as a visualisation tool for data mining applications [8, 11]. The visualisation is achieved by observing the two dimensional clusters of the multi dimensional input data set and identifying the inter and intra cluster proximities and distances. Once such clusters are identified, the data analyst generally develops hypotheses on clusters and the data and can use other methods such as statistics to test such hypotheses. In other instances, attribute values of the clusters can be analysed to identify the *useful* clusters in the data set for further analysis.

As such the feature maps generated by the SOM can be used in the preliminary stage of data mining process where the analyst uses such a map to obtain an *initial idea* about the *nature* of the data. The current usage of feature maps in data mining has been mainly for obtaining an initial unbiased segmentation of the data. Although this is a useful function, the feature maps offer the potential of providing more descriptive information regarding the data. Developing the feature maps to obtain such additional information can be thought of as a step towards developing the feature maps as a complete data mining tool.

Damminda Alahakoon: *Monitoring Shift and Movement in Data using Dynamic Feature Maps,*
Studies in Computational Intelligence (SCI) **4**, 29–41 (2005)
www.springerlink.com   © Springer-Verlag Berlin Heidelberg 2005

The Growing Self Organising Map (GSOM) has been proposed as a dynamically generating neural map which has particular advantages for data mining applications [3, 4, 5]. Several researchers have previously developed incrementally growing SOM models [6, 9, 10]. These models have similarities as well as differences from each other, but all attempt to solve the problem of pre-defined, fixed structure SOMs. A parameter called the Spread Factor (SF) in the GSOM provides the data analyst with control over the spread of the map. This ability is unique to the GSOM and could be manipulated by the data analyst to obtain a progressive clustering of a data set at different levels of detail.

A limitation of feature maps is that they do not have a fixed shape for a given set of input data. This becomes apparent when the same set of data records are presented to the network in changed order (say, sorted by different dimensions or attributes). Although the map would produce a similar clustering, the positioning of the clusters would be different. The factors which contribute to such changed positioning in traditional SOMs are the initial weights of the nodes and the order of data record presentation to the network, while in the GSOM it is only the order of records presented to the network, since the node weights are initialised according to the input data values.

Therefore comparing two maps becomes a non-trivial operation as even similar clusters may appear in different positions and shapes. Due to this problem, the use of feature maps for detecting *changes* in data is restricted. In this article a solution to this problem is proposed with the development of a layer of *summary* nodes as a conceptual model of the clusters in the data. This model is dependent upon the proportions of inter cluster distances and independent of the actual positions of the clusters in the map. The rest of the article is structured as follows. Section 3.2 provides a brief description of the GSOM algorithm and the effect of the spreading out using a parameter. Section 3.3 describes the method used for comparing separate GSOMs. A number of required definitions are presented in this section. Section 3.4 presents some simulation results to demonstrate the techniques and algorithms for GSOM comparision. Section 3.5 provides the concluding remarks for the paper.

3.2 The Growing Self Organising Map

The GSOM is an unsupervised neural network which is initialised with 4 nodes and *grows* nodes to represent the input data [1, 2]. During the node growth, the weight values of the nodes are *self organised* according to a similar method as the SOM.

3.2.1 The GSOM Algorithm

The GSOM process is as follows:

1. Initialisation Phase
   a) Initialise the weight vectors of the starting nodes (4) with random numbers.
   b) Calculate the Growth Threshold (GT) for the given data set according to the user requirements.
2. Growing Phase
   a) Present input to the network.

b) Determine the weight vector that is closest to the input vector mapped to the current feature map (winner), using Euclidean distance (similar to the SOM). This step can be summarised as:
Find q' such that $|v - w_{q'}| \leq |v - w_q| \; \forall q \in \mathcal{N}$ where v, w are the input and weight vectors respectively, q is the position vector for nodes and $\mathcal{N}$ is the set of natural numbers.

c) The weight vector adaptation is applied only to the neighbourhood of the winner and the winner itself. The neighbourhood is a set of neurons around the winner, but in the GSOM the starting neighbourhood selected for weight adaptation is smaller compared to the SOM (localised weight adaptation). The amount of adaptation (learning rate) is also reduced exponentially over the iterations. Even within the neighbourhood weights which are closer to the winner are adapted more than those further away. The weight adaptation can be described by:

$$w_j(k+1) = \begin{cases} w_j(k), & j \notin N_{k+1} \\ w_j(k) + LR(k) \times (x_k - w_j(k)), & j \in N_{k+1} \end{cases}$$

where the learning rate $LR(k), k \in \mathcal{N}$ is a sequence of positive parameters converging to 0 as $k \to \infty$. $w_j(k), w_j(k+1)$ are the weight vectors of the the node j, before and after the adaptation and N_{k+1} is the neighbourhood of the winning neuron at $(k+1)$th iteration. The decreasing of $LR(k)$ in the GSOM, depends on the number of nodes existing in the network at time k.

d) Increase the error value of the winner (error value is the difference between the input vector and the weight vectors).

e) When $TE_i \geq GT$ (where TE is the total error of node i and GT is the growth threshold), Grow nodes if i is a boundary node. Distribute weights to neighbours if i is a non-boundary node.

f) Initialise the new node weight vectors to match the neighbouring node weights.

g) Initialise the learning rate (LR) to its starting value.

h) Repeat steps b..g until all inputs have been presented, and node growth is reduced to a minimum level.

3. Smoothing Phase
   a) Reduce the learning rate and fix a small starting neighbourhood.
   b) Find winner and adapt weights of winner and neighbors, as in the growing phase.

Therefore, instead of the weight adaptation in the original SOM, the GSOM adapts its weights and architecture to represent the input data. Therefore in the GSOM a node has a weight vector and two dimensional coordinates which identify its position in the net, while in the SOM the weight vector is also called the position vector.

3.2.2 The Spread Factor

As described in the algorithm, the GSOM uses a threshold value called the Growth Threshold (GT) to decide when to initiate new node growth. The GT will decide the amount of spread of the feature map to be generated. Therefore if we require only a very abstract picture of the data, a large GT will result in a map with a fewer number of nodes. Similarly a smaller GT will result in the map spreading out more. When using the GSOM for data mining, it might be a good idea to first generate a *smaller* map, only showing the most significant clustering in the data, which will give the data analyst a summarised picture of the inherent clustering in the total data set.

The node growth in the GSOM is initiated when the error value of a node exceeds the GT. The total error value for node i is calculated as:

$$TE_i = \sum_{H_i} \sum_{j=1}^{\mathcal{D}} (x_{i,j} - w_j)^2 \tag{3.1}$$

where H is the number of hits to the node i and $\mathcal{D}$ is the dimension of the data. $x_{i,j}$ and w_j are the input and weight vectors of the node i respectively. For new node growth:

$$TE_i \geq GT \tag{3.2}$$

The GT value has to be experimentally decided depending on our requirements for map growth. As can be seen from (3.1) the dimension of the data set will make a significant impact on the accumulated error (TE) value, and as such will have to be considered when deciding the GT for a given application. The Spread Factor SF was introduced to address this limitation, thus eleminating the need for the data analyst to consider data dimensionality when generating feature maps. Although exact measurement of the effect of the SF is yet to be carried out, it provides an indicator for identifying the level of spread in a feature map (across differant dimensionalities). The derivation of the SF has been described in [4]. The formula used is:

$$GT = -D \times \ln(SF)$$

Therefore, instead of having to provide a GT, which would take different values for different data sets, the data analyst has to provide a value SF, which will be used by the system to calculate the GT value depending on the dimension (D) of the data. This will allow the GSOMs to be identified with their spread factors, and will be a basis for comparision of different maps.

3.3 Monitoring Movement and Change in Data with GSOMs

In this paper the concept of comparing different feature maps is presented. Several *types of differences* that can occur in feature maps are identified. For the purpose of describing the method of *identifying differences* such differences in data are categorised as shown in Fig. 3.1.

Category 1 consists of different data sets selected from the same domain. Therefore the attributes of the data could be the same although the attribute values can be different. Category 2 is the same data set but analysed at different points in time. Therefore the attribute set and the records are the same initially. During a certain period of time, if modifications are done to the data records, the data set may become different from the initial set. As shown in Fig. 3.1 we categorise such differences as:

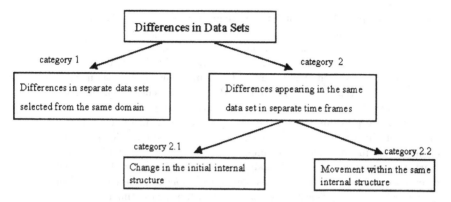

Fig. 3.1. Categorisation of the type of differences in data

1. Category 2.1 → change in data structure
2. Category 2.2 → movement of data while possessing the same structure

A formal definition of *movement* and *change* in data is presented below. Considering a data set $S(t)$ with k clusters Cl_i at time t where $\sum_{i=1}^{k} Cl_i \Longrightarrow S(t)$, and $\bigcap_{i=1}^{k} Cl_i = \emptyset$ categories 2.1 and 2.2 can thus be defined more formally as:

Definition: Change in Structure

Due to additions and modifications of attribute values the groupings existing in a data set may change over time. Therefore we say that the data has *changed* if, at a time t' where $t' > t$, and k and k' are the number of clusters (at similar spread factor) before and after the addition of records respectively, if

$$S(t') \Longrightarrow \sum_{i=1}^{k'} Cl_i(t) \qquad \text{where} \qquad k' \neq k \qquad (3.3)$$

In other words, the number of clusters has been increased or decreased due to changes in the attribute values.

Definition: Movement in Data

Due to the additions and modifications of attribute values the internal *content* of the clusters, *the intra cluster relationships* may change, although the actual number of clustering remain the same. For a given data set $S(t)$, this property can be described as:

$$S(t') \to \sum_{i=1}^{N} Cl_i(t') \qquad (3.4)$$

and $\exists A_{ij}(t')$ such that $A_{ij}(t') \neq A_{ij}(t)$ for some cluster Cl_i.

According to the above categorisation of the type of differences in data, category 1 simply means a difference between two or more data sets. Since the data are from the same domain (same sets of attributes), they can be directly compared.

Such analysis provides insight into the relationships between the functional groupings and the natural groupings of the data. For example, in a customer database, if the data sets have been functionally separated by region, such regions can be separately mapped. Comparing such maps will provide information as to the similarity/differences of functional clusters to natural clusters by region. Such analysis may provide information which can be used to optimise current functional groupings.

Category 2.1 provides information about the *change* in a data set over time. As defined in the previous section the changes will be an increase or decrease in the number of clusters in the map, at the same level of spread. Identification of such change is useful in many data mining applications. For example, in a survey for marketing potential and trend analysis, these would suggest the possibility of the emergence of a new category of customers, with special preferences and needs.

Category 2.2 will identify any *movement* within the existing clusters. It may also be that the additions and modifications to the data has in time resulted in the same number of, but different, clusters being generated. Monitoring such movement will provide an organisation with the advantage of identifying certain shifts in customer buying patterns. Therefore such movement (category 2.2) may be the initial stage of a change (category 2.1) and early identification of such movements towards the change may provide a valuable competitive edge for the organisation.

A further extension to the GSOM called the Attribute Cluster Relationship (ACR) Model has been proposed by developing two further layers of nodes (a layer of nodes representing cluster summaries, and another set of nodes representing data dimensions) on top of the GSOM [3]. The ACR model facilitates the maintenance of cluster information for further analysis and usage. The GSOM with the ACR model provides an automated method of identifying movement and change in data. The summarised and conceptual view of the map provided by the ACR model makes it practical to implement such map comparison. The method of identifying change and movement in data using the GSOM and ACR model is described below.

Figure 3.2 shows two GSOMs and the respective ACR models for a data set S at two instances t_1 and t_2 in time. It can be seen from the maps that significant change has occurred in the data as shown by the increase in the number of clusters. There are several possibilities for the type of change that has occurred and some of them are:

1. An extra cluster has been added while the earlier three clusters have remained the same.
2. The grouping has changed completely. That is, the four clusters in $GMAP_{t_2}$ of Fig. 3.2 are different from the three clusters in $GMAP_{t_1}$.
3. Some clusters may have remained the same while others might have changed, thus creating additional clusters. For example, C_1 and C_2 can be the same as C_4 and C_5 while C_3 has changed. That is, C_3 could have been split into two clusters to generate clusters C_6 and C_7. Another possibility is that C_3 does not exist with the current data and instead the two clusters C_6 and C_7 have been generated due to the new data.

The following terms are defined considering two GSOMs $GMAP_1$ and $GMAP_2$ on which ACR models have been created. These terms are then used in the description of the proposed method.

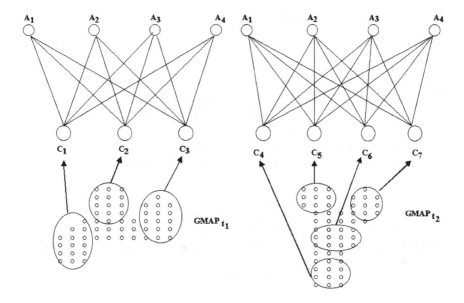

Fig. 3.2. Identification of change in data with the ACR model

Definition: Cluster Error (ERR_{Cl})

A measure called the cluster error (ERR_{Cl}) between two clusters in two GSOMs is defined as:

$$ERR_{Cl}(Cl_j(GMAP_1), Cl_k(GMAP_2))$$
$$= \sum_{i=1}^{D} |A_i(Cl_j) - A_i(Cl_k)| \qquad (3.5)$$

where Cl_j and Cl_k are two clusters belonging to $GMAP_1$ and $GMAP_2$ respectively and $A_i(Cl_j)$, $A_i(Cl_k)$ are the ith attribute of clusters Cl_j and Cl_k.

The ERR_{Cl} value is calculated using the ACR models for the GSOMs. During the calculation of the ERR_{Cl}, if a certain attribute which was considered non-significant to the cluster $Cl_j(GMAP_1)$ is considered as a significant attribute for $Cl_k(GMAP_2)$, then the two clusters are not considered to be similar. Therefore we define a term *significant non-similarity* as shown below.

Definition: Significant Non-Similarity

If $\exists A_i(Cl_j), A_j(Cl_k)$ such that $A_j(Cl_k) = -1$ and $A_i(Cl_j) \neq -1$ then Cl_j and Cl_k are significantly non-similar. The -1 above is considered since the non-significant and non-contributing attribute cluster links (m_{ij}) are assigned -1 values in the ACR model.

Definition: Cluster Similarity

We define two clusters Cl_j and Cl_k as similar when the following conditions are satisfied.

1. Does not satisfy the significant non-similarity condition.
2. $ERR_{Cl}(Cl_j, Cl_k) \leq T_{CE}$ where T_{CE} is the threshold of cluster similarity and has to be provided by the data analyst depending on the level of similarity required. If complete similarity is required, then $T_{CE} = 0$.

We can now derive the range of values for T_{CE} as follows.

Since $0 \leq A_i \leq 1 \; \forall i = 1 \ldots D$, using equation 3.5 we can say

$$0 \leq ERR_{Cl} \leq D \tag{3.6}$$

Since the average ERR_{Cl} value is $D/2$, if $ERR_{Cl} \geq D/2$ the two clusters are more *different* according to attribute values than they are equal. Since we need the threshold value to identify cluster similarity, the maximum value for such a threshold can be $D/2$. Therefore

$$0 < T_{CE} < D/2 \tag{3.7}$$

Definition: Measure of Similarity Indicator

Since the similarity between two clusters depends on the T_{CE} value we define a new indicator called the *measure of similarity* which will indicate the *amount* of similarity when two clusters are considered to be similar. The measure of similarity indicator (I_s) is thus calculated as the fraction of the actual cluster error to the maximum tolerable error for two clusters to be considered similar.

$$I_s = 1 - \frac{ERR_{Cl}(Cl_j, Cl_k)}{\text{Max}(T_{CE})} \tag{3.8}$$

By substituting from (3.7),

$$I_s = 1 - \frac{ERR_{Cl}(Cl_j, Cl_k)}{D/2} \tag{3.9}$$

Considering two GSOMs $GMAP_1$ and $GMAP_2$, the cluster comparison algorithm can now be presented as:

1. Calculate $ERR_{Cl}(Cli, Clj) \; \forall Cl_i \in GMAP_1$ with all $Cl_j \in GMAP_2$.
2. For each $Cl_i \in GMAP_1$, find $ERR_{Cl}(Cl_i, Cl_p)$, $Cl_p \in GMAP_2$, such that

$$ERR_{Cl}(Cl_i, Cl_p) \leq ERR_{Cl}(Cl_i, Cl_j)$$

$$\forall Clj \in GMAP_2, \; p \neq j$$

3. Ensure that the clusters Cl_i, Cl_p satisfy the cluster similarity condition.
4. Assign Cl_p to Cl_i (as similar clusters) with the *amount of similarity* calculated as the measure of similarity value

$$1 - \frac{ERR_{Cl}(Cl_i, Cl_p)}{D/2}$$

5. Identify cluster Cl_i in $GMAP_1$ and Cl_j in $GMAP_2$ which have not been assigned to a cluster in the other GSOM.

The cluster comparison algorithm will provide a measure of the similarity of the clusters. If all the clusters in the two maps being compared have a high measure of similarity values, then the maps are considered equal. The amount of similarity, or difference to be tolerated, will depend on the requirements of the application. If there is one or more clusters in a map (say $GMAP_1$) which do not find a similar cluster in the other map (say $GMAP_2$), the two maps are considered different. The advantage of this comparison algorithm is not only for comparing feature maps for their similarity, but as a data monitoring method. For example, feature maps generated on a transaction data set at different time intervals may identify movement in the clusters or attribute values. The movement may start as a small value initially (the two maps have a high similarity measure) and gradually increase (reduction of the similarity measure) over time. Such movement may indicate an important trend that can be made use of, or the starting of a deviation (problem) which needs immediate corrective action.

3.4 Experimental Results

An animal data set [7] is used to demonstrate the effect of the ACR model for data monitoring. 25 animals out of 99 are selected from the animal data and these are: lion, wolf, cheetah, bear, mole, carp, tuna, pike, piranha, herring, chicken, pheasant, sparrow, lark, wren, gnat, flea, bee, fly, wasp, goat, calf, antelope, elephant and buffalo. The animals for this experiment have been selected such that they equally represent the groups insects, birds, non-meat eating mammals, meat eating mammals and fish. Figure 3.3 shows the GSOM for the 25 selected animals. The clusters are shown with manually drawn boundary lines for easy visualisation.

3.4.1 Identifying the Shift in Data Values

The GSOM of Fig. 3.3 is considered as the map of the initial data and some additional data records (given below) are added to simulate a new data set for identification of any significant shifts in data.

Experiment 1: Adding New Data without the Change in Clusters

In the first experiment 8 new animals are added to the first 25. The new animals are deer, giraffe, leopard, lynx, parakeet, pony, puma and reindeer. These animals were selected such that they belong to groups (clusters) that already exist in the initial data. As such the addition of these new records should not make any difference to the cluster summary nodes in the ACR model. Figure 3.4 shows the GSOM for the new data set with the additional 8 animals. Five clusters C1', C2'...C5' have been identified from $GMAP_2$. Table 3.1 shows cluster error values (calculated according to equation 3.5) between the GSOMs $GMAP_1$ (Fig. 3.3) and $GMAP_2$ (Fig. 3.4). Table 3.2 shows the clusters that have been identified as similar from $GMAP_1$ and $GMAP_2$ with the respective measures of similarity indicator values (I_s). It can be seen that the five clusters have not changed significantly due to the introduction of new data.

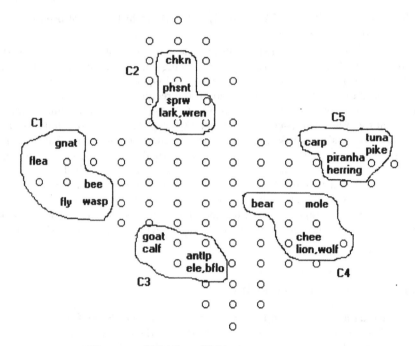

Fig. 3.3. $GMAP_1$ – GSOM for the 25 animals

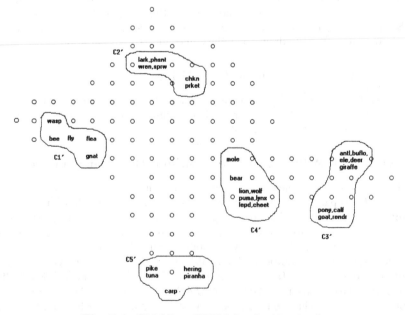

Fig. 3.4. $GMAP_2$ – GSOM for the 33 animals

Table 3.1. The cluster error values calculated for clusters in Fig. 3.4 with clusters in Fig. 3.3

	C1	C2	C3	C4	C5
C1'	0	4.87	8.13	8.73	10.4
C2'	5.00	0.13	7.40	8.67	8.07
C3'	8.17	7.58	0.04	1.84	8.71
C4'	8.88	8.54	1.65	0.15	7.27
C5'	10.4	7.93	8.67	7.27	0

Table 3.2. The measure of similarity indicator values for the similar clusters between $GMAP_1$ and $GMAP_2$

$GMAP_1$	$GMAP_2$	I_s
C1	C1'	1
C2	C2'	0.983
C3	C3'	0.995
C4	C4'	0.981
C5	C5'	1

Experiment 2: Adding New Data with Different Groupings

In the second experiment six different *birds* are added to the original 25 animals. The newly added birds are crow, duck, gull, hawk, swan, vulture. These birds are different from the birds that exist in the original 25 animal data set as some of them are *predators* and others are *aquatic* birds. These sub-categories did not exist in the birds included with the initial 25 animals. The purpose of selecting these additional data is to demonstrate the data movement identification with the ACR model. Figure 3.5 shows the clusters C1', C2' .. C5' identified with the new 31 animal data set.

The cluster error values between $GMAP_1$ and $GMAP_3$ are shown in Table 3.3, and the best matching clusters are given in Table 3.4, with the respective similarity

Table 3.3. The cluster error values calculated for clusters in Fig. 3.4 with clusters in Fig. 3.5

	C1	C2	C3	C4	C5
C1'	0	4.87	8.13	8.73	10.4
C2'	5.79	0.93	8.10	8.15	7.01
C3'	8.13	7.56	0	1.8	8.67
C4'	8.73	8.56	1.8	0	7.27
C5'	10.4	7.93	8.67	7.27	0

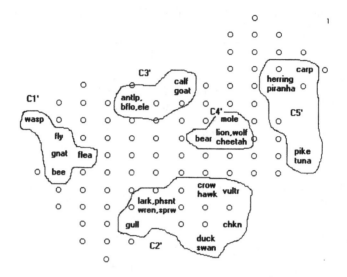

Fig. 3.5. $GMAP_3$ – GSOM for the 31 animals

Table 3.4. The measure of similarity indicator values for the similar clusters between $GMAP_1$ and $GMAP_3$

z $GMAP_1$	$GMAP_3$	I_s
C1	C1'	1
C2	C2'	0.88
C3	C3'	1
C4	C4'	1
C5	C5'	1

indicator (I_s) values. From the I_s values, it can be seen that although clusters C2 and C2' are considered as similar there is a noticeable movement in the cluster. As such the data analyst can focus attention on this cluster to identify the reason for such a change.

3.5 Conclusion

This paper described a method of identifying change in the structure of the clusters within a data set. The need and importance of identifying such change in data was discussed and the advantages and limitations of using SOM based techniques for such change monitoring was highlighted. The paper initially presented several definitions as a foundation framework for building the data change identification model. The GSOM-ACR model has been used as the basic technique. Simple examples have been described to highlight the usefulness of the method. The GSOM can be used to develop clusters at different levels of abstraction, and thus can generate hierarchical levels of clusters. Further work needs to be carried out to highlight the usefulness of comparing hierarchical clusters generated using the GSOM, especially for data mining applications.

References

1. Alahakoon L.D., Halgamuge S.K. and Srinivasan B. A Self Growing Cluster Development Approach to Data Mining. In *Proceedings of the IEEE Conference on Systems Man and Cybernatics*, pp. 2901–2906, 1998.
2. Alahakoon L.D., Halgamuge S.K. and Srinivasan B. A Structure Adapting Feature Map for Optimal Cluster Representation. In *Proceedings of the International Conference on Neural Information Processing*, pp. 809–812, 1998.
3. Alahakoon L.D. *Data Mining with Structure Adapting Neural Networks*. PhD thesis, School of Computer Science and Software Engineering, Monash University, 2000.
4. Alahakoon L.D., Halgamuge S.K. and Srinivasan B. Dynamic Self Organising Maps with Controlled Growth for Knowledge Discovery. *IEEE Transactions on Neural Networks*, 11(3):601–614, 2000.
5. Alahakoon L.D. Controlling the Spread of Dynamic Self Organising Maps. In *Proceedings of the Intelligent Systems, Design and Applications Conference*, 2003.
6. Blackmore J. and Miikkulainen R. Incremental Grid Growing: Encoding High Dimensional Structure into a Two Dimensional Feature Map. In Simpson P., editor, *IEEE Technology Update*. IEEE Press, 1996.
7. Blake C., Keogh E. and Merz C.J. University of California, Irvine, Dept. of Information and Computer Sciences. UCI Repository of machine learning databases, 1998. http://www.ics.uci.edu/~mlearn/MLRepository.html.
8. Deboeck G. *Visual Explorations in Finance*. Springer Verlag, 1998.
9. Fritzke B. Growing Cell Structure: A Self Organising Network for Supervised and Un-supervised Learning. *Neural Networks*, 07:1441–1460, 1994.
10. Martinetz T.M. and Schultan K.J. Topology Representing Networks. *Neural Networks*, 7(3):507–522, 1994.
11. Westphal C. *Data Mining Solutions*. Wiley Computer Publishing, 1998.

4

Serendipity in Text and Audio Information Spaces: Organizing and Exploring High-Dimensional Data with the Growing Hierarchical Self-Organizing Map

Michael Dittenbach[1] and Dieter Merkl[2] and Andreas Rauber[3]

[1] E-Commerce Competence Center – EC3, Donau-City-Straße 1, A–1220 Wien, Austria
   michael.dittenbach@ec3.at
[2] Research Group for Industrial Software Engineering, Technische Universität Wien, Operngasse 9, A–1040 Wien, Austria
   dieter@rise.tuwien.ac.at
[3] Institut für Softwaretechnik, Technische Universität Wien, Favoritenstraße 9–11/188, A–1040 Wien, Austria
   andi@ifs.tuwien.ac.at

Abstract. While tools exist that allow us to search through vast amounts of text within seconds, most systems fail to assist the user in getting an overview of the information available or maintaining orientation within an information space. We present a neural network architecture, i.e. the *Growing Hierarchical Self-Organizing Map*, providing content-based organization of information repositories, facilitating intuitive browsing and serendipitous exploration of the information space. To show the universal potential of this architecture, we present the automatic, content-based organization of two different types of repositories with diverse characteristics, the first being a collection of newspaper articles and the second being a music collection.

Key words: Interactive Exploration, Information Spaces, Growing Hierarchical Self-Organizing Map, GHSOM

4.1 Introduction

Apart from the traditional focus of Information Retrieval (IR) research on optimizing query performance both in terms of speed as well as quality, i.e. recall/precision trade-offs, another aspect of IR is gaining importance, namely serendipity. Traditional query processing requires a well-defined information need combined with the knowledge how and to what degree the answer is contained in an information repository. Serendipity, on the other hand, focuses on the exploration of a repository, facilitating the discovery of information, allowing answers where the question cannot

Michael Dittenbach et al.: *Serendipity in Text and Audio Information Spaces: Organizing and Exploring High-Dimensional Data with the Growing Hierarchical Self-Organizing Map*, Studies in Computational Intelligence (SCI) **4**, 43–60 (2005)
www.springerlink.com

be precisely defined. Methods supporting exploration of data repositories thus provide an important compliment to more traditional approaches. By providing means to explore a data repository, users have a chance to build a mental model of it, to understand its characteristics and to grasp which information is or is not present.

With the methods presented in this paper we address this crucial feature of successful interaction with information by providing a comprehensible mapping of the data space, which can subsequently be visualized and explored in an interactive manner. As the basic tool for providing such an organization based representation we use the *Self-Organizing Map (SOM)* [7, 8], an unsupervised neural network model that provides a topology-preserving mapping from a high-dimensional input space onto a two-dimensional output space. In other words, the similarity between input data items is preserved as faithfully as possible within the representation space of the *SOM*.

Despite the large number of research reports on applications of the *SOM* [6, 15] some deficiencies remained largely untouched. First, the *SOM* uses a static network architecture which has to be defined prior to the start of training, both in terms of number as well as arrangement of neural processing elements. Second, hierarchical relations between input data are rather difficult to detect in the map display. So far, both issues have been addressed separately by means of adaptive architectures, e.g. the *Growing Grid* [5], or hierarchies of independent *SOMs*, e.g. the *Hierarchical Feature Map* [14] or the *Tree-Structured SOM* [10, 11].

We present a neural network model, the *Growing Hierarchical Self-Organizing Map (GHSOM)* [2, 4], that addresses both of the above mentioned deficiencies within one framework. Basically, this neural network model is composed of independent *SOMs*, each of which is allowed to grow in size during the training process until a quality criterion regarding data representation is met. This growth process is further continued to form a layered architecture such that hierarchical relations between input data are further detailed at lower layers of the neural network. We also report on our enhancement of the *GHSOM* where emphasis is directed to the orientation of the various layers during training. Already learned similarities between input data can thus be maintained during the establishment of the hierarchical structure of the *GHSOM*.

Using this approach, we are able to represent data in context, unveiling the relationships between the various items in a repository, thus turning data into information to be explored and understood. In order to demonstrate the capabilities of the *GHSOM* based approach to the exploration of information spaces, we chose two different domains with diverse characteristics, namely collections of text and music.

In the case of text collections, the *GHSOM* allows automatical organization of documents by subject, creating a topic hierarchy according to the characteristics of the data collection. The process is virtually language and domain independent, relying on word histograms for content-based organization. A particularly challenging characteristic of this domain is the high dimensionality of the feature spaces involved in text representation, which quite commonly ranges in the order of several thousand dimensions. The capabilities of this approach have been evaluated in a wide range of document collections within the *SOMLib* Digital Library Project [18, 19].

Regarding music data we aim at creating an organization according to sound similarities of individual titles. A particular challenge in this respect is the extraction of features that allow the computation of similarities according to the sound sensation perceived by users. We address these issues within the *SOM-enhanced*

JukeBox (*SOMeJB*) project [16, 17, 20] relying on a feature representation capturing Rhythm Patterns of music while incorporating psycho-acoustic models of audition.

The remainder of the paper is structured as follows. In Sect. 4.2 we provide a brief overview of the *SOM* followed by a description of the architecture and the training process of the *GHSOM* in Sect. 4.3. Then, we present some results of training the *GHSOM* with real-world text documents in Sect. 4.4 and with music in Sect. 4.5. Finally, we draw some conclusions in Sect. 4.6.

4.2 Self-Organizing Map

The *Self-Organizing Map* is an unsupervised neural network providing a mapping from a high-dimensional input space to a usually two-dimensional output space while preserving topological relations as faithfully as possible. The *SOM* consists of a set of i units arranged in a two-dimensional grid with a weight vector $m_i \in \Re^n$ attached to each unit. Elements from the high-dimensional input space, referred to as input vectors $x \in \Re^n$, are presented to the *SOM* and the activation of each unit for the presented input vector is calculated using an activation function. Commonly, the Euclidean distance between the weight vector of the unit and the input vector serves as the activation function. In the next step the weight vector of the unit showing the highest activation (i.e. the smallest Euclidean distance) is selected as the 'winner' and is modified as to more closely resemble the presented input vector. Pragmatically speaking, the weight vector of the winner is moved towards the presented input signal by a certain fraction of the Euclidean distance as indicated by a time-decreasing learning rate α. Thus, this unit's activation will be even higher the next time the same input signal is presented. Furthermore, the weight vectors of units in the neighborhood of the winner as described by a time-decreasing neighborhood function are modified accordingly, yet to a smaller amount as compared to the winner. This learning procedure finally leads to a topologically ordered mapping of the presented input signals. Consequently, similar input data are mapped onto neighboring regions of the map.

A simple graphical representation of a self-organizing map's architecture and its learning process is provided in Fig. 4.1. In this figure the output space consists of a square of 25 units, depicted as circles, forming a grid of 5 × 5 units. One input

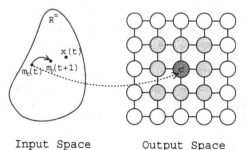

Input Space   Output Space

Fig. 4.1. Adaption of weight vectors during the training process

vector $x(t)$ is randomly chosen and mapped onto the grid of output units. In the second step of the learning process, the winner c showing the highest activation is selected. Consider the winner being the unit depicted as the dark gray unit labeled c in the figure. The weight vector of the winner, $m_c(t)$, is now moved towards the current input vector. This movement is symbolized in the input space in Fig. 4.1. As a consequence of the adaptation, unit c will produce an even higher activation with respect to the input pattern x at the next learning iteration, $t+1$, because the unit's weight vector, $m_c(t+1)$, is now nearer to the input pattern x in terms of the input space. Apart from the winner, adaptation is performed with neighboring units, too. Units that are subject to adaptation are depicted as shaded units in the figure. The shading of the various units corresponds to the amount of adaptation, and thus, to the spatial width of the neighborhood-kernel. Generally, units in close vicinity of the winner are adapted more strongly, and consequently, they are depicted with a darker shade in the figure.

4.3 Growing Hierarchical Self-Organizing Map

4.3.1 Architecture and Training

The key idea of the *GHSOM* is to use a hierarchical structure of multiple layers where each layer consists of a number of independent *SOMs*. One *SOM* is used at the first layer of the hierarchy. For every unit in this map a *SOM* might be added to the next layer of the hierarchy. This principle is repeated with the third and any further layers of the *GHSOM*.

Since one of the shortcomings of *SOM* usage is its fixed network architecture we preferred to use an incrementally growing version of the *SOM* similar to the *Growing Grid*. This relieves us of the burden of predefining the network's size, which is rather determined during the unsupervised training process. We start with a layer 0 consisting of only one single unit. The weight vector of this unit is initialized as the average of all input data. The training process then basically starts with a small map of 2×2 units in layer 1 that is self-organized according to the standard *SOM* training algorithm.

This training process is repeated for a number λ of training iterations. λ is determined by the number of input data to be trained on the map. Ever after λ training iterations the unit with the largest deviation between its weight vector and the input vectors represented by this very unit is selected as the error unit e. In between the error unit e and its most dissimilar neighbor d in terms of the input space either a new row or a new column of units is inserted as shown in Fig. 4.2. The weight vectors of these new units are initialized as the average of their neighbors.

An obvious criterion to guide the training process is the quantization error q_i, calculated as the sum of the distances between the weight vector of a unit i and the input vectors mapped onto this unit. It is used to evaluate the mapping quality of a *SOM* based on the *mean quantization error* (*MQE*) of all units on the map. A map grows until its *MQE* is reduced to a certain fraction τ_1 of q_i of unit i in the preceding layer of the hierarchy. Thus, the map now represents the data mapped onto the higher layer unit i in more detail.

As outlined above, the initial architecture of the *GHSOM* consists of one *SOM*. This architecture is expanded by another layer in case of dissimilar input data being

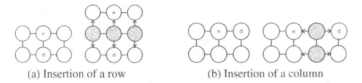

(a) Insertion of a row   (b) Insertion of a column

Fig. 4.2. A row (**a**) or a column (**b**) of units (*shaded gray*) is inserted in between *error unit* e and the neighboring unit d with the largest distance between its weight vector and the weight vector of e in the Euclidean space

mapped on a particular unit. These units are identified by a rather high quantization error q_i which is above a threshold τ_2. This threshold basically indicates the desired granularity of data representation as a fraction of the initial quantization error at layer 0. In such a case, a new map will be added to the hierarchy and the input data mapped on the respective higher layer unit are self-organized in this new map, which again grows until its MQE is reduced to a fraction τ_1 of the respective higher layer unit's quantization error q_i.

A graphical representation of a $GHSOM$ is given in Fig. 4.3. The map in layer 1 consists of 3×2 units and provides a rough organization of the main clusters in the input data. The six independent maps in the second layer offer a more detailed view on the data. Two units from one of the second layer maps have further been expanded into third-layer maps to provide a sufficiently granular data representation.

Depending on the desired fraction τ_1 of MQE reduction, we may end up with either a very deep hierarchy consisting of small maps, a flat structure with large maps, or -- in the most extreme case -- only one large map, which is similar to the *Growing Grid*. The growth of the hierarchy is terminated when no further units are available for expansion. It should be noted that the training process does not necessarily lead to a balanced hierarchy in terms of all branches having the same depth. This is one of the main advantages of the $GHSOM$, because the structure of the hierarchy adapts itself according to the requirements of the input space.

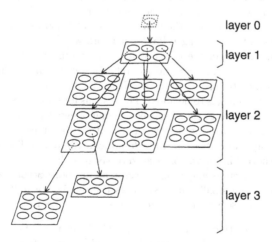

Fig. 4.3. $GHSOM$ reflecting the hierarchical structure of the input data

Therefore, areas in the input space that require more units for appropriate data representation create deeper branches than others.

The growth process of the *GHSOM* is mainly guided by the two parameters τ_1 and τ_2, which merit further consideration. Parameter τ_2 controls the minimum granularity of data representation, i.e. no unit may represent data at a coarser granularity. If the data mapped onto one single unit still has a larger variation, a new map will be added originating from this unit representing this unit's data in more detail at a subsequent layer. This absolute granularity of data representation is specified as a fraction of the inherent dissimilarity of the data collection as such, which is expressed in the *mean quantization error* (mqe^1) of the single unit in layer 0 representing all data points. If we decide after the termination of the training process that a yet more detailed representation would be desirable, it is possible to resume the training process from the respective lower level maps, continuing to both grow them horizontally as well as adding new lower level maps until a stricter quality criterion is satisfied. This parameter thus represents a global termination and quality criterion for the *GHSOM*.

Parameter τ_1 controls the actual growth process of the *GHSOM*. Basically, hierarchical data can be represented in different ways favoring either (a) lower hierarchies with rather detailed refinements presented at each subsequent layer or (b) deeper hierarchies, which provide a stricter separation of the various sub-clusters by assigning separate maps. In the first case we will prefer larger maps in each layer that explain larger portions of the data in their flat representation, allowing less hierarchical structuring. In the second case, however, we will prefer rather small maps, each of which describes only a small portion of the characteristics of the data, and rather emphasizes the detection and representation of hierarchical structure.

Thus, the smaller the parameter τ_1, the larger will be the degree to which the data has to be explained with one single map. This results in larger maps as the map's mean quantization error (MQE) will be lower the more units are available for representing the data. If τ_1 is set to a rather high value, the MQE does not need to fall too far below the mqe of the upper layer's unit it is based upon. Thus, a smaller map will satisfy the stopping criterion for the horizontal growth process, requiring the more detailed representation of the data to be performed in subsequent layers.

In a nutshell we can say, that the smaller the parameter value τ_1, the more shallow the hierarchy, and that the lower the setting of parameter τ_2, the larger the number of units in the resulting *GHSOM* network will be.

Apart from the advantage of automatically determining the number of units required for data representation and the reflection of the hierarchical structure in the data, a considerable speed-up of the *GHSOM* training process as compared to standard *SOM* training has to be noted. The reasons for this are twofold. First, at the transition from one layer to the next, vector components that are (almost) identical for all data items mapped onto a particular unit can be omitted for training of the according next layer map, because they do not contribute to differentiation between them. Hence, shorter input vectors lead directly to reduced training times because of faster winner selection and weight vector adaptation. Secondly, a considerable speed-up results from smaller map sizes, as the number of units that have to be evaluated for winner selection is smaller at each map. This results directly from the

---

[1] Please note the distinction between mqe, which refers to the mean quantization error of a unit, and MQE in capital letters, which refers to a whole map.

fact that the spatial relation of different areas of the input space is maintained by means of the network architecture rather than by means of the training process.

4.3.2 GHSOM Extension: Map Orientation

The hierarchical structuring imposed on the data results in a separation of clusters mapped onto different branches. While this, in principle, is a desirable characteristic helping to understand the cluster structure of the data, it may lead to misinterpretations when large clusters are mapped and expanded on two neighboring, yet different units. Similar input data are thus rather arbitrarily separated in different branches of the hierarchy.

In order to provide a global orientation of the individual maps in the various layers of the hierarchy, their orientation must conform to the orientation of the data distribution on their parents' maps. This can be achieved by creating a coherent initialization of the units of a newly created map, i.e. by adding a fraction of the weight vectors in the neighborhood of the parent unit [3]. This initial orientation of the map is preserved during the training process. By providing a global orientation of all maps in the hierarchy, potential negative effects of splitting a large cluster into two neighboring branches can be avoided, as it is possible to navigate across map boundaries to neighboring maps.

Let unit e be expanded to form a new 2×2 map in the subsequent layer of the hierarchy. This map's four model vectors $e1$ to $e4$ are initialized to mirror the orientation of neighboring units of its parent e. Figure 4.4 provides an illustration of the initialization of new maps and the influence of the parent unit's neighbors. Geometrically speaking, the model vectors of the four corner units are moved in the

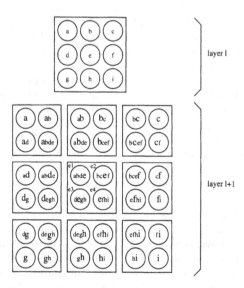

Fig. 4.4. Initialization of units for expanded maps. The letters on the units in layer $l + 1$ denote the vectors that are influential for the respective unit's initialization with the larger size letters denoting the map's corresponding upper-layer unit

data space towards the directions of their respective parent's neighbors by a certain fraction. This initial orientation of the map is preserved during the training process. While new units may be inserted in between, the four corner units will still be most similar to the respective corner units of the maps in neighboring branches. The exact amount by which these corner units are moved in the respective directions does not influence the characteristic of the topology-preserving initialization. We can thus choose to set them to the mean of the parent and its neighbors in the respective directions, e.g. setting $e1 = (e + a + b + d)/4$, $e2 = (e + b + c + f)/4$, etc. In the simplest case, the neighbors' model vectors may be used directly as initial corner positions of the new maps in the data space.

4.4 Experiments: Textual Information Spaces

For the experiments we have used a collection of 11,627 articles from the daily Austrian newspaper *Der Standard* from the second quarter of 1999. These documents cover the broad range of topics typically addressed by the newspaper ranging from national and international politics via sports and economy to culture and special weekend addenda. The experiments presented below as well as experiments using the complete collection of about 50,000 articles from the whole year 1999 are available for interactive exploration at http://www.ifs.tuwien.ac.at/~andi/somlib.

4.4.1 Feature Extraction

In order to use the *SOM* for organizing documents based on their topics, a vector-based description of the content of the documents needs to be created. While manually or semi-automatically extracted content descriptors may be used, research results have shown that a rather simple word frequency based description is sufficient to provide the necessary information in a very stable way [1, 9, 13, 19]. For this word frequency based representation a vector structure is created consisting of all words appearing in the document collection.

Stop words, i.e. words that do not contribute to content representation and topic discrimination between documents, are usually removed from this list of words. Again, while manually crafted stop word lists may be used, simple statistics allow the removal of most stop words in a very convenient and language- and subject-independent way. On the one hand, words appearing in too many documents, say, in more than half of all documents, can be removed without the risk of losing content information, as the content conveyed by these words is too general. On the other hand, words appearing in less than a minimum number of 5 to 10 documents (depending on the total number of documents in the collection) can be omitted for content-based classification, as the resulting sub-topic granularity would be too small to form a topical cluster in its own right. Note that the situation is different in the information retrieval domain, where rather specific terms need to be indexed to facilitate retrieval of a very specific subset of documents. In this respect, content-based organization and browsing of documents constitutes a conceptually different approach to accessing document archives and interacting with them by browsing topical hierarchies. This obviously has to be supplemented by various searching facilities, including information retrieval capabilities as they are currently realized in many systems.

The documents are described by the words they are made up of within the resulting feature space, usually consisting of about 1,000 to 15,000 dimensions, i.e. distinct terms. While basic binary indexing may be used to describe the content of a document by simply stating whether or not a word appears in the document, more sophisticated schemes such as $tf \times idf$, i.e. term frequency times inverse document frequency [21], provide a better content representation. This weighting scheme assigns higher values to terms that appear frequently within a document, i.e. have a high term frequency, yet rarely within the complete collection, i.e. have a low document frequency. Usually, the document vectors are normalized to unit length to make up for length differences of the various documents. For the experiments presented below, we discard terms that appear in more than 7% or in less than 0.56% of the articles, resulting in a vector dimensionality of 3,799 unique terms. These vectors are fed into the *GHSOM* to create a hierarchy of maps for interactive exploration.

4.4.2 GHSOM Atlas

The *GHSOM* training results in a rather shallow hierarchical structure of up to 7 layers. The layer 1 map (cf. Fig. 4.5) grows to a size of 7 × 4 units, all of which are expanded at subsequent layers. The units' labels are obtained by the *LabelSOM* algorithm detailed in [12]. We find the most dominant branches to be, for example, *Sports*, located in the upper right corner of the map, *Internal Affairs* in the lower right corner, *Internet*-related articles on the left hand side of the map or articles about *Commerce* and *banks* in the middle, to name but a few. However, due to the large size of the resulting first layer map, a fine-grained representation of the data is already provided at this layer. This results in some larger clusters to be represented by two neighboring units already at the first layer, rather than being split up in a lower layer of the hierarchy. For example, we find the cluster on *Internal Affairs* to be represented by two neighboring units. One of these, on position $(6/4)^2$, covers solely articles related to the *Freedom Party* and its political leader *Jörg Haider* (characteristic labels are `fpoe`, `joerg` or `haider`), representing one of the most dominant political topics in Austria for some time now, resulting in an accordingly large number of news articles covering this topic. The neighboring unit to the right, i.e. located in the lower right corner on position (7/4), covers other *Internal Affairs*, with one of the main topics being the elections to the *European Parliament*. Figure 4.6 shows these two second-layer maps.

We also find, for example, articles related to the *Freedom Party* on the branch covering the more general *Internal Affairs*, reporting on their role and campaigns for the elections to the *European Parliament*. As might be expected these are closely related to the other articles on the *Freedom Party* which are located in the neighboring branch. Obviously, we would like them to be presented on the left hand side of this map, so as to allow the transition from one map to the next, with a continuous orientation of topics. Due to the initialization of the added maps during the training process, this continuous orientation is preserved as shown in Fig. 4.6. Continuing from the second layer map of unit (6/4) to the right we reach the according second layer map of unit (7/4) where we first find articles focusing, again, on the *Freedom*

---

[2] We refer to a unit located in column x and row y as (x/y) starting with $(0/0)$ in the upper left corner.

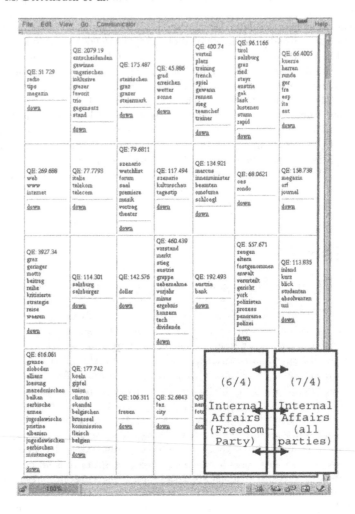

Fig. 4.5. Top-layer map with the topic of Austrian internal affairs being split onto two units in the lower right corner

Party, before moving on to the *Social Democrats*, the *People's Party*, the *Green Party* and the *Liberal Party*.

4.5 Experiments: Music Spaces

In this section, we present some experimental results of our system based on a music archive made up of MP3-compressed files of popular pieces of music from a variety of genres. Specifically, we present in more detail the organization of a small subset of the entire archive, consisting of 77 pieces of music, with a total playing time of about 5 hours, using the *GHSOM*. This subset, due to its limited size offers itself

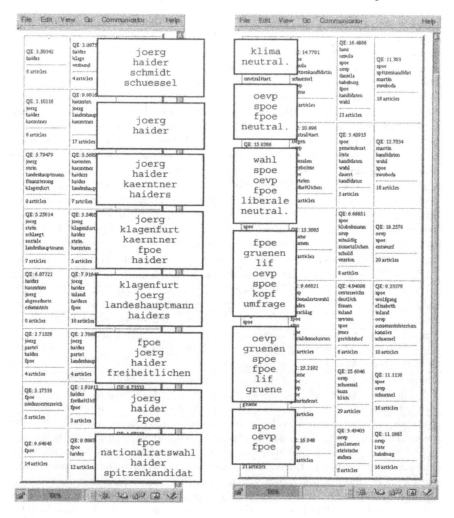

Fig. 4.6. Two neighboring second-layer maps on Internal Affairs

for detailed discussion. We furthermore present results using a larger collection of 359 pieces of music, with a total playing length of about 23 hours. The experiments, including audio samples, are available for interactive exploration at the *SOMeJB* project homepage at http://www.ifs.tuwien.ac.at/~andi/somejb.

4.5.1 Feature Extraction from Audio Signals

The feature extraction process for the Rhythm Patterns is composed of two stages. First, the specific loudness sensation in different frequency bands is computed, which is then transformed into a time-invariant representation based on the modulation frequency. Starting from a standard Pulse-Code-Modulated (PCM) signal, stereo channels are combined into a mono signal, which is further down-sampled to 11 kHz. Furthermore, pieces of music are cut into 6-second segments, removing the first and last two segments to eliminate lead-in and fade-out effects, and retaining only every second segment for further analysis. Using a Fast Fourier Transform (FFT), the raw audio data is further decomposed into frequency ranges using Hanning Windows with 256 samples (corresponding to 23 ms) with 50% overlap, resulting in 129 frequency values (at 43 Hz intervals) every 12 ms. These frequency bands are further grouped into so-called critical bands, also referred to by their unit bark [22], by summing up the values of the power spectrum between the limits of the respective critical band, resulting in 20 critical-band values. A spreading function is applied to account for masking effects, i.e. the masking of simultaneous or subsequent sounds by a given sound. The spread critical-band values are transformed into the logarithmic decibel scale, describing the sound pressure level in relation to the hearing threshold. Since the relationship between the dB-based sound pressure levels and our hearing sensation depends on the frequency of a tone, we calculate loudness levels, referred to as phon, using the equal-loudness contour matrix. From the loudness levels we calculate the specific loudness sensation per critical band, referred to as sone.

To obtain a time-invariant representation, recurring patterns in the individual critical bands resembling rhythm are extracted in the second stage of the feature extraction process. This is achieved by applying another discrete Fourier transform, resulting in amplitude modulations of the loudness in individual critical bands. These amplitude modulations have different effects on our hearing sensation depending on their frequency, the most significant of which, referred to as fluctuation strength, is most intense at 4 Hz and decreasing towards 15 Hz (followed by the sensation of roughness, and then by the sensation of three separately audible tones at around 150 Hz). We thus weight the modulation amplitudes according to the fluctuation strength sensation, resulting in a time-invariant, comparable representation of the rhythmic patterns in the individual critical bands. To emphasize the differences between strongly reoccurring beats at fixed intervals a final gradient filter is applied, paired with subsequent Gaussian smoothing to diminish unnoticeable variations. The resulting 1.200 dimensional feature vectors (20 critical bands times 60 amplitude modulation values) capture beat information up to 10Hz (600 bpm), going significantly beyond what is conventionally considered beat structure in music. They may optionally be reduced down to about 80 dimensions using PCA. These Rhythm Patterns are further used for data signal comparison.

4.5.2 A GHSOM of 77 Pieces of Music

Figure 4.7 depicts a *GHSOM* trained on the music data. On the first level the training process has resulted in the creation of a 3 × 3 map, organizing the collection into 9 major styles of music. The bottom right represents mainly classical music, while the upper left mainly represents a mixture of Hip Hop, Electro, and House by *Bomfunk MCs (bfmc)*. The upper-right, center-right, and upper-center represent mainly disco music such as *Rock DJ* by *Robbie Williams (rockdj)*, *Blue* by *Eiffel 65 (eiffel65-blue)*, or *Frozen* by *Madonna (frozen)*. Please note that the organization does not follow clean "conceptual" genre styles, splitting by definition, e.g. *HipHop* and *House*, but rather reflects the overall sound similarity.

Seven of these 9 first-level categories are further refined on the second level. For example, the bottom right unit representing classical music is divided into 4

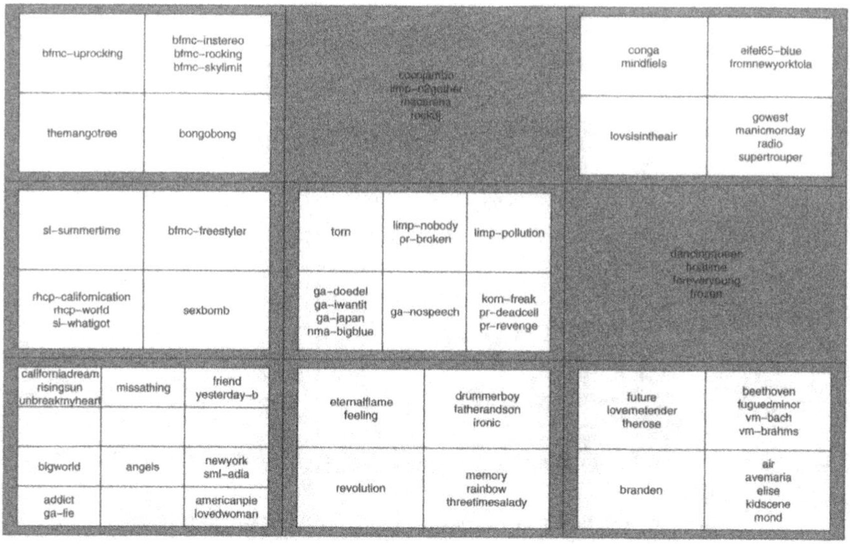

bfmc–uprocking	bfmc–instereo bfmc–rocking bfmc–skylimit		coonjamba bfmc–reggalber mascarena roskjij		conga mindfiels	eifel65–blue fromnewyorktola
themangotree	bongobong				lovsisintheair	gowest manicmonday radio supertrouper
si–summertime	bfmc–freestyler	torn	limp–nobody pr–broken	limp–pollution	dancingqueen hottime forevelryoung frozen	
rhcp–californication rhcp–world si–whatigot	sexbomb	ga–doedel ga–iwantit ga–japan nma–bigblue	ga–nospeech	korn–freak pr–deadcell pr–revenge		
californiadream risingsun unbreakmyheart	missathing	friend yesterday–b	eternalflame feeling	drummerboy fatherandson ironic	future lovemetender therose	beethoven fuguedminor vm–bach vm–brahms
bigworld	angels	newyork sml–adia	revolution	memory rainbow threetimesalady	branden	air avemaria elise kidscene mond
addict ga–lie		americanpie lovedwoman				

Fig. 4.7. GHSOM of music collection

further sub-categories. Of these 4 categories the lower-right represents slow and peaceful music, mainly piano pieces such as *Für Elise (elise)* and *Mondscheinsonate (mond)* by *Beethoven*, or *Fremde Länder und Menschen* by *Schumann (kidscene)*. The upper-right represents, for example, pieces by *Vanessa Mae (vm)*, which, in this case, are more dynamic interpretations of classical pieces played on the violin. In the upper-left orchestral music is located such as the as the end credits of the film *Back to the Future III (future)* and the slow love song *The Rose* by *Bette Midler (therose)*, exhibiting a more intensive sound sensation, whereas the lower right corner unit represents the *Brandenburg Concerts* by *Bach (branden)*.

Generally speaking, we find the softer, more peaceful songs on this second level map located in the lower half of the map, whereas the more dynamic, intensive songs are located in the upper half. This corresponds to the general organization of the map in the first layer, where the unit representing Classic music is located in the lower right corner, having more aggressive music as its upper and left neighbors. This allows us, even on lower-level maps, to move across map boundaries to find similar music on the neighboring map following the same general trends of organization, thus alleviating the common problem of cluster separation in hierarchical organizations.

Some interesting insights into the music collection which the *GHSOM* reveals are, for example, that the song *Freestyler* by *Bomfunk MCs* (center-left) is quite different then the other songs by the same group. *Freestyler* was the group's biggest hit so far and, unlike their other songs, has been appreciated by a broader audience. Generally, the pieces of one group have similar sound characteristics and thus are located within the same categories. This applies, for example, to the songs of *Guano Apes (ga)* and *Papa Roach (pr)*, which are located in the center of the 9 first-level categories together with other aggressive rock songs. However, another exception is *Living in a Lie* by *Guano Apes (ga-lie)*, located in the lower-left. Listening to this

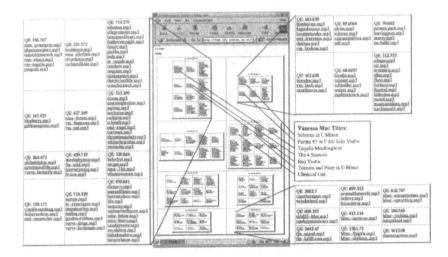

Fig. 4.8. GHSOM of the larger music collection (359 pieces)

piece reveals that it is much slower than the other pieces of the group, and that this song matches very well to, for example, *Addict* by *K's Choice*.

4.5.3 A GHSOM of 359 Pieces of Music

In this section we present results from using the *SOMeJB* system to structure a larger collection of 359 pieces of music. Due to space constraints we cannot display or discuss the full hierarchy in detail. We will thus pick a few examples to show the characteristics of the resulting hierarchy, inviting the reader to explore and evaluate the complete hierarchy via the project homepage.

The resulting *GHSOM* has grown to a size of 2×4 units on the top layer map. All 8 top-layer units were expanded onto a second layer in the hierarchy, from which 25 units out of 64 units total on this layer were further expanded into a third layer. None of the branches required expansion into a fourth layer at the required level-of-detail setting. An integrated view of the two top-layers of the map is depicted in Fig. 4.8. We will now take a closer look at some branches of this map, and compare them to the respective areas in the *GHSOM* of the smaller data collection depicted in Fig. 4.7

Generally, we find pieces of soft classical music in the upper right corner, with the music becoming gradually more dynamic and aggressive as we move towards the bottom left corner of the map. Due to the characteristics of the training process of the *GHSOM* we can find the same general tendency at the respective lower-layer maps. The overall orientation of the map hierarchy is rotated when compared to the smaller *GHSOM*, where the classical titles were located in the bottom right corner, with the more aggressive titles placed on the upper left area of the map. This rotation is due to the unsupervised nature of the *GHSOM* training process. It can, however, be avoided by using specific initialization techniques if a specific orientation of the map were required.

The unit in the upper right corner of the top-layer map, representing the softest classical pieces of music, is expanded onto a 3×2 map in the second layer (expanded to the upper right in Fig. 4.8). Here we again find the softest, most peaceful pieces in the upper right corner, namely part of the soundtrack of the movie *Jurassic Park*, next to *Leaving Port* by *James Horner*, *The Merry Peasants* by *Schumann*, and *Canon* by *Pachelbel*. Below this unit we find further soft titles, yet somewhat more dynamic. We basically find all titles that were mapped together in the bottom right corner unit of the *GHSOM* of the smaller collection depicted in Fig. 4.7 on this unit, i.e. *Air*, *Ave Maria*, *Für Elise*, *Fremde Länder und Menschen (kidscene)* and the *Mondscheinsonate*. Furthermore, a few additional titles of the larger collection have been mapped onto this unit, the most famous of which probably are *Die kleine Nachtmusik* by *Mozart*, the *Funeral March* by *Chopin* or the *Adagio* from the Clarinet Concert by *Mozart*.

Let us now take a look at the titles that were mapped onto the neighboring units in the previously presented smaller data collection. The *Brandenburgische Konzerte*, located on the neighboring unit to the right in the first example, can be found in the lower left corner of this map, together with, for example, *Also sprach Zarathustra* by *Richard Strauß*. Mapped onto the upper neighboring unit in the smaller *GHSOM* we had titles like the *First Movement of the 5th Symphony* by *Beethoven*, or the *Toccata and Fugue in D Minor* by *Bach*. We find these two titles in the upper left corner of the 2-layer map of this *GHSOM*, together with two of the three titles mapped onto the diagonally neighboring unit in the first *GHSOM*, i.e. *Love me Tender* by *Elvis Presley*, and *The Rose* by *Bette Midler*, which are again soft, mellow, but a bit more dynamic. The third title mapped onto this unit in the smaller *GHSOM*, i.e. the soundtrack of the movie *Back to the Future III* is not mapped into this branch of this *GHSOM* anymore. When we listen to this title we find it to have mainly strong orchestral parts, which have a different, more intense sound than the soft pieces mapped onto this branch, which is more specific with respect to very soft classical titles as more of them are available in the larger data collection. Instead, we can find this title on the upper right corner in the neighboring branch to the left, originating from the upper left corner unit of the top-layer map. There it is mapped together with *The Beauty and the Beast* and other orchestral pieces, such as *Allegro Molto* by *Brahms*. We thus find this branch of the *GHSOM* to be more or less identical to the overall organization of the smaller *GHSOM* in so far as the titles present in both collections are mapped in similar relative positions to each other.

Due to the topology preservation provided by the *GHSOM* we can move from the soft classical cluster map to the left to find somewhat more dynamic classical pieces of music on the neighboring map (expanded to the left in Fig. 4.8). Thus, a typical disadvantage of hierarchical clustering and structuring of datasets, namely the fact that a cluster that might be considered conceptually very similar is subdivided into two distinct branches, is alleviated in the *GHSOM* concept, because these data points are typically located in the close neighborhood. We thus find, on the right border of the neighboring map, the more peaceful titles of this branch, yet more dynamic than the classical pieces on the neighboring right branch discussed above.

Rather than continuing to discuss the individual units we shall now take a look at the titles of a specific artist and its distribution in this hierarchy. In total, there are 7 titles by *Vanessa Mae* in this collection, all violin interpretations, yet of a distinctly different style. Her most "conventional" classical interpretations, such as

Brahms' *Scherzo in C Minor (vm-brahms)* or Bach's *Partita #3 in E for Solo Violin (vm-bach)* are located in the classical cluster in the upper right corner branch on two neighboring units on the left side of the second-layer map. These are definitely the most "classical" of her interpretations in the given collection, yet exhibiting strong dynamics. Further 3 pieces of Vanessa Mae (*The 4 Seasons* by Vivaldi, *Red Violin* in its symphonic version, and *Tequila Mockingbird*) are found in the neighboring branch to the left, the former two mapped together with *Western Dream* by *New Model Army*. All of these titles are very dynamic violin pieces with strong orchestral parts and percussion.

When we look for the remaining 2 titles by *Vanessa Mae*, we find them on the unit expanded below the top right corner unit, thus also neighboring the classical cluster. On the top-left corner unit of this sub-map we find *Classical Gas*, which starts in a classical, symphonic version, and gradually has more intensive percussion being added, exhibiting a quite intense beat. Also on this map, on the one-but-next unit to the right, we find another interpretation of the *Toccata and Fuge in D Minor* by Bach, this time in the classical interpretation of *Vanessa Mae*, also with a very intense beat. The more "conventional" organ interpretation of this title, as we have seen, is located in the classic cluster discussed before. Although both are the same titles, the interpretations are very different in their sound characteristics, with *Vanessa Mae*'s interpretation definitely being more pop-like than the typical classical interpretation of this title. Thus, two identical titles, yet played in different styles, end up in their respective stylistic branches of the *SOMeJB* system. We furthermore find that the system does not organize all titles by a single artist into the same branch, but actually assigns them according to their sound characteristics, which makes it particularly suitable for localizing pieces according to one's liking, independent of the typical assignment of an artist to any category, or to the conventional assignment of titles to specific genres.

Further units depicted in more detail in Fig. 4.8 are the bottom right unit representing the more aggressive, dynamic titles. We leave it to the reader to analyze this sub-map and compare the titles with the ones mapped onto the upper left corner map in Fig. 4.7.

4.6 Conclusions

In this paper we have presented an approach to supporting serendipity in information repositories. Data is transformed into an information space, revealing the context of individual data items, their relationships, as well as the extent and limits of the information available. The resulting organization offers itself for interactive exploration and navigation, allowing the creation of a mental model of the information available.

At its core, the approach is based on the *Growing Hierarchical Self-Organizing Map*, a structure adapting neural network model. The major features of this neural network are its hierarchical architecture, where the depth of the hierarchy is determined during the unsupervised training process. Each layer in the hierarchy consists of a number of independent *SOMs* which determine their size and arrangement of units during training. Thus, this model is especially well suited for applications which require hierarchical clustering of the input data. We have shown that significant improvement in data representation can be achieved by directing particular emphasis at the orientation of the various *SOMs* constituting the different branches of the hierarchy. Maps of neighboring branches now show the same orientation as the map they are derived from. Therefore, navigation between different branches of the hierarchy is facilitated.

We have demonstrated the capabilities of this model in two application domains, namely (1) for the organization of textual document collections within the *SOMLib* digital library system, as well as (2) for the creation of a musical landscape as part of the *SOMeJB* system. Both systems allow the interactive exploration of the corresponding data repositories, be it navigating topical clusters in the case of text collections, or groups of music of similar styles for music repositories.

References

1. Chen, H, Schuffels, C, and Orwig, R (1996) Internet categorization and search: A self-organizing approach. *Journal of Visual Communication and Image Representation*, 7(1):88–102.
2. Dittenbach, M, Merkl, D, and Rauber, A (2000) The growing hierarchical self-organizing map. In *Proc Int'l Joint Conf Neural Networks*, Como, Italy.
3. Dittenbach, M, Rauber, A, and Merkl, D (2001) Recent advances with the growing hierarchical self-organizing map. In *Advances in Self-Organizing Maps*, pp. 140–145, Lincoln, GB.
4. Dittenbach, M, Rauber, A, and Merkl, D (2002) Uncovering the hierarchical structure in data using the growing hierarchical self-organizing map. *Neurocomputing*, 48(1–4):199–216, November.
5. Fritzke, B (1995) Growing Grid – A self-organizing network with constant neighborhood range and adaption strength. *Neural Processing Letters*, 2(5).
6. Kaski, S, Kangas, J, and Kohonen, T (1998) Bibliography of self-organizing map (SOM) papers: 1981–1997. *Neural Computing Surveys*, 1(3&4):1–176.
7. Kohonen, T (1982) Self-organized formation of topologically correct feature maps. *Biological Cybernetics*, 43.
8. Kohonen, T (1995) *Self-organizing maps*. Springer, Berlin.
9. Kohonen, T, Kaski, S, Lagus, K, Salojärvi, J, Honkela, J, Paatero V, and Saarela, A (2000) Self-organization of a massive document collection. *IEEE Transactions on Neural Networks*, 11(3):574–585, May.
10. Koikkalainen, P (1995) Fast deterministic self-organizing maps. In *Proc Int'l Conf Neural Networks*, Paris, France.
11. Koikkalainen, P and Oja, E (1990) Self-organizing hierarchical feature maps. In *Proc Int'l Joint Conf Neural Networks*, San Diego, CA.
12. Merkl, D and Rauber, A (1999) Automatic labeling of self-organizing maps for information retrieval. In *Proc Int'l Conf on Neural Information Processing*, Perth, Australia.
13. Merkl, D and Rauber, A (2000) Document classification with unsupervised neural networks. In Crestani, F and Pasi, G, editors, *Soft Computing in Information Retrieval*, pp. 102–121. Physica Verlag.
14. Miikkulainen, R (1990) Script recognition with hierarchical feature maps. *Connection Science*, 2.
15. Oja, M, Kaski, S, and Kohonen, T (2003) Bibliography of self-organizing map (SOM) papers: 1998-2001 addendum. *Neural Computing Surveys*, 3:1–156.

16. Pampalk, E, Rauber, A, and Merkl, D (2002) Content-based organization and visualization of music archives. In *Proceedings of ACM Multimedia 2002*, pp. 570–579, Juan-les-Pins, France, December 1–6. ACM.

17. Rauber, A and Frühwirth, M (2001) Automatically analyzing and organizing music archives. In *Proceedings of the 5th European Conference on Research and Advanced Technology for Digital Libraries (ECDL 2001)*, Springer Lecture Notes in Computer Science, Darmstadt, Germany, Sept. 4–8. Springer.

18. Rauber, A and Merkl, D (1999) The SOMLib Digital Library System. In Abiteboul, S and Vercoustre, AM, editors, *Proceedings of the 3rd European Conference on Research and Advanced Technology for Digital Libraries (ECDL99)*, number LNCS 1696 in Lecture Notes in Computer Science, pp. 323–342, Paris, France, September 22–24. Springer.

19. Rauber, A and Merkl, D (2003) Text mining in the SOMLib digital library system: The representation of topics and genres. *Applied Intelligence*, 18(3):271–293, May-June.

20. Rauber, A, Pampalk, E, and Merkl, D (2003) The SOM-enhanced JukeBox: Organization and visualization of music collections based on perceptual models. *Journal of New Music Research*. to appear.

21. Salton, G (1989) *Automatic Text Processing: The Transformation, Analysis, and Retrieval of Information by Computer*. Addison-Wesley, Reading, MA.

22. Zwicker, E and Fastl, H (1999) *Psychoacoustics, Facts and Models*. Springer.

5

D-GridMST: Clustering Large Distributed Spatial Databases

Ji Zhang and Han Liu

Department of Computer Science, University of Toronto, Toronto, Ontario, M5S 3G4, Canada
{jzhang, hanliu}@cs.toronto.edu

Abstract. In this paper, we will propose a novel distributable clustering algorithm, called Distributed-GridMST (D-GridMST for short), which deals with large distributed spatial databases. D-GridMST employs the notion of a grid to partition the data space involved and uses density criteria to extract representative points from spatial databases, on which a global MST of representatives is constructed. Such an MST is partitioned according to users' clustering specification and used to label data points in the respective distributed spatial database thereafter. D-GridMST is characterized by fast speed, low space requirement and small network transferring overhead. Experimental results show that D-GridMST is effective since it is able to produce exactly the same clustering result as that produced in the centralized paradigm, making D-GridMST a promising tool for clustering large distributed spatial databases.

Key words: Databases, Data Mining, Clustering and Distributed Spatial Databases

5.1 Introduction

With the rapid development of techniques in data acquisition and storage, spatial databases store an increasing amount of space-related data such as satellite maps, remotely sensed images and medical images. These data, if analyzed, can reveal useful patterns and knowledge to human users. Clustering is a process whereby a set of objects are divided into several clusters in which each of the members is in some way similar and is different from the members of other clusters [11]. Spatial data clustering, aiming to identify clusters, or densely populated regions in a large spatial dataset, serves as an important task of spatial data mining. Though a large number of spatial clustering algorithms have been proposed in literature so far, most of them assume the data to be clustered are locally resident in centralized scenario, making them unable to work with inherently distributed spatial data sources.

Ji Zhang and Han Liu: *D-GridMST: Clustering Large Distributed Spatial Databases*, Studies in Computational Intelligence (SCI) **4**, 61–72 (2005)
www.springerlink.com

Reference [2] presents a distributed data clustering algorithm that is based on a classical clustering algorithm PAM and a spanning tree clustering algorithm, called Clusterize. Reference [3] proposes an approach to deal with clustering data emanating from different sites. It operates in three major steps: (i) find the local clusters of data in each site; (ii) find (high) clusters from the union of the distribution data sets at the central site; (iii) finally compute the associations between the two sets of clusters. This method is limited in that the local clustering results generally differ significantly from the global clustering result. Thus the union of local clusters is not only expensive but also inaccurate. In [6], three classical clustering methods, namely K-means, K-Harmonic Means (KHM) and Maximum Expectation (EM) are modified to parallelise the clustering of distributed data sources. However, the distributed version of these methods cannot produce completely identical clustering results as those of their centralized version. More recently, a parallel implementation of K-means based on the message-passing model is presented [5]. To deal with heterogeneity of data across distributed sites, [9] presents a Collective Hierarchical Clustering (CHC) algorithm for analyzing distributed and heterogeneous data. This method first generates local cluster models and then combines them to generate the global cluster model of the data.

In this paper, we will propose a distributable clustering algorithm, called *Distributed-GridMST (D-GridMST)*, that deals with large distributed spatial databases. D-GridMST employs the notions of grid to partition the data space involved and uses density criteria to extract representative points from spatial databases, on which a global MST of representatives is constructed. Such an MST is partitioned according to users' clustering specification and used to label data points in the respective distributed spatial database thereafter. Since only the compact information of the distributed spatial databases is transferred via network, D-GridMST is characterized by small network transferring overhead. Experimental results show that the distributed (D-GridMST) version of our clustering technique is efficient and effective.

The remainder of this paper is organized as follows. In Sect. 5.2, we will present GridMST, our clustering technique for clustering centralized spatial databases. Section 5.3 gives the details of D-GridMST. Experimental results are reported in Sect. 5.4. The final section concludes this paper.

5.2 GridMST

GridMST is a new clustering technique that is fast, scalable, robust to noise, and effective in accomplishing multi-resolution clustering. Figure 5.1 gives an overview of GridMST. It consists of three major parts. The first part deals with scaling the algorithm for very large spatial databases. The second part deals with extracting the necessary information to build a summary structure for multi-resolution clustering. The final part deals with how the multi-resolution clustering makes use of the summary structures to perform analysis.

5.2.1 Sampling Large Spatial Databases

Similar to BIRCH, CURE, and C2P, GridMST handles the problem of very large spatial databases by sampling the databases. Reference [7] derives a theorem to

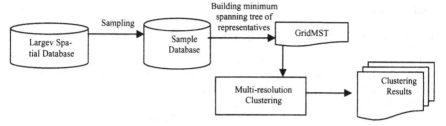

Fig. 5.1. Overview of GridMST

determine the minimum sample size required to ensure that a fraction of the cluster is always included in the sample with probability δ. That is, for a cluster u, if the sample size s satisfies

$$s \geq fN + \frac{N}{|u|} \log\left(\frac{1}{\delta}\right) + \frac{N}{|u|} \sqrt{\left(\log\left(\frac{1}{\delta}\right)\right)^2 + 2f|u| \log\left(\frac{1}{\delta}\right)}$$

then the probability that the sample contains fewer than $f|u|$ points belonging to cluster u is less than δ, where N is the size of the dataset, $|u|$ is the size of the cluster $u, 0 \leq f \leq 1, 0 \leq \delta \leq 1$.

GridMST uses this theorem to determine the sample size and performs uniform sampling on the large spatial database to obtain a sample database. We observe that this sample database could still be too large to fit entirely into the main memory. In this case, GridMST will divide the sample database into several smaller partitions, each of which can be loaded into the main memory. The partitions are read in one at a time and processed for their density information. When all the partitions have been scanned, the grid cells occupied by the whole sampling dataset are obtained (the grid structure will be discussed later in this paper). The size of occupied grid cell is small enough to be stored in the main memory. Representative points can now be generated based on the density information of these grid cells. Note that these representative points are for the entire sample database. This makes GridMST flexible and yet effective in handling samples of all sizes.

5.2.2 Constructing the R-MST

In GridMST, a number of representative points of the database are picked using the density criterion. A minimum spanning tree of these representative points, denoted as R-MST, is built. GridMST constructs R-MST in a number of steps. First, a grid data structure is constructed whereby each point in the dataset is assigned to one and only one cell in the grid. The density of each grid cell is then computed. If the density of a grid cell exceeds some user-specified threshold, then the cell is considered to be dense and the centroid of the points in the dense cell is selected as the representative point of this cell. Once the representative points have been selected, a graph-theoretic based algorithm is used to build the R-MST.

Definitions

Definition 1: Relative Density (RD) of a Grid Cell

Let g be some cell in a grid structure G. Let n be the number of points in g and avg be the average number of points in a cell in G. Then, the Relative Density of g, denoted as $RD(g)$, is defined as the ratio n/avg.

A grid cell is a neighbor of some grid cell g if it is directly connected to g. Hence, a center grid cell will have 8 neighboring grid cells, an edge grid cell will have 5 neighboring grid cells, and a corner grid cell will have only 3 neighboring grid cells.

Definition 2: Neighborhood Density (ND) of a Grid Cell

Let g be some cell in a grid structure G and *Neighbor* be the set of neighboring grid cells of g. The Neighborhood Density of g, denoted as $ND(g)$, is defined as the average of the densities of g and its neighboring grid cells. $ND(g)$ is given by the following formula:

$$ND(g) = \frac{1}{t} * (RD(g) + \sum_{g_i \in Neighborg_{(g)}} RD(g_i))$$

where $(t - 1)$ is the number of neighboring grid cells of g. Specifically, for a 2-dimensional spatial database, we have

$$t = \begin{cases} 6 & \text{if } g \text{ is an edge grid cell} \\ 4 & \text{if } g \text{ is a corner grid cell} \\ 9 & \text{otherwide} \end{cases}$$

Definition 3: Dense vs. Non-dense Grid Cells

Let g be some cell in a grid structure G. g is a dense grid cell if $ND(g)$ is greater than or equal to some user specified density threshold, Td, otherwise g is a non-dense grid cell.

Suppose X and Y are the horizontal and vertical number of grid cells in grid structure and N is the size of the dataset. Then the average number of points in the cells of the grid is given by N/XY. Figure 5.2 shows the densities of the non-empty grid cells in the grid structure. Note that from Definition 3, a grid cell is considered dense if its neighborhood density, ND, exceeds some user-specified threshold. The reason for using ND rather than RD to determine the denseness of a grid cell is that ND measures not only its own density, but it also reflects the density of its neighboring area. This actually compensates for the effect of outliers.

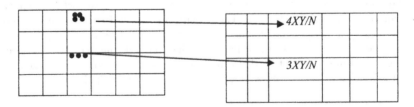

Fig. 5.2. Relative density of grid cells

Definition 4: R-MST

Suppose $A = \langle V, E \rangle$ is an undirected graph, where V is the set of vertices denoting the set of representative points and E is the set of edges connecting these representative points. R-MST is a connected acyclic sub-graph of A that has the smallest total cost (or length), which is measured as the sum of the costs of its edges.

Generating Representative Points

To generate the representative points, grid cell mapping of data points is perform, i.e. assign each data point into one and only one grid cell. Once a point has been assigned, the density of its corresponding cell is incremented by 1. Dense cells are identified based on some pre-specified density threshold, Td. A cell is a dense grid cell if its Neighborhood Density is greater than or equal to Td, otherwise it is a non-dense grid cell. The representative points are generated using the centroids of all the points in the dense cells and are added to the list of representative points. The steps of the algorithm are shown in Fig. 5.3.

Constructing the R-MST

The algorithm for constructing the R-MST is as follows:

1. Compute all the pair-wise Euclidean distances of representatives and sort them in ascending order. This forms a pool of potential R-MST edges;
2. Pick edges from the pool of potential edges obtained in (1), starting with the shortest edge. Only those edges that do not create a cycle with the edges that have already been chosen are picked. The process terminates when the number of such edges is equal to $(N_r - 1)$, where N_r is the number of representative points.

Procedure Repr_Generation (Dataset D)
Begin
1. For each point p in D Do
2. {
3.   Cell (p)=**Map**(p);
4. j = **Hash** (Cell(p));
5.   **Count** $[j]$++;
6. }
7. For each cell i in the hash table do
8. {
9.   ND=**Neighborhood_Den**(i);
10.   If (ND $>=T_d$) Then {
11.   Cell i is a dense cell;
12   Repr_List=Repr_List$\cup$**CentroidOfPoints**(i);
13.   } }
End

Fig. 5.3. Algorithm to extract representative points

5.2.3 Using R-MST for Spatial Clustering

After the R-MST has been constructed, multi-resolution clustering can be easily achieved. Suppose a user wants to find k clusters. A graph search through the R-MST is initiated, starting from the largest cost edge, to the lowest cost edge. As an edge is traversed, it is marked as deleted from the R-MST. The number of partitions resulting from the deletion is computed. The process stops when the number of partitions reaches k. Any change in the value of k simply implies re-initiating the search-and-marked procedure on the R-MST. Once the R-MST has been divided into k partitions, we can now propagate this information to the original dataset so that each point in the dataset is assigned to one and only one partition/cluster. A naive approach is to go through the dataset once, and compute the distance between each point to all the representative points. The data point is then assigned to the cluster whose representative point it is closest to. However, we observed that if a data point falls into a dense cell, say dc, then the nearest representative point is the representative point of dc. Thus, we can immediately assign the data point the same cluster label as the representative point of cell dc. For those data points that fall into non-dense cells, we use an indexing structure for high dimensional data, the X-tree [1], to speed up the search for the closest representative point. Once the closest representative point is found, the corresponding data point will be assigned the same cluster label as its closest representative point.

5.3 D-GridMST (Distributed GridMST)

After discussing the technique of GridMST that is mainly applicable in clustering spatial database in the centralized paradigm, we, in this section, will present D-GridMST, the distributed version of GridMST that works with distributed spatial databases. The adaptation from GridMST to D-GridMST mainly involves generation of global data model by combining local data models in centralized site. In order to produce the clustering result of these distributed databases that is comparable to result of a centralized database in D-GridMST, globalization of local data model is entailed to obtain global data model that captures the cluster features for the whole dataset. A coordinator in the system, called the *central site*, is in charge of globalizing local data models and broadcasting the global data model. The data clustering analysis is still performed locally in each distributed site. Specifically, the globalization of local data model in D-GridMST involves:

(1) Globalize Range of Every Dimension of Data in Each Distributed Site

Global range of every dimension of the data is required to construct a structure of global grid. Here, we assume the all the spatial data reside in the distributed databases are homogenous in nature, which ensures that such combination can be performed meaningfully. The objective of obtaining a global range of every dimension is to ensure that the grid constructed is able to encapsulate all the data points stored in the distributed sites. To this end, all distributed sites are required to provide the central site with the information regarding the maximum and minimum values, i.e. the range of every dimension of local data points. Upon the receipt of such information from all distributed sites, the central site will commence to produce the global range. Specifically, Let S denote the number of distributed sites, d be number of dimensions of the dataset (typically we have $d = 2$ in most spatial databases), $L_{\max}(i, j)$ and $L_{\min}(i, j)$ be the local maximum and minimum values of jth dimension of the ith distributed site. The global maximum and minimum values of dimension j, denoted as $G_{\max}(j)$ and $G_{\min}(j)$, are produced as follows:

$$G_{\min}(j) = \min(L_{\min}(1, j), L_{\min}(2, j), \ldots L_{\min}(S, j))$$
$$G_{\max}(j) = \max(L_{\max}(1, j), L_{\max}(2, j), \ldots L_{\max}(S, j))$$

The range of ith dimension, denoted as $R(i)$ can be computed as

$$R(i) = G_{\max}(i) - G_{\min}(i) \ (1 \le i \le d)$$

(2) Globalize Locally Occupied Cells in Each Distributed Site

Here, the occupied cells refer to the cells that occupied by the data points in the database. In other words, the occupied cells are those cells whose density is at least 1. Locally occupied cells are those cells occupied by the local data points in the distributed site, and globally occupied cells are those cells occupied by all the data points. The globally occupied cells serve as the potential pool for the selection of dense cells: the dense cells are only the occupied cells whose neighborhood density exceeds some threshold. The globally occupied cells are simply the union of local occupied cells. Suppose there are S distributed sites and $LOC(i)$ denotes the local occupied cells of the ith distributed site. The global occupied cells, denoted by GOC, can be generated as follows:

$$GOC = LOC(1) \cup LOC(2) \ldots \cup LOC(S)$$

5.3.1 D-GridMST Algorithm

In this section, we will give the algorithm of D-GridMST that performs clustering of distributed spatial databases in Table 5.1. Table 5.2 gives the annotation of the notations used in the *Transfer/location* field in Table 5.1.

Table 5.1. Algorithm of D-GridMST

Step	Transfer/Location	Operation
1	DS → CS	Transfer local range of every dimension of data
2	CS	Globalize local range to global range
3	CS → DS	Transfer global range and create global grid G
4	DS	Assign local points into G and compute the density
5	DS → CS	Transfer locally occupied cells and their densities
6	CS	Globalize locally occupied cells of the grid
7	CS	Generate representative points and construct MST
8	CS	Perform multi-resolution clustering using MST
9	CS → DS	Transfer clustering result of representative points
10	DS	Label local data points

Table 5.2. Annotations of the **Transfer/location** field in Table 5.1

Value	Meaning
CS	Clustering operations in centralized site
DS	Clustering operations in all the distributed sites
CS → DS	Data are transferred from centralized site to all the distributed sites
DS → CS	Data are transferred from all distributed sites to the centralized site

5.3.2 The Complexity of D-GridMST

The complexity analysis in the subsection includes the analysis of its computational complexity, space complexity and transfer overhead. Let N be the total size of spatial databases in all distributed sites, d be the number of dimension of each spatial database, S be the total number of the distributed sites, N_c be the number of cells in the grid and N_r be the number of the global representative points.

(a) Computational Complexity: The computational workload for D-GridMST involves Step 2, 4, 6, 7, 8 and 10 of the algorithm. Globalizing the local range to the global one in Step 2 requires a complexity of $O(d*S)$. Assigning local points into the grid G and compute the density in Step 4 require a complexity of $O(N)$, where $N = N_1 + N_2 + \cdots + N_d$. In Step 6, globalizing locally occupied cells of the grid involves the union of occupied cells in the grid of all distributed sites, whose complexity is at most $O(S^* N_c)$. Generation of representative points and construction of MST require $O(N_c + N_r^2)$. Using MST to cluster the representatives only requires $O(N_r)$. Finally, labeling all points has a complexity of $O(N)$. In sum, the computational complexity is $O(d*S + N + S*N_c + N_c + N_r^2 + N_r + N)$. Given $d \ll N, N_c \ll N$ and $N_r \ll N$, thus the computational complexity of D-GridMST is approximately $O(N)$.

(b) Space Complexity: The storage requirements for the centralized and distributed sites are different in D-GridMST. The centralized site has to store the range of each dimension of all distributed sites ($O(d*S)$), the occupied cells of all distributed sites ($O(N_c * S)$) and the MST generated ($O(N_r)$). Because $d \ll N_c$ and $N_r \ll N_c$ for most spatial datasets, thus the space complexity of D-GridMST is $O(N_c * S)$. As for each distributed site, it will have to store the global range of data across all the sites and clustering result of representative points in addition to the original data in this site, therefore the storage requirement for distributed site S_i is $O(N_i + \mathrm{Max}(d*S, N_r))$ $(1 \le i \le S)$.

(c) Transferring Overhead: The data transferring between the centralized and distributed sites occurs in Step 1, 3, 5 and 9, respectively. The overhead of transferring the local range of each dimension of distributed data and the global data range between the centralized and distributed sites is $O(d^*S)$. Transferring local occupied cells and their densities from distributed sites to centralized site has an overhead of $O(S^* N_c)$ at most. Finally, the overhead of transferring the clustering result of representative points from centralized site to distributed sites is $O(N_r)$. Therefore, the total transferring overhead is $O(d*S + S*N_c + N_r)$.

Remarks. From the above analysis, we can see that D-GridMST is promising in the sense that: (i) It is very efficient because that the computational time is approximately linear with respect to the total size of the distributed data, therefore it is well scalable to large spatial databases; (ii) It is space economic because D-GridMST only imposes small space requirements on both centralized and distributed sites; and (iii) It has small transferring overhead since the overhead of $O(d * S + S * N_c + N_r)$ is much smaller than the space required by the centralized algorithm, which having a transferring overhead of $O(N * d)$ at least.

5.4 Experimental Results

In this part, we will study the effect of the size of spatial databases in each distribute site and the number of distributed sites on the efficiency of D-GridMST. Also, we will propose a metric, termed *CluSim*, and use it for measuring the similarity between the clustering results produced by using the centralized and distributed versions of the same clustering algorithm, respectively.

5.4.1 Efficiency of Clustering Distributed Databases using D-GridMST

We will first study the efficiency of D-GridMST on distributed databases. We will investigate the effect of the size of spatial databases in each distributed site and the number of distributed sites on the efficiency of D-GridMST. For the sake of simplicity, we set the spatial databases in each distributed site in this experiment to be an equal size. The size of the database in each distributed site ranges from 100,000 to 500,000 and the number of distributed sites varies from 10 to 30 in the experiments. The results are shown in Figs. 5.4 and 5.5. The two basic findings of the experiments are: (i) The execution time of clustering is approximately linear with respect

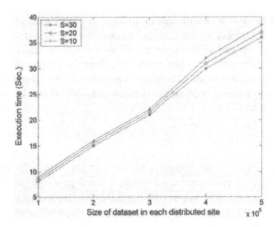

Fig. 5.4. Execution time of D-GridMST on varying dataset sizes in each distributed site

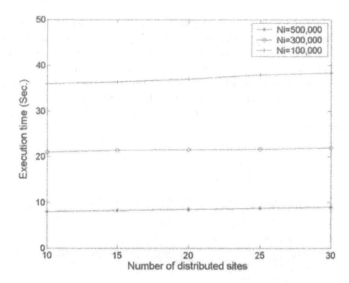

Fig. 5.5. Execution time of D-GridMST on varying number of the distributed sites

to the size of databases residing in the distributed sites. This is because the assignment of each point in the database into the grid and the labeling of all the points in the database will dominate the execution time of D-GridMST, thus the complexity of D-GridMST is linear with respect to the total size of databases it works on; (ii) The execution time is nearly not affected by the factor of S. This is because that the local clustering in each distributed site can be inherently paralleled.

5.4.2 Effectiveness of Clustering Distributed Databases using D-GridMST

As far as the effectiveness of a distributed clustering algorithm is concerned, we are mainly interested in whether the clustering result of the distributed algorithm is consistent with the result of the centralized version of the algorithm. To measure this, we devise a metric that, to some extent, reflects the closeness of the clustering results of the centralized and distributed versions of the algorithm. This metric, termed *Clustering Similarity* (*CluSim* for short), is defined as follows:

$$CluSim = \frac{1}{N} \sum_{i=1}^{N} (label_{\text{cen}}(p_i) = label_{\text{dis}}(p_i)) * 100\%$$

where $label_{\text{cen}}(p_i)$ and $label_{\text{dis}}(p_i)$ denote the cluster label of the point p_i using the centralized and distributed versions of clustering algorithm, respectively. $label_{\text{cen}}(p_i) = label_{\text{dis}}(p_i)$ returns 1 when point p_i is assigned the same cluster label under the two algorithms and returns 0 otherwise. Given that the cluster labels of two algorithms might be inconsistent (the same cluster may have different cluster labels in the two clustering results). Therefore, we have to first make the cluster labeling of the two algorithms consistent before *CluSim* can be computed. This can be achieved by searching the best-matched pairs of clusters in the two results through computing the distance between the centroids of the two clusters. Two clusters are regarded as a best matched pair if they have the minimum inter-cluster distance among all possible pair of clusters in the result. The points in the two clusters that are a best-matched pair have the same clustering label. Using the above formula, we compute *CluSim* of D-GridMST and two exiting distributed clustering algorithms, distributed K-means and distributed EM. The *CluSim* results of the three methods are D-GridMST (100%), K-means (93%) and EM (90%). Clearly, D-GridMST outperforms the other two algorithms and is able to produce exactly the same clustering result as that produced by the centralized version of the algorithm.

5.5 Conclusion

In this paper, we proposed a distributable clustering algorithm, called D-GridMST, which deals with large distributed spatial databases. D-GridMST employs the notions of grid to partition the data space involved and uses density criteria to extract representative points from spatial databases, on which a global MST of representatives is constructed. Such a MST is partitioned according to users' clustering specification and used to label data points in the respective distributed spatial database thereafter. D-GridMST is fast and requires low space and small network transferring overhead. Experimental results show that D-GridMST is effective since it is able to produce exactly the same clustering result as that produced in the centralized paradigm.

References

1. S. Berchtold, D. A. Keim and H. Kriegel. The X-tree: An Index Structure for High-Dimensional Data. *Proc. 22nd International Conference on Very Large Data Base (VLDB'96)*, Mumbai, India, 1996.
2. D. K. Bhattacharyya and A. Das. A New Distributed Algorithm for Large Data Clustering. *IDEAL'2000*, pp. 29–34, 2000.
3. A. Bouchachia. Distributed Data Clustering. *CAiSE*, 2003.
4. M. Charikar, C. Chekuri, T. Feder, and R. Motwani. Incremental Clustering and Dynamic Information Retrieval. *ACM Symposium on Theory of Computing*, 1997.
5. I. S. Dhillon and D. S. Modha. A Data-clustering Algorithm on Distributed Memory Multiprocessors. *Large-Scale Parallel Data Mining*, pp. 245–260, 2002.
6. G. Forman and B. Zhang. Distributed Data Clustering Can be Efficient and Exact. *SIGKDD Explorations*, Vol. 2 Issue 2, pp. 34–38, 2000.
7. V. Ganti, R. Ramakrishnan, J. Gehrke, A. Powell, and J. French. Clustering Large Datasets in Arbitrary Metric Spaces. *Proc. 15th International Conference on Data Engineering (ICDE'99)*, Sydney, Australia, 1999.
8. J. He, A. H. Tan, C. L. Tan, and S. Y. Sung. On Quantitative Evaluation of Clustering Systems. *Information Retrieval and Clustering*, Kluwer Academic Publishers, 2002.
9. E. L. Johnson and H. Kargupta. Collective Clustering from Distributed Heterogeneous Data. *Large-Scale Parallel Data Mining*, pp. 221–244, 2000.

10. J. MacQueen. Some methods for classification and analysis of multivariate observations. *Proc. 5th Berkeley Symposium on Math, Statistics and Probability,* 1, pp. 281–297, 1967.
11. M. Zait and H. Messatfa. A Comparative Study of Clustering Methods. *Future Generation Computer Systems,* Vol. 13, pp. 149–159, 1997.

6

A Probabilistic Approach
to Mining Fuzzy Frequent Patterns

Attila Gyenesei and Jukka Teuhola

Turku Centre for Computer Science (TUCS), Department of Information
Technology, University of Turku, Lemminkäisenkatu 14A, FIN-20500 Turku,
Finland
{gyenesei,teuhola}@it.utu.fi

Abstract. Deriving association rules is a typical task in data mining. The problem
was originally defined for transactions of discrete items, but it was soon extended to
include quantitative and fuzzy items. Since the fuzzy data types include non-fuzzy
and discrete types as special cases, we shall concentrate on fuzzy items. Finding
frequent combinations of items is the major task in mining association rules. The
traditional algorithms (e.g. Apriori) for finding frequent itemsets typically apply a
breadth-first approach, making several passes through the dataset, but depth-first
techniques have also been presented. Here we suggest a probabilistic approach which
iteratively creates candidates for frequent combinations of fuzzy items by computing
estimates of frequency on the basis of repeated counting. The number of iterations
is usually much lower than with the Apriori method. The number of candidates is
typically 1–3 times the number of frequent fuzzy itemsets. Though probabilistic, the
algorithm finds the precise result to the problem.

Key words: data mining, frequent itemsets, association rules, fuzzy items, probabilistic algorithms

6.1 Introduction

We study the generalized *frequent itemset mining* problem, where items can be *fuzzy*
intervals. Originally, frequent itemset mining was defined as a subtask of discovering
association rules among discrete attributes within transactions [1]. The standard example is the market basket dataset, where each transaction consists of items bought
at one time by a single customer. A typical basket itemset could be {milk, diapers,
beer}. It is called *frequent*, if its *support* is not less than a given threshold. The
support is defined as the number of transactions containing all the items in the set.
An association rule means an implication of the type {milk, diapers} → {beer}.
The interestingness of the rule $X \to Y$ can be defined in various ways [15], but the
traditional conditions set lower bounds to the support of $X \cup Y$ and *confidence*
$Y|X$, defined as support($X \cup Y$)/support(X).

Attila Gyenesei and Jukka Teuhola: *A Probabilistic Approach to Mining Fuzzy Frequent Patterns*, Studies in Computational Intelligence (SCI) **4**, 73–89 (2005)
www.springerlink.com

Quantitative attributes can be handled by partitioning the range of values into intervals, and treating the intervals as items [23]. E.g. for a person dataset we might have attributes like height: {0–100, 101–150, 151–180, 181–250} and shoe size: {0–6, 6.5–8, 8.5–10, 10.5–15}. For these we might discover a frequent combination: "30% of all people have height 181–250 and shoe size 10.5–15. This observation is considered interesting if the threshold was set to ≤30%. A corresponding association rule might say that "60% of people with height 180–250 have shoe size 10.5–15". Again, this is interesting, if the confidence threshold is at most 60%. An essential distinction from the market basket is that the intervals are mutually exclusive, and usually one of them must occur per attribute in each transaction. The basic itemset mining algorithm works inefficiently if this property is not considered.

Fuzzy intervals of quantitative attributes generalize the item concept further. They were introduced in [18] to represent intervals with soft boundaries and thereby to enable more expressive rules, called *fuzzy association rules*. In recent years, many algorithms have been proposed to discover them, see [3, 7, 8, 16, 20, 27]. In fuzzy set theory [25], the membership of an element in the set is not boolean (yes/no), but expressed as a real value within [0, 1], representing the degree of membership. It is customary to give meaningful names to fuzzy intervals, e.g. the length of a person might be "short", "average" or "tall". The membership of value 180 in the set "tall" might be 0.4 and in the set "average" 0.6. Because of the naming, fuzzy intervals are called also *fuzzy items*. A frequent fuzzy itemset could be {Height: tall, Shoe: big}, and a related fuzzy association rule {Height: tall} → {Shoe: big}. Notice that a "fuzzy itemset" means a set of fuzzy items, not a fuzzy set of items.

A separate problem in mining fuzzy itemsets is deciding the number of fuzzy items (intervals) per attribute, and determining their membership functions. This is a clustering problem, usually solved before actual mining, though the quality of the resulting itemsets can be improved by careful choices. In the mining algorithm to be presented, we assume that the fuzzy items and their membership functions have been defined for each attribute before starting the actual mining.

The best-known frequent itemset mining algorithm, called Apriori, was independently discovered by Agrawal et al. [2], and Mannila et al. [19]. It was soon extended to quantitative items [23] and fuzzy items [16, 18]. The Apriori method is an iterative, breadth-first algorithm, based on generating stepwise longer *candidate* itemsets, and clever pruning of non-frequent itemsets. Pruning takes advantage of the *monotonicity* principle of frequent itemsets: all subsets of a frequent itemset must also be frequent. Each candidate generation step is followed by a counting step where the frequencies of candidates are checked and non-frequent ones deleted. Generation and counting alternate, until at some step all generated candidates turn out to be non-frequent.

The drawback of Apriori is that each counting step requires scanning through the dataset. It is possible to reduce this by creating more (longer) candidates in one pass, at the cost of an increased total number of candidates. The other main approach to itemset generation is *depth-first* order. The best examples of this approach, in the case of discrete attributes, are *Eclat* [26] and *FP-growth* [14]. These are in many cases faster than Apriori, but it must be emphasized that this is not a universal observation; the superiority depends on the density of the dataset, the correlations between items, and the selected support threshold [28]. Extensions of these methods to fuzzy items have not, to our knowledge, been presented, and in fact they might be non-trivial.

The majority of existing algorithms for frequent itemset mining can be characterized as *analytical*, because they are based on logical rules to create and check the candidate itemsets. Our approach is *probabilistic*, based on observed frequencies of fuzzy items in the dataset. Using these, we can compute estimates for the frequencies of candidate fuzzy itemsets, i.e. combinations of fuzzy items. As in Apriori, we also have to check the actual frequencies of itemsets, by doing a count through the dataset. Based on this, we can compute an estimate of the still missing frequent itemsets. As in Apriori, generation and testing phases alternate iteratively, but now the number of iterations is typically much smaller. The number of candidates is in most cases moderate which, together with the simplicity of the algorithm, results in quite fast implementation.

The method to be presented is based on our *PIE* algorithm ("Probabilistic Iterative Expansion of Candidate Itemsets") [12], which was applicable only to discrete items. Now we generalize the method to fuzzy items, and call the new method *F-PIE* (Fuzzy-PIE). We have earlier presented another probabilistic method for fuzzy itemset mining, called *FFP* [10], but we consider F-PIE to be simpler. In Sect. 6.2, we define the fuzzy concepts and notations to be used in describing the method in Sect. 6.3. Section 6.4 reports experimental results for different types of datasets, using the fuzzy Apriori as a comparison method. Some conclusions are drawn in Sect. 6.5.

6.2 Deriving Fuzzy Items

As a preliminary step before mining, we have to split the range of each quantitative attribute into intervals. The basic principle is that the intervals constitute a *partition* of the range. For the crisp case, this means that the intervals are continuous, mutually disjointed and together cover the whole range. In the fuzzy case we demand that for each point, the sum of its membership values in different fuzzy intervals is equal to one. As mentioned above, the fuzzy intervals are also called *fuzzy items*.

The partitioning task can be solved either as an independent step, regardless of the mining process that follows, or it can be connected, see [4, 24]. The other division is into *univariate* and *multivariate* partitioning. The former divides all attribute ranges independently of each other. Thus, any one-dimensional clustering technique can be applied; for a survey, see [17]. Multivariate methods solve the task as a whole, and in principle can take the mining step into consideration [4]. A third classification is into *top-down* and *bottom-up* methods. The former take the whole range and split it repeatedly into subranges [11], while the latter start from some initial "atomic" intervals and merge them stepwise [24]. In the experiments (Sect. 6.4) we will use our multivariate, top-down partitioning method [11], which tries to create a good starting point for mining, but is not directly connected to the mining algorithm itself.

In this article we concentrate on quantitative attributes, but discrete ones can be processed as special cases, as will be exemplified in Sect. 6.4. We denote our dataset with $T = \{t_1, \ldots, t_n\}$ where each t_i is called a *tuple* of attributes. Thus, we can consider T as a relation of quantitative values. For each attribute a_i, the partitioning algorithm produces fuzzy intervals whose membership functions are piecewise linear, exemplified in Fig. 6.1 for 3 intervals (with fuzzy items "low", "middle", and "high"). The points with membership 0.5 represent the corresponding crisp bounds, and

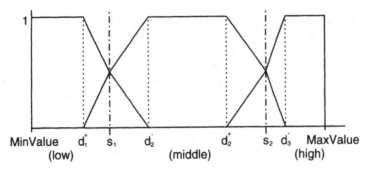

Fig. 6.1. Graphical view of the membership functions

are called 0.5-*bounds* of the fuzzy interval. The shape of the membership function is controlled by an *overlap parameter p*, defining the percentage of the 0.5-bound interval, for which the membership value of the current fuzzy item is between 0.5 and 1, and for which the membership of the neighbouring items is between 0 and 0.5. This p% area consists of two symmetric $(p/2)$% areas in the lower and upper parts of the interval. For the rest $(100-p)$% of the 0.5-bound interval, the membership is 1 for the current fuzzy item and 0 for the others.

More formally, we define the lower and upper 0- and 1-bounds as follows:

Definition 1. The **upper 1-bound**, denoted d_i^+ for fuzzy interval i, is given by:

$$d_i^+ = s_i - 0.5(s_i - s_{i-1})p/100 \,,$$

where p is the overlap parameter in %, and $s_{i-1}(s_i)$ is the left (right) 0.5-bound of fuzzy interval $i, i = \{1, 2, \ldots, m\}$. d_i^+ is also the **lower 0-bound** of interval $i+1$.

Definition 2. The **lower 1-bound**, denoted d_j^- for fuzzy interval j, is as follows:

$$d_j^- = s_{j-1} + 0.5(s_j - s_{j-1})p/100 \,,$$

where p is the overlap parameter in %, and $s_{j-1}(s_j)$ is the left (right) 0.5-bound of fuzzy interval $j, j = \{2, 3, \ldots, m+1\}$. d_j^- is also the **upper 0-bound** of interval $j-1$.

We shall use the notation $\langle a_i : x_i \rangle$ for an arbitrary item x_i of attribute a_i[1]. For itemsets, we use notations like $\langle A : X \rangle$ to denote an ordered set A of attributes and a corresponding set X of some items, one per attribute, i.e. $\langle A : X \rangle = [\langle a_1 : x_1 \rangle, \langle a_2 : x_2 \rangle, \ldots, \langle a_q : x_q \rangle]$.

For the mining task, we have to define the concept *fuzzy support* of $\langle A : X \rangle$, because that will be used as a criterion in deciding whether an itemset is frequent or not. Let $f_{\langle a_i : x_i \rangle}$ be the membership function of item $\langle a_i : x_i \rangle$. Assuming that tuple t_k of the dataset T contains value $t_k(a_i)$ for attribute a_i, then the *fuzzy support* of $\langle A : X \rangle$ with respect to T is defined as

$$FS_T(A : X) = \sum_{k=1}^{n} \Pi_{i=1}^{m} f_{\langle a_i : x_i \rangle}(t_k(a_i)) \,. \tag{6.1}$$

---

[1] The angle brackets can be omitted within parentheses.

The relative fuzzy support, also called fuzzy probability, is now

$$FP_T(A:X) = \frac{FS_T(A:X)}{n} . \qquad (6.2)$$

By item "probability" we mean the expected membership value for the item. We use the product rule to combine the membership values within a tuple, instead of other basic t-norms like minimum, Lukasiewicz, etc. (see [5]), because we treat the membership values as probabilities. For example, if the tuple [Age: 18, Height: 180] has membership values [0.8, 0.7] in the fuzzy itemset [⟨Age: young⟩, ⟨Height: tall⟩], then the tuple contributes by 0.56 to the fuzzy support of [⟨Age: young⟩, ⟨Height: tall⟩]. Our task is now to find all such fuzzy itemsets, for which $FS_T(A:X)$ is not less than a predefined minimum support threshold σ. In probabilistic terms, we may also use a relative threshold σ_{rel}, being equal to σ/n.

6.3 The Mining Algorithm

In this section we generalize our PIE algorithm [12] to the fuzzy case. The core of the method is a generate-and-test loop, such as in the well-known Apriori method [2]. However, our candidate generation is based on probabilistic estimates of the supports of fuzzy itemsets. The testing phase includes counting the supports of candidates, but also additional book-keeping for the next iteration.

6.3.1 Relative Item Supports

Probabilistic estimation is based on the relative fuzzy support $FP_T(a:x)$ of each item x of each attribute a, separately. Therefore, the initial step of the algorithm is to determine these elementary supports. For example, for items "young", "middle" and "old" of attribute "Age", we might get relative fuzzy support values 0.3, 0.5, 0.2, respectively, using formulas (6.1) and (6.2). Notice that, in spite of overlaps, the sum of relative item supports (probabilities) within an attribute must still be 1, because of the fuzzy partitioning principle: every age-value contributes altogether 1 to the supports of different fuzzy age items.

The items not satisfying the threshold condition can be discarded immediately because they cannot be part of any frequent itemset. The monotonicity property holds also for fuzzy itemsets: All subsets of a frequent fuzzy itemset must also be frequent.

6.3.2 Data Structure

Candidates for frequent itemsets will be represented as a *trie* (prefix tree) structure, which is typical in this context, see e.g. [13]. Figure 6.2 shows the complete trie for the attributes Age, Height and Shoe size. For space reasons, each attribute has now only 2 fuzzy items. Structuring of the trie is based on some fixed order of attributes and items, so the depths are decreasing from left to right. In practice, the whole trie, representing all item combinations, is of course huge, so only a part of it (marked with solid lines in Fig. 6.2) will actually be materialized during the

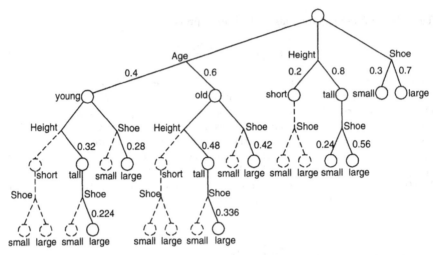

Fig. 6.2. An initial trie for three attributes where each of them has two fuzzy items. The minimum support threshold σ_{rel} was set to 0.2. The virtual nodes with relative supports less than 0.2 are drawn using *dashed lines*. Only the relative support values of real nodes are shown

mining; the rest (marked with dashed lines) are called *virtual* nodes. Each node is labelled by one item, and the labels on the path from the root to a node define the itemset related to the node. An important generalization from the discrete case is that sibling nodes are grouped so that items of the same attribute (like "young" and "old" of Age) belong to the same group. In Fig. 6.2 this is expressed by joining the parent-child links between nodes. In the implementation, there are no separate nodes for attributes, so for example the trie in Fig. 6.2 has in practice three levels, plus the root which represents an empty itemset. Another important observation is that a path can contain at most one item per attribute. In principle, due to fuzzy overlap, we might define combinations like [⟨Age: young⟩, ⟨Age: old⟩] having non-zero support, but we do not consider these meaningful for the mining task.

6.3.3 Initial Candidates for Frequent Itemsets

The first approximation of candidate itemsets is obtained by computing estimates of relative supports of various item combinations, based on the relative supports of elementary items. Here we make a straightforward approximation, as if the items were *independent*: For an itemset $[⟨A : X⟩ = [⟨a_1 : x_1⟩, ⟨a_2 : x_2⟩, \ldots, ⟨a_q : x_q⟩]$ we compute the estimated relative support (probability) as

$$EP(A : X) = FP(a_1 : x_1) \cdot FP(a_2 : x_2) \cdot \ldots \cdot FP(a_q : x_q) . \qquad (6.3)$$

From this point on, the default dataset T is omitted in the formulas. The calculation of *EP*-values for the nodes in the trie of Fig. 6.2 is shown in Table 6.1.

The support estimates are monotonically decreasing on the way down. Thus they need be computed on each path of the trie only until a value $< \sigma_{rel}$ is met. The

Table 6.1. Relative fuzzy supports (probabilities) of fuzzy itemsets

Fuzzy Itemset	Relative Fuzzy Support	Virtual Node
⟨Age: young⟩	0.4	no
⟨Age: old⟩	0.6	no
⟨Height: short⟩	0.2	no
⟨Height: tall⟩	0.8	no
⟨Shoe: small⟩	0.3	no
⟨Shoe: large⟩	0.7	no
[⟨Age: young⟩, ⟨Height: short⟩]	$0.4 \cdot 0.2 = 0.08$	yes
[⟨Age: young⟩, ⟨Height: tall⟩]	$0.4 \cdot 0.8 = 0.32$	no
. . .	. . .	. . .
[⟨Height: tall⟩, ⟨Shoe: small⟩]	$0.8 \cdot 0.3 = 0.24$	no
[⟨Height: tall⟩, ⟨Shoe: large⟩]	$0.8 \cdot 0.7 = 0.56$	no
[⟨Age:young⟩, ⟨Height:short⟩, ⟨Shoe:small⟩]	$0.4 \cdot 0.2 \cdot 0.3 = 0.024$	yes
[⟨Age:young⟩, ⟨Height:short⟩, ⟨Shoe:large⟩]	$0.4 \cdot 0.2 \cdot 0.7 = 0.056$	yes
. . .	. . .	. . .
[⟨Age: old⟩, ⟨Height: tall⟩, ⟨Shoe: small⟩]	$0.6 \cdot 0.8 \cdot 0.3 = 0.144$	yes
[⟨Age: old⟩, ⟨Height: tall⟩, ⟨Shoe: large⟩]	$0.6 \cdot 0.8 \cdot 0.7 = 0.336$	no

subtrees below will then be initially virtual. Hence, we do not need to compute the whole of Table 6.1.

6.3.4 Counting the Supports

The next step is to read the tuples from the dataset and count the true (cumulative) support values $FS(p)$ for each real node p (i.e. the support related to the path of items to that node), using the earlier mentioned product rule to combine values of membership functions for items on the path. This is done for each tuple, and the products are summed to $FS(p)$. Note that a node name represents the related ⟨attribute: item⟩ pair.

Another value computed for each visited node is the so called *pending support* (PS), which represents an upper bound to the sum of supports of virtual children, accumulated for all tuples. For each tuple t_k, matching the ancestor path of node p, we increment $PS(p)$ by $FS_{t_k}(p)$ times the maximal relative support $FS_{t_k}(v)$ among virtual children v of p:

$$PS(p) = \sum_{k=1}^{n} \left[FS_{t_k}(p) \cdot \max_{v \in virtual\_children(p)} FS_{t_k}(v) \right] . \tag{6.4}$$

The pending support will be our criterion for expanding the trie by converting some virtual descendants into real ones: If $PS(p) \geq \sigma$, then it is possible (but not certain) that one or more of the virtual children of p are frequent, and real nodes must be created for them. If there are no such nodes, the process is ready and the frequent fuzzy itemsets can be extracted from the tree as paths ending at any node p having $FS(p) \geq \sigma$.

6.3.5 Expanding the Candidate Set

Trie expansion starts a new cycle, and iteration continues until no new candidates need to be generated. A more detailed stopping rule will be formulated in Subsect. 6.3.6. The expansion must be performed very carefully, in order to prevent "explosion" of the trie. Which virtual nodes should be materialized, in order to approach the final solution fast but avoiding too many useless nodes? Here we resort to elementary probabilities of fuzzy items and the computed pending supports. Suppose that we have a node $p = \langle a_i : x_i \rangle$ with observed fuzzy support $FS(p)$ and observed pending support $PS(p) \geq \sigma$. Node p has children (either real or virtual), representing items of attributes $a_{i+1}, a_{i+2}, \ldots$ The pending support should now be divided among the virtual children, but it must be noticed that they are not mutually exclusive, so a more elaborated analysis is needed. First denote the existing items of attribute a_j $(j > i)$ among children of p by $Real(a_j)$, and the non-existing items by $Virtual(a_j)$. Now we can derive the *local probability* of the $Virtual(a_j)$ as

$$LP(Virtual(a_j)) = \frac{FP(Virtual(a_j))}{1 - FP(Real(a_{i+1}) \cdot FP(Real(a_{i+2})) \cdot \ldots} . \tag{6.5}$$

Here the product $FP(Real(a_{i+1})) \cdot FP(Real(a_{i+2})) \cdot \ldots$ represents the estimated probability of the combination $[Real(a_{i+1})), Real(a_{i+2}), \ldots]$, whereas $1 - FS$ $(Real(a_{i+1})) \cdot FS(Real(a_{i+2})) \cdot \ldots$ represents the estimated probability of combinations having at least one virtual component. This is just the situation where pending support was incremented, so $LP(Virtual(a_j))$ represents a *conditional probability* of the virtual items of a_j. This is a direct generalization of the conditional probability that we used in the discrete case in [12]. The estimate is *local* because we did not yet consider the path from root to p. On the basis of the pending support of p we obtain the estimated support of $Virtual(a_j)$ as

$$ES(Virtual(a_j)) = LP(Virtual(a_j)) \cdot PS(p) . \tag{6.6}$$

Finally, to get estimates for the individual item supports we divide the support of $Virtual(a_j)$ in proportion to the item supports, because they are mutually exclusive:

$$ES(a_j : x_j) = \frac{FP(a_j : x_j)}{FP(Virtual(a_j))} \cdot ES(Virtual(a_j)) . \tag{6.7}$$

If $ES(a_j : x_j) \geq \sigma$ then we conclude that $\langle a_j : x_j \rangle$ is expected to be frequent, and we generate a real node for that. However, in order to guarantee a finite number of iterations in the worst case, we have to somewhat relax this condition. Because the true distribution may be extremely skewed, almost the whole pending support may actually "belong" to only one virtual child. To ensure that the iteration converges, we add a tuning factor to the support condition:

$$ES(a_j : x_j) \geq \alpha^r \sigma \tag{6.8}$$

with some constant α between 0 and 1 and r being the iteration number ($r = 1, 2, \ldots$). Sooner or later this will result in expansion, so that we get rid of PS-values greater or equal to the threshold. In our tests we used the data-dependent value $\alpha =$ average fuzzy probability (FP) of frequent fuzzy items. The reasoning behind this is that it accelerates the local expansion growth by r levels, on the

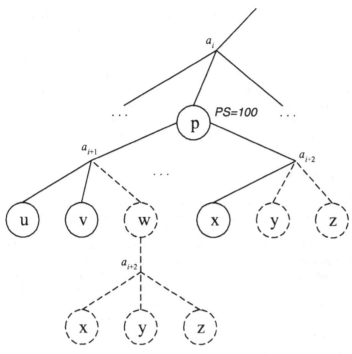

Fig. 6.3. An example of expansion for probabilities $FP(u) = 0.5$, $FP(v) = 0.3$, $FP(w) = 0.2$, $FP(x) = 0.6$, $FP(y) = 0.3$ and $FP(z) = 0.1$

average, at iteration r. The largest extensions are applied only to the skewest parts of the trie, so the total size remains tolerable.

When a virtual node has been materialized, we immediately check its expansion, based on its ES-value, recursively. In recursion the formulas (6.5)–(6.7) can in principle be applied, but they degenerate, because there are no real children, yet. Therefore, the estimated support of an item $\langle a_k: x_k \rangle$, a potential child of $\langle a_j : x_j \rangle$ is just the product $FP(a_k : x_k) \cdot ES(a_j : x_j)$.

Example. Figure 6.3 shows a numeric example of trie expansion, assuming the minimum support threshold $\sigma = 50$ ($n = 1000$, $\sigma_{rel} = 0.05$) and $\alpha = 0.4$. Note that for space reasons, only a generic part of a larger trie is presented. The fuzzy probabilities are assumed to be $FP(u) = 0.5$, $FP(v) = 0.3$, $FP(w) = 0.2$, $FP(x) = 0.6$, $FP(y) = 0.3$ and $FP(z) = 0.1$. Node p has a pending support 100, related to its virtual children w, y and z, so we have to test for expansion. Using our formulas (6.5)–(6.7), the estimated supports are

$$ES(w) = \frac{0.2}{0.2} \cdot \frac{0.2}{1 - 0.8 \cdot 0.6} \cdot 100 = 38.46 \geq \alpha \cdot \sigma = 20$$

$$ES(y) = \frac{0.3}{0.4} \cdot \frac{0.4}{1 - 0.8 \cdot 0.6} \cdot 100 = 57.69 \geq \alpha \cdot \sigma$$

$$ES(z) = \frac{0.1}{0.4} \cdot \frac{0.4}{1 - 0.8 \cdot 0.6} \cdot 100 = 19.23 < \alpha \cdot \sigma$$

Nodes w and y are materialized, but z is not. Then, we immediately test the expansion of materialized nodes, based on their ES-values. However, a_{i+2} is the last attribute, so the expansion test continues down only from w. The estimated supports are: $ES(x) = 38.46 \cdot 0.6 = 23.07$, $ES(y) = 38.46 \cdot 0.3 = 11.54$ and $ES(z) = 38.46 \cdot 0.1 = 3.84$. We expand x, since its estimated support is $> \alpha \cdot \sigma = 20$.

The expansion phase is continued by a new counting phase, and new values for node supports and pending supports are calculated. The two phases alternate, until all pending supports are less than σ. The name of our method, F-PIE, stands for this probabilistic iterative expansion of the candidate trie for fuzzy itemsets.

6.3.6 Stopping Rule

From efficiency viewpoint, we should restrict the work done, especially in the counting phase, because that is the most expensive part, making repeated scans through the dataset and extensive walks in the trie. In later iterations, fewer and fewer of the nodes are still "active", in the sense that there is no need to create new children for them. We call such nodes "ready". Readiness can be checked easily with a recursive process: A leaf with observed fuzzy support less than σ is ready. A non-leaf node is ready if all its real children are ready, and its pending support is $< \sigma$. In counting, we process one tuple at a time, and scan each subset of items down the tree, but only until the first ready node on each path. Also the expansion can of course be skipped for ready nodes and their children. When the root becomes ready, we can stop the whole procedure.

6.3.7 Pseudocode

We gather the description of the F-PIE method into the following high-level algorithm.

Algorithm F-PIE – Probabilistic iterative expansion of candidates
in fuzzy frequent itemset mining

```
Input: A relational dataset T, the minimum support threshold σ.
Output: The complete set of fuzzy frequent itemsets.

1. // Initial steps.
2. scan T and collect the set F of fuzzy frequent items of all
        attributes;
3. α := average probability of fuzzy items in F;
4. iter := 0;
5. // The first generation of candidates, based on fuzzy
   // item probabilities.
6. create an F-PIE-trie P so that it contains all such
        ordered subsets S ⊆ F for which Π(Prob(s∈S)) · |T| ≥ σ
        and which do not contain items from the same attribute;
7. set the status of all nodes of P to not-ready;
8. // The main loop: alternating count, test and expand.
9. loop
10.     // Scan the dataset and check readiness.
```

```
11.     scan T and count the fuzzy support and pending
        support values for non-ready nodes in P;
12.     iter := iter + 1;
13.     for each node p ∈ P do
14.         if pending_support(p) < σ then
15.             if p is a leaf then set p ready
16.             else if the children of p are ready then
17.                 set p ready;
18.         if root(P) is ready then exit loop;
19.     // Expansion phase: Creation of subtries on the
        // basis of observed pending supports.
20.     for each non-ready node p in P do
21.         if pending_support(p) ≥ σ then
22.             for each virtual child <a: v> of p do
23.                 compute estim_support(<a: v>) by formulas
                    (6.5)-(6.7)
24.                 if estim_support(<a: v>) ≥ α^{iter}σ then
25.                     create node v as the child of p;
26.                     add such ordered subsets S ⊆ F\{1..v} as
                        descendant paths of v, for which
                            Π(Prob(s∈S)) · estim_support(v) ≥ α^{iter}σ
                        and which do not contain items from the
                        same attribute;
27. // Gather up results from the trie
28. return the paths for nodes p in P such that
    support(p) ≥ σ;
29. end
```

6.4 Experimental Results

We examined the effectiveness of our F-PIE algorithm by experimenting with four real-life datasets. They represent rather different kinds of domains, and we wanted to include dense and non-dense, strongly correlated and weakly correlated datasets, as well as various numbers of tuples and attributes. The first dataset (Family) comes from a research by the U.S. Census Bureau [6], the second (Biology) is a study of the population biology of abalone by Nash et al. [21], the third one (Medical) was collected by Chiba University Hospital for laboratory examinations and given by the PKDD'99 Conference Organizers [22] and the last one (Mushroom) includes descriptions of hypothetical samples corresponding to 23 species of gilled mushrooms in the Agaricus and Lepiota Family [9]. Table 6.2 summarizes the dataset information.

The first three datasets had quantitative attributes, which were divided into fuzzy intervals using our proposed multidimensional fuzzy partitioning algorithm [11]. As an example, Table 6.3 shows the results of partitioning for dataset "Family" with overlap parameter values 0%, 20%, 40% and 60%. To analyze the different overlap values, we calculated the average information amount ("interestingness") of frequent itemsets [11]. It can be noticed that the partitioning method gives similar average information values for the four different overlap parameters. We chose overlap = 40% to be used in the rest of the experiments.

Table 6.2. Test dataset description

Dataset	#Tuples	#Attributes	#Fuzzy Items
Family	60607	5	17
Biology	4177	8	31
Medical	23176	10	40
Mushroom	8124	120	–

Table 6.3. The number of fuzzy items per attribute, the number of frequent itemsets and their average information for different overlap values of dataset "Family" with minimum support threshold $\sigma_{\mathrm{rel}} = 0.01$

Overlap	Age	FamPers	EdHead	IncHead	IncFam	#Frequent Itemsets	Average inf.
0%	3	4	3	5	4	452	6958
20%	3	4	3	5	3	430	6990
40%	3	4	3	4	3	446	7093
60%	3	4	3	4	3	452	7194

It is important to notice that the fuzzy frequent itemset mining algorithms can be applied to the discrete (non-fuzzy) case, as well. For that purpose, the highly dense "Mushroom" dataset was used during the efficiency evaluations.

Table 6.4 shows interesting statistics from the proposed F-PIE method for dataset "Medical", with different values of relative minimum support threshold σ_{rel}. The data to be collected were the number of candidates, depth of the trie, and the number of iterations. The table also shows the number of frequent items and frequent itemsets, to enable comparison with the number of candidates. For this medium dense dataset, the number of candidates varies between 1–2 times the number of frequent itemsets. For non-dense datasets the ratio is usually larger.

The values of the "security parameter" α (average probability of frequent items) are also given in Table 6.4. Considering I/O performance, we can see that the number of iteration cycles (=number of file scans) is quite small, compared to the fuzzy

Table 6.4. Statistics from the F-PIE algorithm for dataset "Medical"

σ_{rel}	#Freq. Items	#Freq. Itemsets	Alpha	#Cand	Trie Depth	#Iter.	#Apriori's Iterations
0.11	18	1906	0.528	3743	9	3	8
0.09	20	2468	0.485	4811	9	4	9
0.07	20	3209	0.485	6021	10	3	9
0.05	22	4403	0.446	8102	10	3	10
0.03	23	6741	0.428	11333	10	3	10
0.01	27	14491	0.368	24066	10	4	10

Table 6.5. Development of the trie for dataset "Medical", with three different values of σ_{rel}

σ_{rel}	Iteration	#Frequent Itemsets Found	#Nodes	#Ready Nodes
0.09	1	2175	2268	1123
	2	2440	4414	4218
	3	2468	4805	4794
	4	2468	4811	4811
0.05	1	3949	4306	2432
	2	4377	7639	7418
	3	4403	8102	8102
0.01	1	13560	15778	11807
	2	14425	23025	22648
	3	14479	23941	23935
	4	14491	24066	24066

version of the Apriori method, for which the largest frequent itemset dictates the number of iterations. This is roughly the same as the trie depth in the F-PIE method.

The F-PIE method can also be characterized by describing the development of the trie during the iterations. The most interesting figures are the number of nodes and the number of ready nodes, given in Table 6.5. Especially the number of ready nodes implies that even though we have many candidates (=nodes in the trie), large parts of them are not touched in the later iterations.

For comparison of speed, we chose our fuzzy Apriori implementation, which has been used in previous data mining research [10]. The experiments were performed on a 1.5 GHz Pentium 4 PC machine with 512 MB main memory. Both algorithms were coded in C. The results for the four test datasets and for different relative minimum support thresholds are shown in Table 6.6. The execution times are also illustrated graphically in Fig. 6.4.

The results show that F-PIE is almost always more efficient than fuzzy Apriori. It is faster for sparse and weakly correlated datasets, such as "Family" and "Medical", but less superior for dense (Mushroom) and very correlated (Biology) datasets, especially with smaller minimum support thresholds. This is probably a general observation: the performance of most frequent itemset mining algorithms is highly dependent on the dataset and threshold.

6.5 Conclusion

We have studied the problem of mining frequent fuzzy itemsets, and suggested a related mining algorithm. It applies to datasets consisting of tuples of quantitative attributes, the ranges of which have been partitioned into fuzzy intervals, called items. In the experiments, the performance of the new algorithm was in most cases better than the fuzzy version of the Apriori method. The algorithm also fits to the crisp case, where the "fuzzyness" is set to zero. Even the discrete attributes can be

Table 6.6. Comparison of execution times (in seconds) of F-Apriori and F-PIE algorithms for four datasets

	(a) Family				(b) Biology		
σ_{rel}	#Freq. Itemsets	Fuzzy Apriori	F-PIE	σ_{rel}	#Freq. Itemsets	Fuzzy Apriori	F-PIE
0.11	77	2.89	1.70	0.11	563	0.609	0.605
0.09	92	2.94	1.74	0.09	773	0.657	0.625
0.07	116	3.00	2.08	0.07	1067	0.734	0.687
0.05	156	2.95	1.78	0.05	1385	0.891	0.891
0.03	212	3.02	1.77	0.03	2030	1.060	0.921
0.01	452	3.09	1.77	0.01	3700	1.380	1.300
	(c) Medical				(d) Mushroom		
σ_{rel}	#Freq. Itemsets	Fuzzy Apriori	F-PIE	σ_{rel}	#Freq. Itemsets	Fuzzy Apriori	F-PIE
0.11	1906	12.3	11.3	0.50	153	1.22	0.70
0.09	2468	13.8	13.5	0.45	329	1.61	0.98
0.07	3209	17.6	16.0	0.40	565	2.05	1.42
0.05	4403	25.7	21.0	0.35	1189	3.72	3.41
0.03	6741	32.2	22.8	0.30	2735	9.89	8.86
0.01	14491	47.8	29.1	0.25	5545	21.20	25.00

handled by regarding them directly as items. In spite of the generality, the efficiency of the algorithm is almost the same as that of an implementation tailored to the discrete case.

The mining method differs from most other approaches to itemset mining in that it is probability-based. This makes it rather robust and easy to program, needing in most cases only a small number of iterations to obtain the result. In the estimation of frequency for fuzzy itemsets, we used only the observed values of item probabilities, and combined them as if the items were independent. This is a rough approximation, and might be improved by deriving more statistics about item combinations, at the cost of increased complexity in the program logic.

Our implementation was simple also in the sense that we did not preprocess or reorganize the dataset in any way. One option that could be tried is the so called vertical layout, so that for each node, the tuples with non-zero support are extracted and constitute the basis for subsequent counting in the related subtree. This means that the dataset is distributed to the nodes, and duplicated to a greater or lesser extent.

Another simplification made in our implementation was assuming the trie would fit in the main memory. While we anticipate this to hold for most practical datasets, in extreme cases (lots of attributes with lots of frequent fuzzy items and/or very low support thresholds) the performance will degrade due to virtual memory swapping. Making the trie external would require partitioned representation, and processing in a depth-first rather than breadth-first manner. However, we leave the details of the external version to future work.

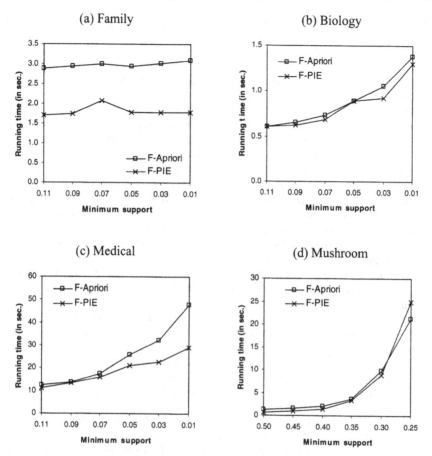

Fig. 6.4. Running times for different minimum support thresholds

References

1. R. Agrawal, T. Imielinski and A. Swami, Mining association rules between sets of items in large databases, in Proceedings of ACM SIGMOD International Conference on Management of Data, 1993, pp. 207–216.
2. R. Agrawal and R. Srikant, Fast algorithm for mining association rules in large databases, in Proceedings of the 20th International Conference on Very Large Data Bases, 1994, pp. 487–499.
3. W.-H. Au and K.C.C. Chan, An effective algorithm for discovering fuzzy rules in relational databases, in Proceedings of the 7th IEEE International Conference on Fuzzy Systems, 1998, pp. 1314–1319.
4. S.D. Bay, Multivariate discretization for set mining, Knowledge and Information Systems 3(4), 2001, pp. 491–512.
5. C. Carlsson and R. Fullér, Fuzzy reasoning in decision making and optimization, Physica-Verlag, 2002.
6. U.S. Census Bureau, http://www.census.gov/.

7. G. Chen, Q. Wei and E. Kerre, Fuzzy data mining: Discovery of fuzzy generalized association rules, Recent Issues on Fuzzy Databases, 2000, pp. 45–66.

8. G. Chen and Q. Wei, Fuzzy association rules and the extended mining algorithms, Information Sciences 147, 2002, pp. 201–228.

9. Frequent Itemset Mining Implementations (FIMI'03) Workshop website, http://fimi.cs.helsinki.fi, 2003.

10. A. Gyenesei and J. Teuhola, Mining fuzzy frequent patterns without repeated candidate generation, in Proceedings of the 1st International Conference on Fuzzy Systems and Knowledge Discovery, 2002, pp. 374–380.

11. A. Gyenesei and J. Teuhola, Multidimensional fuzzy partitioning of attribute ranges for mining quantitative data, in Proceedings of the International Conference on Fuzzy Information Processing, 2003, pp. 413–422.

12. A. Gyenesei and J. Teuhola, Probabilistic iterative expansion of candidates in mining frequent itemsets, in Proceedings of the IEEE ICDM Workshop on Frequent Itemset Mining Implementations, 2003, pp. 49–56.

13. B. Goethals, Efficient frequent pattern mining, PhD thesis, University of Limburg, Belgium, Dec. 2002.

14. J. Han, J. Pei, and Y. Yin, Mining frequent patterns without candidate generation, in Proceedings of ACM SIGMOD International Conference on Management of Data, 2000, pp. 1–12.

15. R.J. Hilderman and H.J. Hamilton, Knowledge discovery and interestingness measures: a survey, Technical Report CS 99–04, University of Regina, Canada, 1999.

16. T-P. Hong, C-S. Kuo and S-C. Chi, Mining association rules from quantitative data, Intelligent Data Analysis, 3(5), 1999, pp. 363–376.

17. A.K. Jain, M.N. Murty and P.J. Flynn, Data clustering: a review, ACM Computing Surveys, 31(3), 1999, pp. 264–323.

18. C.M. Kuok, A. Fu and M.H. Wong, Fuzzy association rules in databases, ACM SIGMOD Record, 27 (1), 1998, pp. 41–46.

19. H. Mannila, H. Toivonen, and A.I. Verkamo, Efficient algorithms for discovering association rules, in Proceedings of the AAAI Workshop on Knowledge Discovery in Databases, 1994, pp. 181–192.

20. L.J. Mazlack, Granulation of quantitative association rules, International Journal of Fuzzy Systems, 3(2), 2001, pp. 400–408.

21. W.J. Nash, T.L. Sellers, S.R. Talbot, A.J. Cawthorn and W.B. Ford, The population biology of abalone in Tasmania, Sea Fisheries Division, Technical Report No. 48, 1994.

22. Principles and Practice of Knowledge Discovery in Databases (PKDD'99) Conference website, http://lisp.vse.cz/pkdd99/, 1999.

23. R. Srikant and R. Agrawal, Mining quantitative association rules in large relation tables, in Proceedings of ACM SIGMOD International Conference on Management of Data, 1996, pp. 1–12.

24. K. Wang, S.H.W. Tay and B. Liu, Interestingness-based interval merger for numeric association rules, in Proceedings of the Fourth International Conference on Knowledge Discovery and Data Mining, 1998, pp. 121–147.

25. L.A. Zadeh, Fuzzy sets, Information and Control, 8 (1965) 338–353.

26. M.J. Zaki, Scalable algorithms for association mining, IEEE Transactions on Knowledge and Data Engineering 12 (3), 2000, pp. 372–390.

27. W. Zhang, Mining fuzzy quantitative association rules, IEEE International Conference on Tools and Artificial Intelligence, 1999, pp. 99–102.
28. Z. Zheng, R. Kohavi and L. Mason, Real world performance of association rule algorithms, in F. Provost and R. Srikant (eds.), in Proceedings of the Seventh ACM SIGKDD International Conference on Knowledge Discovery and Data Mining, 2001, pp. 401–406.

Identifying Interesting Patterns
in Multidatabases

Chengqi Zhang and Jeffrey Xu Yu and Shichao Zhang

Department of Systems Engineering and Engineering Management, Chinese
University of Hong Kong, Shatin, New Territories, Hong Kong
{cqzhang, yu}@se.cuhk.edu.hk; zhangsc@it.uts.edu

Abstract. In this chapter we develop a new technique for mining multidatabases.
The new mining algorithm, by comparing to traditional multidatabase mining strategies that have been focused on identifying mono-database-mining-like patterns, is
able to identify both the *commonality* and *individuality* among the local patterns in
branches within a company. While the commonality is important in terms of global
decision-making, exceptional patterns often present as more glamorous than commonality patterns in such areas as marketing, science discovery, and information
safety. We evaluated the proposed technique, and our experimental results demonstrate that the approach is efficient and promising.

Key words: multidatabase mining, local pattern analysis, high-vote pattern, exceptional pattern

7.1 Introduction

On-going work proposed in [22] explores techniques for mining association rules in
multidatabases. The increasing use of multidatabase technology, such as computer
communication networks, distributed database systems, federated database systems,
multidatabase language systems, and homogeneous multidatabase language systems,
has led to the development of many multidatabase systems in real-world applications. Many organizations need to mine multiple databases, which are distributed
across their branches, for the purpose of decision-making. On the other hand, there
are essential differences between mono- and multi-database mining. Because they are
fascinated with mono-database mining techniques, traditional multidatabase mining
techniques are not adequate for discovering patterns such as "85% of the branches
within a company agreed that a customer usually purchases sugar if he or she purchases coffee". Therefore, developing effective and efficient techniques for mining
association rules in multidatabases is very important. And we have developed a new
strategy for analyzing local patterns in [20], referred to *local pattern analysis*.

Within a company, each branch, large or small, has an equal power to vote for
patterns for global decision-making. Some patterns can receive votes from most of
the branches. These patterns are referred to *high-vote patterns*. High-vote patterns

Chengqi Zhang et al.: *Identifying Interesting Patterns in Multidatabases*, Studies in Computational Intelligence (SCI) **4**, 91–112 (2005)
www.springerlink.com

represent the commonness of the branches. Therefore, these patterns may be far more important in terms of global decision-making within the company.

Unlike high-vote patterns, however, exceptional patterns can be hidden in local patterns. Exceptional patterns reflect the "individuality" of, say, branches of an interstate company. In real-world applications, exceptional patterns often present as more glamorous than high-vote patterns in such areas as marketing, science discovery, and information safety.

This chapter designs a new algorithm for identifying both high-vote and exceptional patterns from multidatabases. We begin with presenting the problem statement in Sect. 7.2. Section 7.3 proposes a procedure for identifying high-vote patterns. In Sect. 7.4, a model for identifying exceptional patterns is proposed. Section 7.5 evaluates the proposed approach by the use of experiments. Finally, this chapter is summarized in Sect. 7.6.

7.2 Problem Statement

For description, this section states multidatabase mining problem in a simple way.

7.2.1 Related Work

Data mining techniques (see [22] and [14]) have been successfully used in many diverse applications. Developed techniques are oriented towards mono-databases.

Multidatabase mining has been recently recognized as an important research topic in the KDD community. Yao and Liu have proposed a means of searching for interesting knowledge in multiple databases according to a user query [23]. The process involves selecting all interesting information from many databases by retrieval. Mining only works on the selected data.

Liu, Lu and Yao have proposed another mining technique in which relevant databases are identified [8]. Their work has focused on the first step in multidatabase mining, which is the identification of databases that are most relevant to an application. A relevance measure was proposed to identify relevant databases for mining with an objective to find patterns or regularity within certain attributes. This can overcome the drawbacks that are the result of forcedly joining all databases into a single very large database upon which existing data mining techniques or tools are applied. However, this database classification is typically database-dependent. Therefore, Zhang and Zhang have proposed a database-independent database classification in [21], which is useful for general-purpose multidatabase mining.

Zhong et al. [24] proposed a method of mining peculiarity rules from multiple statistical and transaction databases based on previous work. A peculiarity rule is discovered from peculiar data by searching the relevance among the peculiar data. Roughly speaking, data is peculiar if it represents a peculiar case described by a relatively small number of objects and is very different from other objects in a data set. Although it appears to be similar to the exception rule from the viewpoint of describing a relatively small number of objects, the peculiarity rule represents the well-known fact with common sense, which is a feature of the general rule.

Other related research projects are now briefly described. Wu and Zhang advocated an approach for identifying patterns in multidatabase by weighting [17].

Reference [12] described a way of extending the INLEN system for multidatabase mining by incorporating primary and foreign keys as well as developing and processing knowledge segments. Wrobel [16] extended the concept of foreign keys to include foreign links since multidatabase mining also involves accessing non-key attributes. Aronis et al. [2] introduced a system called WoRLD that uses spreading activation to enable inductive learning from multiple tables in multiple databases spread across the network. Existing parallel mining techniques can also be used to deal with multidatabases [4, 5, 10, 11, 13].

The above efforts provide a good insight into multidatabase mining. However, there are still some limitations in traditional multidatabase mining that are discussed in next subsection.

7.2.2 Limitations of Previous Multidatabase Mining

As we have seen that traditional multidatabase mining is fascinated with mono-database mining techniques. It consists of a two-step approach. The first step is to select the databases most relevant to an application. All the data is then pooled together from these databases to amass a huge dataset for discovery upon which mono-database mining techniques can be used. However, there are still some limitations discussed below.

1. Putting all the data from relevant databases into a single database can destroy some important information that reflect the distributions of patterns.

   In some contexts, each branch of an interstate company, large or small, has equal power in voting patterns for global decisions. For global applications, it is natural for the company headquarters to be interested in the patterns voted for by most of the branches. It is therefore inadequate in multidatabase mining to utilize existing techniques for mono-databases mining.

2. Collecting all data from multidatabases can amass a huge database for centralized processing using parallel mining techniques.

   It may be an unrealistic proposition to collect data from different branches for centralized processing because of the huge data volume. For example, different branches of Wal-Mart receive 20 million transactions a day. This is more than the rate at which data can be feasibly collected and analyzed using today's computing power.

3. Because of data privacy and related issues, it is possible that some databases of an organization may share their patterns but not their original databases.

   Privacy is a very sensitive issue, and safeguarding its protection in a multi-database is of extreme importance. Most multidatabase designers take privacy very seriously, and allow some protection facility. For source sharing in real-world applications, sharing patterns is a feasible way of achieving this.

   From the above observations, it is clear that traditional multidatabase mining is inadequate to serve two-level applications of an interstate company. This prompts the need to develop new techniques for multidatabase mining.

7.2.3 Our Approach

Based on the above analysis, the problem for our research can be formulated as follows.

> Let $D_1, D_2, \ldots, D_m$ be m databases in the m branches $B_1, B_2, \ldots, B_m$ of a company, respectively; and LI_i be the set of patterns (local patterns) from D_i ($i = 1, 2, \ldots, m$). We are interested in the development of new techniques for identifying both high-vote and exceptional patterns of interest in the local patterns.

Our model in this chapter is the first research effort in this direction because traditional multidatabase mining, which puts all the data together from relevant databases to amass a huge dataset for discovery, cannot identify both high-vote and exceptional patterns in multidatabases.

7.2.4 Local Patterns

Local pattern analysis is a strategy for identifying laws, rules, and useful patterns from a set of local patterns from multidatabases [20]. This subsection presents the definition of local pattern.

Strictly speaking, a *local pattern* is a pattern that has been identified in the local database of a branch. A local pattern may be a frequent itemset, an association rule, causal rule, dependency, or some other expression. For description purposes, this book sometimes takes frequent itemsets, sometimes association rules, and sometimes both frequent itemsets and association rules, as local patterns. Example 1 illustrates local patterns.

Example 1. Consider a company that has five branches with five databases D_1, $D_2, \ldots, D_5$ as follows.

$$D_1 = \{(A, B, C, D); (B, C); (A, B, C); (A, C)\}$$
$$D_2 = \{(A, B); (A, C); (A, B, C); (B, C); (A, B, D); (A, C, D)\}$$
$$D_3 = \{(B, C, D); (A, B, D); (B, C); (A, B, D); (A, B)\}$$
$$D_4 = \{(A, C, D); (A, B, C); (A, C); (A, D); (D, C)\}$$
$$D_5 = \{(A, B, C); (A, B); (A, C); (A, D)\}$$

where each database has several transactions, separated by a semicolon; each transaction contains several items, separated by a comma.

When $minsupp = 0.5$. Local frequent itemsets in D_1, D_2, D_3, D_4, D_5 are listed in Tables 7.1, 7.2, 7.3, 7.4, and 7.5, respectively.

In Tables 7.1, 7.2, 7.3, 7.4, and 7.5, "XY" stands for the conjunction of X and Y. All of the local frequent itemsets discovered in $D_1, D_2, \ldots, D_5$ refer to local patterns from the five branches. All association rules generated from the local frequent itemsets are also referred to as local patterns.

In a multidatabase environment, a pattern has attributes: the name of the pattern, the rate voted for by branches, and supports (and confidences for a rule) in

Table 7.1. Local frequent itemsets in database D_1

Itemsets	Support	$\geq minsupp$
A	0.75	y
B	0.75	y
C	1.0	y
AB	0.5	y
AC	0.75	y
BC	0.75	y
ABC	0.5	y

Table 7.2. Local frequent itemsets in database D_2

Itemsets	Support	$\geq minsupp$
A	0.833	y
B	0.667	y
C	0.667	y
AB	0.5	y
AC	0.5	y

Table 7.3. Local frequent itemsets in database D_3

Itemsets	Support	$\geq minsupp$
A	0.6	y
B	1.0	y
D	0.6	y
AB	0.6	y
BD	0.6	y

Table 7.4. Local frequent itemsets in database D_4

Itemsets	Support	$\geq minsupp$
A	0.8	y
C	0.8	y
D	0.6	y
AC	0.6	y

branches that vote for the pattern (For details, please see [20]). In other words, a pattern is a super-point of the form

$$P(name, vote, vsupp, vconf) \qquad (7.1)$$

where,

name is the dimension of the name of the pattern;
vote is the dimension of the voted rate of the pattern;

Table 7.5. Local frequent itemsets in database D_5

Itemsets	Support	$\geq minsupp$
A	1.0	y
B	0.5	y
C	0.5	y
AB	0.5.	y
AC	0.5	y

$vsupp$ is a vector that indicates the m dimensions of supports in m branches (local pattern sets), referred to as the support dimensions; and

$vconf$ is a vector that indicates the m dimensions of confidences in m branches (local pattern sets), referred to as the confidence dimensions.

7.2.5 Distribution of Local Patterns

An interstate company must face two-level decisions: the company decisions (global applications) and the branch decisions (local applications). To avoid re-mining multiple databases, local patterns, including local frequent itemsets and local association rules mined in branches, are reused.

For huge number of local patterns, a company's headquarters often selects predictive patterns based on the lift in the top 5%, 10%, or 15%. And these patterns would generally be voted for by most of their branches. Hence, reusing local patterns is important, and sometimes necessary. The techniques in this section focus on identifying high-vote patterns from local patterns, which are regarded as a novel pattern in multidatabases.

Where graphical methods usually attempt to present a set of local patterns in pictorial form so as to give users an adequate visual description. Therefore, we use figures below to illustrate in a concrete form which patterns in local patterns are of interest.

Consider four databases $DB1$, $DB2$, $DB3$, and $DB4$ of a company. In Fig. 7.1, the distribution of patterns in branches (databases) of the company is depicted, where each local pattern in the databases $DB1$, $DB2$, $DB3$, $DB4$ is a point marked by "r".

Figure 7.1 displays the distribution of local patterns (patterns) in the four databases. Intuitively, the patterns distributed at the intersection of the four databases would be of the most interest to company headquarters, as they have been voted for by all branches. Figure 7.2 highlights the intersections of interest.

In Fig. 7.2, the shaded portion is the intersection of $DB1$, $DB2$, $DB3$, and $DB4$. There are also three other parts: $T1$, T_2 and T_3, which may be of interest, where T_1 is the intersection of $DB1$, $DB2$ and $DB3$; $T2$ is the intersection of $DB1$, $DB3$ and $DB4$; and $T3$ is the intersection of $DB2$, $DB3$ and $DB4$.

In real applications, it is common sense that company headquarters would be interested in patterns that are voted for by the majority of their branches. In Fig. 7.2, the local patterns that occur in the labelled areas $T1$, $T2$, and $T3$ are of interest. This includes the shaded area. These areas are referred to as *high-vote-pattern area*, in which the patterns have been voted for by most of the branches of the company.

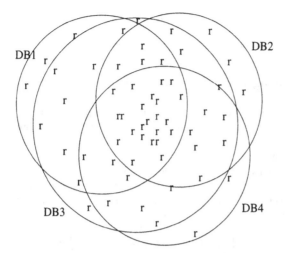

Fig. 7.1. The intersection of four databases

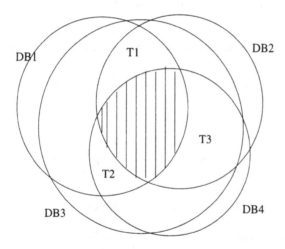

Fig. 7.2. The shaded part of the intersection

7.3 Identifying High-Vote Patterns

High-vote patterns can grasp the distribution of patterns in local patterns, and reflect the "commonness" of branches in their voting. High-vote patterns are useful for global applications of interstate companies. This section presents techniques for identifying this kind of pattern from local patterns.

To identify high-vote patterns from local patterns, the projection of $P(name, vote, vsupp, vconf)$ on $name$ and $vote$ is considered in this chapter. That is, the projection

$$P(name, vote) \tag{7.2}$$

is used to search for high-vote patterns of interest.

7.3.1 Measuring High-Vote Patterns

In some contexts, each branch, large or small, has equal power to vote for their local patterns when global decisions are being made by their company. Thus, we may expect to obtain some commonness from local patterns. This commonness could be of interest to company headquarters.

We now formally define the patterns below.

Let $D_1, D_2, \ldots, D_m$ be m databases in the m branches $B_1, B_2, \ldots, B_m$ of a company, respectively; and LI_i be the set of patterns (local patterns) from D_i ($i = 1, 2, \ldots, m$), and

$$LI = \{r_j | r_j \in LI_1 \cup LI_2 \cup \cdots \cup LI_m, 1 \le j \le n\}$$

where $n = |LI_1 \cup LI_2 \cup \cdots \cup LI_m|$.

The table of frequency of patterns voted for by the branches of an interstate company for local patterns is listed in Table 7.6.

Table 7.6. Frequencies of patterns voted for by branches of an interstate company

	r_1	r_2	$\cdots$	r_n
B_1	$a_{1,1}$	$a_{1,2}$	$\cdots$	$a_{1,n}$
B_2	$a_{2,1}$	$a_{2,2}$	$\cdots$	$a_{2,n}$
$\cdots$	$\cdots$	$\cdots$		$\cdots$
B_m	$a_{m,1}$	$a_{m,2}$	$\cdots$	$a_{m,n}$
$Voted\ Number$	$voting_1$	$voting_2$	$\cdots$	$voting_n$

In Table 7.6, B_i is the ith branch of an interstate company ($1 \le i \le m$); $a_{i,j} = 1$ means that branch B_i votes for pattern r_j (where, r_j is a valid pattern in branch B_i), $a_{i,j} = 0$ means that branch B_i does not vote for pattern r_j (where, r_j is not a valid pattern in branch B_i) ($1 \le i \le m$ and $1 \le j \le n$); and $voting_i$ is the number of branches that vote for the ith patterns ($1 \le i \le m$).

From Table 7.6, the average voting rate can be obtained as

$$AverageVR = \frac{voting(r_1) + voting(r_2) + \cdots + voting(r_n)}{n} \tag{7.3}$$

where $voting(r_i) = voting_i / m$, which is the voting ratio of r_i.

The voting rate of a pattern is *high* if it is greater than $AverageVR$. By using $AverageVR$, these patterns can be classified into four classes which are depicted in Fig. 7.3.

In Fig. 7.3, $votingrate = AverageVR$ is a reference line for measuring the interest of the pattern r_i. Certainly, the further the voting rate is from the line, the more the interest. If the voting rate of a pattern is in $[x_1, 1]$, the pattern is referred

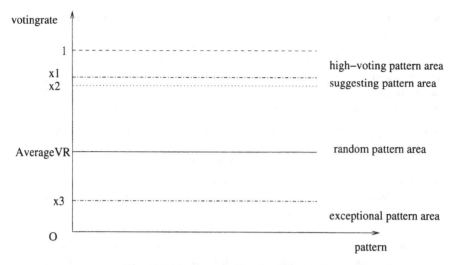

Fig. 7.3. Patterns in local pattern sets

to as a "high-vote pattern", and $[x_1, 1]$ refers to as a *high-vote pattern area*. If the voting rate of a pattern is in $[x_2, x_1)$, the pattern is referred to as a "suggested pattern", and $[x_2, x_1)$ refers to as a *suggested pattern area*. If the voting rate of a pattern is in (x_3, x_2), the pattern is referred to as a "random pattern", and (x_3, x_2) refers to as a *random pattern area*. Random patterns are of no interest to company headquarters. Nor are they considered in this book. If the voting rate of a pattern is in $[0, x_3]$, the pattern is referred to as an "exceptional pattern" and $[0, x_3]$ refers to as an *exceptional pattern area*. For any pattern r in the exceptional pattern area, r is of interest if it satisfies the conditions given in next section.

To measure the interest of a high-vote pattern r_i, the relationship between the voting rate, $voting(r_i)$, and the average voting rate, $AverageVR$, are considered. Certainly, if $voting(r_i) > AverageVR$, the pattern r_i refers to a high-vote pattern. Therefore, $voting(r_i) - AverageVR$ satisfies:

$$0 < voting(r_i) - AverageVR \le 1 - AverageVR$$

In particular, we have

$$0 < \frac{voting(r_i) - AverageVR}{1 - AverageVR} \le 1$$

Certainly, the bigger the ratio $(voting(r_i) - AverageVR)/(1 - AverageVR)$, the more interesting the pattern.

Consequently, the interest measure $LPI(r_i)$ of a pattern r_i is defined according to the deviation of the voting rate $voting(r_i)$ from the average voting rate $AverageVR$, and

$$LPI(r_i) = \frac{voting(r_i) - AverageVR}{1 - AverageVR} \tag{7.4}$$

for $1 - AverageVR \neq 0$, where $LPI(r_i)$ is referred to as the interest of r_i, given $AverageVR$.

From this interest measure, $LPI(r_i)$ is positively related to the real voting ratio of the pattern r_i. It is highest if the real voting ratio is 1. A pattern r_i in local patterns is an interesting high-vote pattern if its interest measure LPI_i is equal to, or greater than, a threshold – the minimum interest degree $(minVR)$ given by users or experts. Obviously, for $LPI(r_i)$:

- if $voting(r_i) = AverageVR$, r_i is of no interest, and the interest measure of the pattern is as
$$LPI(r_i) = 0$$

- if $voting(r_i) - AverageVR > 0$, r_i is a high-vote pattern. When $voting(r_i) = 1$ is the strongest voting rate, r_i is a high-vote pattern of interest, and the interest measure of the pattern is as
$$LPI(r_i) = 1$$

- again, if $voting(r_i) - AverageVR < 0$, r_i is a low-vote pattern. When $voting(r_i) = 0$ is the weakest voting rate: r_i is a low-vote pattern of interest. The interest measure of this pattern will be defined in Chap. 8.

The problem of finding high-vote patterns can now be stated as follows.

Given a set of local pattern sets, $LIset$, find all patterns for which interest degrees are equal to, or greater than, $minVR$.

7.3.2 Algorithm Design

High-vote patterns reflect the commonness of branches within an interstate company. Finding high-vote patterns in multiple databases is a procedure that identifies all patterns for which interest degrees are equal to, or greater than, $minVR$. This subsection presents the algorithm *highvotingPatterns*.

Procedure 1 *highvotingPatterns*
begin
Input: LI_i $(1 \leq i \leq M)$: *sets of local patterns, minVR: threshold value that is the minimum interest voting ratio;*
Output: *FPattern: set of high-vote patterns;*
(1) **call** the procedure *GoodClass* to generate a classification $class^\alpha$ for M local patterns LI_i;
(2) **if** $\alpha = 1$ **then**
    begin
        input please suggest a value for α;
        call the procedure *GreedyClass* to generate a classification
            $class^\alpha$;
    end;
(3) **let** $FPattern \leftarrow \{\}$;
(4) **for** each $class_i$ in $class^\alpha$ **do**
    begin
        (4.1) **let** $FPattern_i \leftarrow \{\}$;
        let pattern set $P \leftarrow \emptyset$;

(4.2) **for** a local pattern set LI in $class_i$ **do**
      for each pattern r in LI **do**
          begin
            if $r \in P$ **then**
              let $r_f \leftarrow r_f + 1$;
            else
              begin
                let $P \leftarrow P \cup \{r\}$;
                let $r_f \leftarrow 1$;
              end
          end
(4.3) **for** each pattern r in P **do**
      if $LPI(r_f) \geq minVR$ **then**
          let $FPattern_i \leftarrow FPattern_i \cup \{r\}$;
(4.4) **let** $FPattern \leftarrow FPattern \cup \{FPattern_i\}$;
    end
(5) **output** the high-vote patterns in $FPattern$;
end;

The procedure *highvotingPattens* above searches for all high-vote patterns from M given local pattern sets.

Step (1) finds a good classification $class^{\alpha}$ for the M local pattern sets (i.e., from M databases) using the algorithm in [18]. For $class^{\alpha}$, we need to search for high-vote patterns one by one with respect to the classes. The set of high-vote patterns found in a class is taken as an element of $FPattern$. Step (3) allows the set $FPattern$ to be an empty set.

Step (4) analyzes the patterns for each class, $class_i$, in $class^{\alpha}$. Step (4.1) initializes the set variable $FPattern_i$ that is used to save all high-vote patterns in $class_i$, and the set variable P that is used to save all patterns in $class_i$. Step (4.2) sums up the frequency for each pattern r in $class_i$. Step (4.3) generates all high-vote patterns from set P and saves them in $FPattern_i$, where each pattern r in $FPattern_i$ has a voting ratio $LPI(r_f) \geq minVR$. Step (4.4) appends $FPattern_i$ into $FPattern$.

Step (5) outputs all high-vote patterns in the given local pattern sets class by class.

7.3.3 Searching High-Vote Patterns: An Example

We now demonstrate the use of the algorithm *highvotingPatterns*.

Example 2. Consider the seven local pattern sets LI_1, $LI_2, \ldots, LI_7$ which are obtained from branches B_1, $B_2, \ldots, B_7$, respectively.

$$LI_1 = \{(A, 0.4); (B, 0.5); (C, 0.7); (D, 0.6); (E, 0.9)\}$$
$$LI_2 = \{(A, 0.5); (B, 0.45); (C, 0.5)\}$$
$$LI_3 = \{(A, 0.55); (B, 0.42)\}$$
$$LI_4 = \{(A, 0.51); (C, 0.4); (D, 0.44)\}$$
$$LI_5 = \{(E, 0.38); (F, 0.5); (G, 0.85); (I, 0.44)\}$$
$$LI_6 = \{(E, 0.47); (I, 0.34); (J, 0.6)\}$$
$$LI_7 = \{(E, 0.55); (F, 0.35)\}$$

Here each local pattern set has several patterns, separated by a semicolon, and each pattern consists of its name and its support in a branch, separated by a comma. Let the minimal voting ratio $minVR$ equal 0.25.

Firstly, the algorithm $highvotingPatterns$ searches for all high-vote patterns from the seven local pattern sets. After calling $GoodClass$, there is a good classification $class^{0.0375}$ for the seven local pattern sets, and $class^{0.0375}$ has two elements as follows:

$$class_1 = \{LI_1, LI_2, LI_3, LI_4\}$$
$$class_2 = \{LI_5, LI_6, LI_7\}$$

For $class_1$ and $class_2$, the votes of patterns are summed up, as listed in Tables 7.7 and 7.8.

Table 7.7. Frequency of patterns in $class_1$ voted for by branches

	A	B	C	D	E
B_1	1	1	1	1	1
B_2	1	1	1	0	0
B_3	1	1	0	0	0
B_4	1	0	1	1	0
$Voted\ Number$	4	3	3	2	1

Table 7.8. Frequency of patterns in $class_2$ voted for by branches

	E	F	G	I	J
B_5	1	1	1	1	0
B_6	1	0	0	1	1
B_7	1	1	0	0	0
$Voted\ Number$	3	2	1	2	1

From the voting ratios of patterns, we can get the interest measurements of all patterns. For $class_1$, $AverageVR = 0.65$. Because the voting rates of D and E are less than $AverageVR$, they are not high-vote patterns. For A, B and C,

$$LPI(A) = \frac{voting(A) - AverageVR}{1 - AverageVR} = \frac{1 - 0.65}{1 - 0.65} = 1$$

$$LPI(B) = \frac{voting(B) - AverageVR}{1 - AverageVR} = \frac{0.75 - 0.65}{1 - 0.65} \approx 0.286$$

$$LPI(c) = \frac{voting(C) - AverageVR}{1 - AverageVR} = \frac{0.75 - 0.65}{1 - 0.65} \approx 0.286$$

For $class_2$, $AverageVR = 0.6$. Because the voting rates of G and J are less than $AverageVR$, they are not high-vote patterns. For E, F and I,

$$LPI(E) = \frac{voting(E) - AverageVR}{1 - AverageVR} = \frac{1 - 0.6}{1 - 0.6} = 1$$

$$LPI(F) = \frac{voting(F) - AverageVR}{1 - AverageVR} = \frac{0.667 - 0.6}{1 - 0.6} \approx 0.167$$

$$LPI(I) = \frac{voting(I) - AverageVR}{1 - AverageVR} = \frac{0.667 - 0.6}{1 - 0.6} \approx 0.167$$

When $minVR = 0.25$, high-vote patterns in $FPattern_1$ and $FPattern_2$ are as follows.

$$FPattern_1 = \{(A, 1); (B, 0.75); (C, 0.75)\}$$
$$FPattern_2 = \{(E, 1)\}$$

Here each set has several patterns, separated by a semicolon, and each pattern consists of its name and its voting ratio by the branches in a class, separated by a comma.

We can represent high-vote patterns in natural language. For example, let B stand for "$i_1 \rightarrow i_2$". Then the high-vote pattern $(B, 0.75)$ can be represented as "75% of the branches agreed that if i_1 then i_2".

7.4 Identifying Exceptional Patterns of Interest

Exceptional patterns often present as more glamorous than high-vote patterns in such areas as marketing, science discovery, and information safety. For example, "20% of 10 toy branches strongly supported the new toy "Mulan" which was purchased with rather high frequency". These local patterns can be used to analyze the possible purchasing trends, although "Mulan" has a low-vote rate.

7.4.1 Measuring Exceptional Patterns

As we have said, exceptional patterns can grasp the individuality of branches. This subsection presents models for measuring the interest of such patterns.

To identify exceptional patterns of interest from local patterns, the projection of $P(name, vote, vsupp, vconf)$ on $name$, $vote$ and $vsupp$ is considered, i.e., the projection

$$P(name, vote, vsupp) \tag{7.5}$$

is considered. The table of frequencies of patterns voted for by the branches of an interstate company for local patterns is the same as that in Table 7.6.

If the voting rate of a pattern is less than $AverageVR$, the pattern might be an exceptional pattern. This means that interesting exceptional patterns are hidden in low-vote patterns. To measure the interest of an exceptional pattern, r_j, its voting rate and its supports in branches must be considered. In order to reflect the factors, two metrics for the interest are constructed below.

The first metric considers the relationship between the voting rate $voting(r_j)$, and the average voting rate $AverageVR$. If $voting(r_j) < AverageVR$, the pattern r_j refers to a low-vote pattern. In this case, $voting(r_j) - AverageVR$ satisfies:

$$-AverageVR \leq voting(r_j) - AverageVR < 0$$

In particular, we have

$$0 < \frac{voting(r_j) - AverageVR}{-AverageVR} \leq 1$$

Certainly, the bigger the ratio $(voting(r_j) - AverageVR)/(-AverageVR)$, the more interesting the pattern.

Consequently, one of the interest measures, $EPI(r_j)$, of a pattern, r_j, is defined as the deviation of the voting rate, $voting(r_j)$, from the average voting rate, $AverageVR$.

$$EPI(r_j) = \frac{voting(r_j) - AverageVR}{-AverageVR} \qquad (7.6)$$

for $AverageVR \neq 0$, while $EPI(r_j)$ is referred to as the interest of r_j, given $AverageVR$.

From the above interest measure, $EPI(r_j)$ is negatively related to the voting ratio of the pattern r_j. It is highest if the voting ratio is 0. A pattern r_j in local patterns is an exceptional pattern if its interest measure $EPI(r_j)$ is greater than or equal to the threshold minimum interest degree ($minEP$) given by users or experts.

Because an exceptional pattern is a low-vote pattern, it cannot be a valid pattern in the whole dataset that is the union of a class of databases. However, exceptional patterns are those that are strongly supported by a few of branches. That is, exceptional patterns reflect the individuality of such branches. Consequently, the second metric considers the support dimensions of a pattern.

From local patterns, the supports of patterns in branches of an interstate company are listed in Table 7.9.

Table 7.9. Supports of patterns in branches of an interstate company

	r_1	r_2	$\cdots$	r_n	$minsupp$
B_1	$supp_{1,1}$	$supp_{1,2}$	$\cdots$	$supp_{1,n}$	$minsupp_1$
B_2	$supp_{2,1}$	$supp_{2,2}$	$\cdots$	$supp_{2,n}$	$minsupp_2$
$\cdots$	$\cdots$	$\cdots$		$\cdots$	$\cdots$
B_m	$supp_{m,1}$	$supp_{m,2}$	$\cdots$	$supp_{m,n}$	$minsupp_m$

In Table 7.9, B_i is the ith branch of the interstate company ($1 \leq i \leq m$), and $supp_{i,j} \neq 0$ stands for the support of pattern r_j in branch B_i, when r_j is a valid pattern in B_i. If $supp_{i,j} = 0$, then r_j is not a valid pattern in branch B_i ($1 \leq i \leq m$ and $1 \leq j \leq n$). Meanwhile, $minsupp_i$ is the minimum support used to identify local patterns within the ith branch.

A pattern, r_j, in local patterns is exceptional if its supports in a few of branches are pretty high. How large can a support be referred to as *high*? In the case of 0.5 and 0.2, 0.5 is obviously higher than 0.2. However, the supports of a pattern are collected from different branches, and local databases in branches may differ in size. For example, let 100 and 100000 be the number of transactions in databases D_1 and

D_2 respectively, and 0.5 and 0.2 be the supports of r_j in D_1 and D_2 respectively. For these supports of r_j, the support 0.5 of r_j in D_1 cannot be simply regarded as high, and the support 0.2 of r_j in D_2 cannot be simply regarded as low.

Consequently, a reasonable measure for high support should be constructed. Though it is difficult to deal with raw data in local databases, the minimum supports in branches are suitable references for determining which of the supports of a pattern are high.

Generally, the higher the support of a pattern in a branch, the more interesting the pattern. Referenced with the minimum support $minsupp_i$ in branch B_i, the interest of a pattern r_j in B_i is defined as

$$RI_i(r_j) = \frac{supp_{i,j} - minsupp_i}{minsupp_i} \qquad (7.7)$$

where $supp_{i,j}$ is the support of r_j in branch B_i.

From the above measure, $RI_i(r_j)$ is positively related to the support of r_j in branch B_i. It is highest if the support is 1. The pattern r_j in B_i is of interest if RI_i is greater than or equal to a threshold: the minimum interest degree ($minEPsup$) given by users or experts.

Example 3. Consider the supports of patterns in four branches given in Table 7.10. Pattern r_1 is a valid pattern in branches B_1 and B_4 and not a valid pattern in branches B_2 and B_3. The interesting degrees $RI_1(r_1)$ and $RI_4(r_1)$ of r_1 in B_1 and B_4 are as follows

$$RI_1(r_1) = \frac{supp_{1,1} - minsupp_1}{minsupp_1} = \frac{0.021 - 0.01}{0.01} = 1.1 \,,$$

$$RI_4(r_1) = \frac{supp_{4,1} - minsupp_4}{minsupp_4} = \frac{0.7 - 0.5}{0.5} = 0.4$$

This means that the interest of r_1 in B_1 is higher than that in B_4 though $supp_{1,1} = 0.021 < supp_{4,1} = 0.7$.

For other patterns, their interesting degrees are shown in Table 7.11.

In Table 7.11, "−" stands for that a pattern is not voted for by a branch.

From the above discussion, an exceptional pattern r is of interest if

(C1) $EPI(r) \geq minEP$; and
(C2) $RI_i(r) \geq minEPsup$ for branches that vote for r.

The problem of finding exceptional patterns can now be stated as follows: given a set of local pattern sets, $LIset$, find all patterns that satisfy both the conditions (C1) and (C2).

Table 7.10. Supports of patterns in four branches of an interstate company

	r_1	r_2	r_3	r_4	$minsupp$
B_1	0.021	0.012	0.018	0	0.01
B_2	0	0.6	0	0.3	0.25
B_3	0	0	0.8	0	0.4
B_4	0.7	1	0	0	0.5

Table 7.11. Interest of patterns in branches

	r_1	r_2	r_3	r_4	$minsupp$
B_1	1.1	0.2	0.8	–	0.01
B_2	–	1.4	–	0.2	0.25
B_3	–	–	1	–	0.4
B_4	0.4	1	–	–	0.5

7.4.2 Algorithm Designing

Below, we design an algorithm, $Exceptional Patterns$, for identifying interesting exceptional patterns from local patterns. Then, in the next section, an example will be used to illustrate how to search exceptional patterns from local patterns by way of algorithms.

Procedure 2 $ExceptionalPatterns$
begin
    Input: LI_i $(1 \le i \le M)$: sets of local patterns,
    $minEP$: threshold value that is the minimal interest degree for the voting ratios of exceptional patterns;
    $minEPsupp$: threshold value that is the minimal interest degree for the supports of exceptional patterns in branches;
    Output: $EPattern$: set of exceptional patterns of interest;
(1) **generate** a classification $class$ for M local patterns LI_i;
(2) **let** $EPattern \leftarrow \{\}$;
(3) **for** each $class_i$ in $class$ **do**
    begin
     (3.1) **let** $EPattern_i \leftarrow \{\}$; $giveup_i \leftarrow \{\}$;
        let pattern set $E \leftarrow \emptyset$;
     (3.2) **for** a local pattern set LI in $class_i$ **do**
        for each pattern r in LI **do**
            if $(r \notin giveup_i)$ and $(RI_{LI}(r) < minEPsup)$ **then**
            if $r \in E$ **then**
                let $E \leftarrow E - \{r\}$; $giveup_i \leftarrow giveup_i \cup \{r\}$;
            else
              if $r \in E$ **then**
                //$r.count$ is the counter of r.
                let $r.count \leftarrow r.count + 1$;
              else
                begin
                    let $E \leftarrow E \cup \{r\}$;
                    let $r.count \leftarrow 1$;
                end
     (3.3) **for** each pattern r in E **do**
        if $EPI(r) \ge minEP$ **then**
            let $EPattern_i \leftarrow EPattern_i \cup \{r\}$;
     (3.4) **let** $EPattern \leftarrow EPattern \cup \{EPattern_i\}$;
    end

(4) **output** the exceptional patterns in $EPattern$;
  end;

The procedure $ExceptionalPatterns$ is to search all interesting exceptional patterns from given M local pattern sets.

Step (1) finds a classification $class$ for the M local pattern sets (from M databases) using our database clustering in [21]. For the above $class$, we need to search exceptional patterns for classes one by one. The set of the interesting exceptional patterns found in a class is taken as an element of $EPattern$. Step (2) initializes the set $EPattern$ to be an empty set.

Step (3) is to handle the patterns for each class $class_i$ in $class$. Step (3.1) initializes the set variables $EPattern_i$, $giveup_i$ and E. $EPattern_i$ is used to save all exceptional patterns in $class_i$, and $giveup_i$ is used to save the patterns that are given up. E is used to save all patterns in $class_i$. Step (3.2) is to sum up the frequency for a pattern r that satisfies both $r \notin giveup_i$ and $RI_{LI}(r) \geq minEPsup$ in the local pattern set of $class_i$. Step (3.3) generates all interesting exceptional patterns from the set E and saves them in $EPattern_i$, where each pattern r in $EPattern_i$ is with $EPI(r) \geq minEP$ in the local pattern set of $class_i$. Step (3.4) is to append $EPattern_i$ into $EPattern$.

Step (4) outputs all exceptional patterns in the given local pattern sets class by class.

7.4.3 An Example

Consider the five local pattern sets $LI_1, LI_2, \ldots, LI_5$ below identified from branches $B_1, B_2, \ldots, B_5$, respectively.

$$LI_1 = \{(A, 0.34); (B, 0.45); (C, 0.47); (D, 0.56); (F, 0.8)\}$$
$$LI_2 = \{(A, 0.35); (B, 0.45); (C, 0.46)\}$$
$$LI_3 = \{(A, 0.05); (B, 0.02); (AB, 0.02)\}$$
$$LI_4 = \{(A, 0.38); (B, 0.45); (C, 0.65); (F, 0.68)\}$$
$$LI_5 = \{(A, 0.22); (C, 0.3); (D, 0.61)\}$$

where each local pattern set has several patterns, separated by a semicolon, and each pattern consists of its name and its support in a branch, separated by a comma.

Let $minsupp_1 = 0.3$, $minsupp_2 = 0.3$, $minsupp_3 = 0.008$, $minsupp_4 = 0.25$, $minsupp_5 = 0.2$, $minEP = 0.5$, and $minEPsup = 0.75$.

We now use the algorithm $ExceptionalPatterns$ to search all exceptional patterns from the five local pattern sets. There is one class $class_1 = \{LI_1, LI_2, LI_3, LI_4, LI_5\}$ for the given local pattern sets.

For $class_1$, the interesting degrees of patterns are shown in Table 7.12.

Because $minEPsup = 0.75$, patterns D, F, and AB are selected according to Step (3.2) in the algorithm $ExceptionalPatterns$. For $class_1$, $AverageVR = 0.6$, and for D, F, and AB

Table 7.12. Interest of patterns in $class_1$

	A	B	C	D	F	AB	$minsupp$
LI_1	0.133	0.5	0.567	0.867	1.667	–	0.3
LI_2	0.167	0.5	0.533	–	–	–	0.3
LI_3	5.25	1.5	–	–	–	1.5	0.008
LI_4	0.52	0.8	1.6	–	1.72	–	0.25
LI_5	0.1	–	0.5	2.05	–	–	0.2

$$EPI(D) = \frac{voting(D) - AverageVR}{-AverageVR} = \frac{0.4 - 0.6}{-0.6} \approx 0.333$$

$$EPI(F) = \frac{voting(F) - AverageVR}{-AverageVR} = \frac{0.4 - 0.6}{-0.6} \approx 0.333$$

$$EPI(AB) = \frac{voting(AB) - AverageVR}{-AverageVR} = \frac{0.2 - 0.6}{-0.6} \approx 0.667$$

According to $minEP = 0.5$, we can obtain all exceptional patterns in the set $EPattern$. And $EPattern$ consists of one element: $EPattern_1 = \{(AB, 0.2, (0, 0.25, 0, 0, 0))\}$. There is only one interesting exceptional pattern $P(AB, 0.2, (0, 0.25, 0, 0, 0))$ in the given local pattern sets.

7.5 Experiments

To study the effectiveness of the approach proposed in this chapter, a set of experiments was carried out on Dell, using Java. The experiments were designed to check whether the proposed approach can effectively identify interesting patterns (high-vote and exceptional patterns). From the proposed algorithms and our experiments, they are the same at efficiency and performance. Therefore, without loss of generality, in this section we only illustrate the experiments for examining the algorithm: $ExceptionalPatterns$.

From dataset selection, it is hoped that some interesting exceptional patterns will be found to exist in classes of databases. This can be done by changing the names of some high-vote patterns in some local pattern sets. For example, let A be a high-vote pattern that is voted for by 28 of 30 local pattern sets. A is changed to X in 25 of the 28 local pattern sets such that the supports of A in the remaining 3 local pattern sets are very high.

To obtain multiple, possibly relevant, databases, the techniques in [8, 23] were adopted. That is, a database was vertically partitioned into a number of subsets, each containing a certain number of attributes. It was hoped that some databases obtained are relevant to each other in classes. In this set of experiments, the databases are generated by Synthetic Classification Data Sets (http://www.kdnuggets.com/). One of the experiments is demonstrated in the followings, in which four databases were generated from the web site. The main properties of the four databases are as follows. There are $|R| = 1000$ attributes, and the average number T of attributes per row is 6. The number $|r|$ of rows is approximately 300000. The average size I of

Table 7.13. Synthetic data set characteristics

Data Set Name	$|R|$	T	$|r|$	Number of Subset
T6I4N1	1000	6	300256	30
T6I4N2	1000	6	299080	30
T6I4N3	1000	6	298904	30
T6I4N4	1000	6	301005	30

maximal frequent sets is 4. The databases are vertically partitioned into 30 datasets. Table 7.13 summarizes the parameters for the databases.

To evaluate the proposed approach, the algorithm *ExceptionalPatterns* is used to generated all exceptional frequent-itemsets (patterns). The effectiveness and efficiency of the algorithm *ExceptionalPatterns* for identifying exceptional patterns are as follows. Table 7.14 shows the results for generating exceptional patterns.

Table 7.14. Exceptional frequent-itemsets ($minEP = 0.40$)

Group Name	$minsupp = 0.0015$	$minsupp = 0.001$
N1	5	5
N2	8	8
N3	15	15
N4	10	10

In Table 7.14, "0.0015" is the minimum support used to measure frequent-itemsets when the datasets are mined. "Ni" is the name of the ith group of datasets that are obtained from the ith database T6I4Ni. "5" is the number of exceptional frequent-itemsets in "N1" when $minsupp = 0.0015$ and $minEP = 0.4$, and "5" is the number of exceptional frequent-itemsets in "N1" when $minsupp = 0.001$ and $minEP = 0.4$.

Figure 7.4 depicts the distribution of exceptional frequent-itemsets in the four groups of datasets.

Table 7.15 shows the running time of *ExceptionalPatterns* when $minEP = 0.4$ for $minsupp = 0.0015$ and $minsupp = 0.001$.

Table 7.15. Running time ($minEP = 0.4$)

Group Name	$minsupp = 0.0015$	$minsupp = 0.001$
N1	13	15
N2	19	22
N3	15	18
N4	16	17

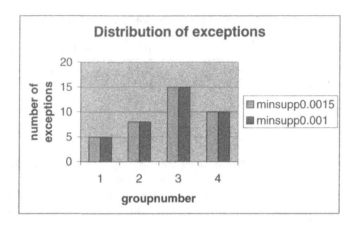

Fig. 7.4. Distribution of exceptional frequent-itemsets in the four groups of datasets when $minEP = 0.4$

Figure 7.5 illustrates the running time for searching exceptional frequent-itemsets in the four groups of datasets when $minEP = 0.4$.

The results from the proposed approach for identifying exceptional patterns are promising. It is also evident from the experimental results that the proposed approach is very efficient due to the fact that it focuses only on local patterns.

As we have previously stated, traditional multidatabase mining techniques put all the data from relevant databases into a single database for pattern discovery. This can destroy some important information such as the above exceptions. In other words, our multidatabase mining technique can identify two new kinds of patterns: exceptions and high-vote patterns.

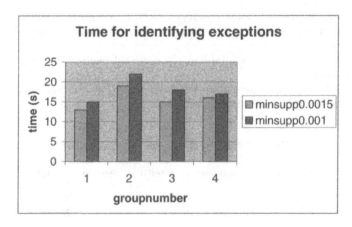

Fig. 7.5. Running time for identifying exceptional frequent-itemsets when $minEP = 0.4$

7.6 Summary

Because putting all the data from (a class of) multidatabases into a single dataset can destroy important information, such as "10% of branches strongly support that the fact that a customer is likely to purchase pork if he/she purchase chili", this chapter develops techniques for identifying novel patterns, known as high-vote and exceptional patterns, in multidatabases by analyzing local patterns. The main contributions of this chapter are as follows.

(1) Designed an algorithm for identifying high-vote patterns.
(2) Constructed two measurements for the interest of exceptional patterns.
(3) Evaluated the proposed techniques by experiments.

Commonality patterns (high-vote patterns), such as "80% of 15 supermarket branches reported that customers like to buy the products of Sunshine", are useful in market forecasting, customer behavior prediction, and global decision-making of an interstate company. Because traditional multidatabase mining techniques cannot identify commonality patterns, they are regarded as relatively novel patterns that describe the distribution of patterns within branches.

Unlike high-vote patterns, exceptional patterns (Individuality patterns) are hidden in local patterns. In real-world applications, exceptional patterns often present as more glamorous than high-vote patterns in such areas as marketing, science discovery, and information safety.

However, traditional multidatabase mining puts all the data together from relevant databases to amass a huge dataset for discovery upon mono-database mining techniques that can be used. This means that traditional multidatabase mining techniques cannot identify high-vote and exceptional patterns in multidatabases. Therefore, our models in this chapter is the first research effort in this direction.

References

1. J. Albert, Theoretical Foundations of Schema Restructuring in Heterogeneous Multidatabase Systems. In: *Proceedings of CIKM*, 2000: 461–470.
2. J. Aronis et al, The WoRLD: Knowledge discovery from multiple distributed databases. *Proceedings of 10th International Florida AI Research Symposium*, 1997: 337–341.
3. P. Chan, An Extensible Meta-Learning Approach for Scalable and Accurate Inductive Learning. *PhD Dissertation*, Dept of Computer Science, Columbia University, New York, 1996.
4. J. Chattratichat, etc., Large scale data mining: challenges and responses. In: *Proceedings of KDD*, 1997: 143–146.
5. D. Cheung, V. Ng, A. Fu and Y. Fu, Efficient Mining of Association Rules in Distributed Databases, *IEEE Trans. on Knowledge and Data Engg.*, 8(1996), 6: 911–922.
6. E. Han, G. Karypis and V. Kumar, Scalable Parallel Data Mining for association rules. In: *Proceedings of ACM SIGMOD*, 1997: 277–288.
7. A. Hurson, M. Bright, and S. Pakzad, *Multidatabase systems: an advanced solution for global information sharing*. IEEEComputer Society Press, 1994.

8. H. Liu, H. Lu, and J. Yao, Identifying Relevant Databases for Multidatabase Mining. In: *Proceedings of Pacific-Asia Conference on Knowledge Discovery and Data Mining*, 1998: 210–221.

9. J. Park, M. Chen, P. Yu: Efficient Parallel and Data Mining for Association Rules. In: *Proceedings of CIKM*, 1995: 31–36.

10. A. Prodromidis, S. Stolfo. Pruning meta-classifiers in a distributed data mining system. In: *Proc. of the First National Conference on New Information Technologies*, 1998: 151–160.

11. A. Prodromidis, P. Chan, and S. Stolfo, Meta-learning in distributed data mining systems: Issues and approaches, In *Advances in Distributed and Parallel Knowledge Discovery*, H. Kargupta and P. Chan (editors), AAAI/MIT Press, 2000.

12. J. Ribeiro, K. Kaufman, and L. Kerschberg, Knowledge discovery from multiple databases. In: *Proceedings of KDD95*. 1995: 240–245.

13. T. Shintani and M. Kitsuregawa, Parallel mining algorithms for generalized association rules with classification hierarchy. In: *Proceedings of ACM SIGMOD*, 1998: 25–36.

14. G. Webb, Efficient search for association rules. In: *Proceedings of ACM SIGKDD*, 2000: 99–107.

15. D.H. Wolpert, Stacked Generalization. *Neural Networks*, 5(1992): 241–259.

16. S. Wrobel, An algorithm for multi-relational discovery of subgroups. In: J. Komorowski and J. Zytkow (eds.) *Principles of Data Mining and Knowledge Discovery*, 1997: 367–375.

17. Xindong Wu and Shichao Zhang, Synthesizing High-Frequency Rules from Different Data Sources, *IEEE Transactions on Knowledge and Data Engineering*, Vol. 15, No. 2, March/April 2003: 353–367.

18. Xindong Wu, Chengqi Zhang and Shichao Zhang, Database Classification for multidatabase Mining. Information Systems, accepted, forthcoming.

19. Shichao Zhang, *Knowledge Discovery in multidatabases by Analyzing Local Instances*, PhD thesis, Deakin University, 31 October 2001.

20. Shichao Zhang, Xindong Wu and Chengqi Zhang, multidatabase Mining. *IEEE Computational Intelligence Bulletin*, Vol. 2, No. 1, June 2003: 5–13.

21. Chengqi Zhang and Shichao Zhang, Database Clustering for Mining multidatabases. In: *Proceedings of the 11th IEEE International Conference on Fuzzy Systems*, Honolulu, Hawaii, USA, May 2002.

22. Chengqi Zhang and Shichao Zhang, *Association Rules Mining: Models and Algorithms*. Springer-Verlag Publishers in Lecture Notes on Computer Science, Volume 2307, p. 243, 2002.

23. J. Yao and H. Liu, Searching Multiple Databases for Interesting Complexes. In: *Proc. of PAKDD*, 1997: 198–210.

24. N. Zhong, Y. Yao, and S. Ohsuga, Peculiarity oriented multidatabase mining. In: *Proceedings of PKDD*, 1999: 136–146.

8

Comparison Between Five Classifiers for Automatic Scoring of Human Sleep Recordings

Guillaume Becq[1], Sylvie Charbonnier[2], Florian Chapotot[1], Alain Buguet[4], and Lionel Bourdon[1] and Pierre Baconnier[3]

[1] Centre de Recherches du Service de Santé des Armées, 24 Avenue des Maquis du Grésivaudan, BP 87, 38702 La Tronche cedex, France
guillaume.becq@crssa.net
[2] Laboratoire d'Automatique de Grenoble, Ecole Nationale Supérieure d'Ingénieurs Electriciens de Grenoble, rue de la Piscine, BP 46, 38402 Saint Martin d'Hères, France
Sylvie.Charbonnier@lag.ensieg.inpg.fr
[3] Laboratoire du Traitement de l'Image de la Modélisation et de la Cognition, faculté de médecine, 38700 La Tronche, France
[4] Institut de Médecine Tropicale du Service de Santé des Armées, 13998 Marseille Armées, France

Abstract. The aim of this work is to compare the performances of 5 classifiers (linear and quadratic classifiers, k nearest neighbors, Parzen kernels and neural network) to score a set of 8 biological features extracted from EEG and EMG, in six classes corresponding to different sleep stages as to automatically elaborate an hypnogram and help the physician diagnosticate sleep disorders. The data base is composed of 17265 epochs of 20 s recorded from 4 patients. Each epoch has been classified by an expert into one of the six sleep stages. In order to evaluate the classifiers, learning and testing sets of fixed size are randomly drawn and are used to train and test the classifiers. After several trials, an estimation of the misclassification percentage and its variability is obtained (optimistically and pessimistically). Data transformations toward normal distribution are explored as an approach to deal with extreme values. It is shown that these transformations improve significantly the results of the classifiers based on data proximity.

Key words: Bayesian Classifiers, Error Estimation, Neural Networks, Normalization, Polysomnography, Representation, Sleep Staging

8.1 Introduction

In biology, taxonomy has been the source of numerous studies and still remains one of the predominant fields of research (genome studies). The development of multidimensional exploratory analyses, computing power and numerical solutions

Guillaume Becq et al.: *Comparison Between Five Classifiers for Automatic Scoring of Human Sleep Recordings*, Studies in Computational Intelligence (SCI) **4**, 113–127 (2005)
www.springerlink.com

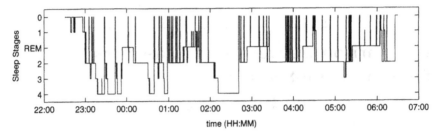

Fig. 8.1. A human hypnogram. Sleep-wake stage scoring has been realized by an expert into 6 different sleep stages from 22 h 30 min to 06 h 30 min over epochs of 20 s: 0-Wake and Movement Time, 1-stage 1 (transition from waking to sleeping), 2-stage 2, 3-stage 3, 4-stage 4 (stage 2, 3 and 4 are part of the orthodox sleep with more and more slow waves observed on the recording), REM–Rapid Eye Movement (or paradoxical) sleep (with rapid brain activity with or without rapid eye movements and muscle atonia)

can explain the growth of such studies. However, in the case of time series, one notices that relative few works have been developed to deal with clustering or classification techniques. One interesting source of such studies is the study of sleep, where several classification techniques have been tested to determine structures on real temporal data [24, 27].

The starting point of sleep studies has been the observation of the electrical activity of the brain measured by electrodes fixed on the scalp, during all night recordings. First observations showed that several patterns were similar from one individual to another, their distributions fluctuating throughout the night. Originally (about 1940) [20, 21], analog signals were plotted on pages of paper. At that time, sleep recordings consisted of huge blocks of paper. With the first discoveries and the evidence of different phases of electroencephalic activity during the night [2, 6, 7, 15], several techniques for electrodes placement were applied and various practices for classifying these activities sprang up. In order to extract the different patterns of such recordings, one expert was assigned to read signals page by page, and give a score corresponding to specific processes of the sleep activity of the brain. The result of this reading has been called an hypnogram and consists in a succession of stages through the night (see Fig. 8.1).

The advantages of working with hypnograms are: an extraction of information from raw data generated by polysomnographic (PSG: multi-channel sleep) recordings, an easier interpretation of the architecture of the night and a better vision of the organization of long term biological processes. It is then easier to discriminate strange charts from normal ones. Therefore, the hypnogram becomes a powerful tool for the diagnosis of sleep pathologies. Besides, the hypnogram, as a summary of the night, considerably reduces the storage of data and allows different laboratories to exchange results and share their knowledge. For that reason, a consensus for a standardization of the rules used to score PSG recordings was held in 1968, bringing together the different leaders in electroencephalography. It led to the creation of the manual by Rechtschaffen and Kales (R&K) [25] currently applied in the different sleep laboratories where pathologies, sleep disorders and untypical hypnograms are studied.

Since 1970, and the growth of computerized methods, interests have been initiated in order to score automatically polysomnographic recordings [11, 29, 30], allowing the expert to avoid spending too much time on this time-consuming work. But studies are still in progress and improvements have to be made. For a complete review of the history of sleep, the reader is referred to [14] where the author, speaking about automatic sleep staging, notes: "The task turned out to be much more difficult because of ambiguities, artifacts and variations in human scoring".

This study has been developed in order to understand the different difficulties encountered with real biological data, while comparing expert practices and machine learning algorithms. For that purpose, a comparative study of five classifiers for automatic analysis of human sleep recordings is presented where temporal data coming from different individuals are mixed together. The interest of transformations toward normal distribution is emphasized since they lead to homogeneous representations for the different selected features. In the first paragraph, the database, the different classifiers, the method chosen to evaluate the performances of the classifiers and the transformations toward normal distribution are presented. In the second paragraph, the results obtained are discussed.

8.2 Materials and Methods

8.2.1 Presentation of the Database

The study has been realized over $N = 11$ polysomnographic recordings available in our database (from 4 healthy subjects). Features were extracted from one EEG (electro-encephalogram, differential lead C3–A2) and one bipolar EMG (electromyogram, position chin), sampled at 200 Hz. The choice of these features has been made in accord with experts in an effort to test a minimal set of electrodes considered necessary for the scoring of sleep.

Eight features thought to represent important physiological processes calculated over epochs of 20 s have been considered and are reported in Table 8.1, where σ denotes the standard deviation and $P_{\rm rel}$ the relative power in a given frequency band. The different bands are: δ (0.5–4.5 Hz), θ (4.5–8.5 Hz), α (8.5–11.5 Hz), σ (11.5–15.5 Hz), β (15.5–22.0 Hz), γ (22.0–45.0 Hz) and corresponds to the ones generally employed in sleep and waking EEG spectral studies [3].

During these epochs of fixed temporal intervals ($\Delta t = 20$ s), EEG can be considered approximately stationary [22]. This assumption is fundamental for the estimation of the different retained features, both in time domain and in the spectral domain. In each epoch, a score has been attributed by an expert. This score is assigned from a set constituted of $K = 6$ classes representing the 6 different stages encountered during human sleep defined in regards with the conventional criteria of R&K [25]: 0-Wake and Movement Time, 1-stage 1, 2-stage 2, 3-stage 3, 4-stage 4, 5-Rapid Eye Movement sleep (or Paradoxical sleep). The different aspects of EEG and EMG signals are represented Fig. 8.2, in order to appreciate the variations of the different signals throughout human sleep.

Once all the signals have been segmented into epochs, preprocessed and their features extracted, we can represent any observation **x** by a state representation in an R^d space ($d = 8$ for our study) where $(.)^t$ denotes the transpose of the vector:

Table 8.1. Description of the features used in the study and their statistical values for a) raw data, b) with z-score normalisation and c) with transformations toward normal distribution

Feature		μ	σ	min	max	min	max	Transform.	min	max
F_1	$\sigma(EEG)$	16.87	13.67	4.69	227.31	-0.89	15.40	$\log(1+x)$	-1.97	5.38
F_2	$P_{\rm rel}(EEG,\delta)$	0.69	0.16	0.01	0.99	-4.27	1.88	$\arcsin(\sqrt{x})$	-4.94	2.58
F_3	$P_{\rm rel}(EEG,\theta)$	0.14	0.07	0.00	0.68	-1.92	7.46	$\arcsin(\sqrt{x})$	-2.93	5.60
F_4	$P_{\rm rel}(EEG,\alpha)$	0.05	0.04	0.00	0.46	-1.34	10.65	$\log(\frac{x}{1-x})$	-4.91	2.95
F_5	$P_{\rm rel}(EEG,\sigma)$	0.05	0.04	0.00	0.50	-1.20	11.54	$\log(\frac{x}{1-x})$	-4.53	2.87
F_6	$P_{\rm rel}(EEG,\beta)$	0.04	0.04	0.00	0.94	-0.93	23.31	$\log(\frac{x}{1-x})$	-3.62	2.95
F_7	$P_{\rm rel}(EEG,\gamma)$	0.06	0.10	0.00	1.35	-0.59	13.44	$\log(\frac{x}{1-x})$	-2.95	2.45
F_8	$\sigma(EMG)$	21.42	39.54	0.00	394.97	-0.54	9.45	$\log(1+x)$	-2.10	3.25

$$\mathbf{x} = (F_1, F_2, \ldots, F_d)^t \tag{8.1}$$

When regrouping both the temporal instants and the different individuals, we obtain an array of observations or statistical units [19] representing the database over which the classification study is done:

$$M = \begin{pmatrix} F_1(t_0(1)) & F_2(t_0(1)) & \cdots & F_d(t_0(1)) & C(t_0(1)) \\ \cdots & \cdots & \cdots & \cdots & \cdots \\ F_1(t_f(1)) & F_2(t_f(1)) & \cdots & F_d(t_f(1)) & C(t_f(1)) \\ \vdots & & \vdots & & \vdots \\ F_1(t_k(i)) & F_2(t_k(i)) & \cdots & F_d(t_k(i)) & C(t_k(i)) \\ \vdots & & \vdots & & \vdots \\ F_1(t_1(N)) & F_2(t_1(N)) & \cdots & F_d(t_1(N)) & C(t_1(N)) \\ \cdots & \cdots & \cdots & \cdots & \cdots \\ F_1(t_f(N)) & F_2(t_f(N)) & \cdots & F_d(t_f(N)) & C(t_f(N)) \end{pmatrix} \tag{8.2}$$

where $t_k(i) = t_0(i) + k\Delta t$ for individual i.

The database was then constituted of 17265 observations over 8 parameters in which we introduced into the last column the expert's classification. A visual display of such a database is given in Fig. 8.3, with features transformed into normal distribution (detailed in 8.2.4) for a better homogeneity of representation.

8.2.2 Learning and Testing Sets

In machine learning, the supervised learning approach tries to learn rules, statistics, mathematical models, with a computer, from a desired result. A database containing both the different features used to solve the problem and the corresponding desired results are used. The aim is to find the model that minimizes a criteria which is a function of the difference between the results calculated by the machine and the desired results.

For this reason, it is common to separate the database into 2 sets: the first is used to induce the machine in a so called learning (or training) phase; the second

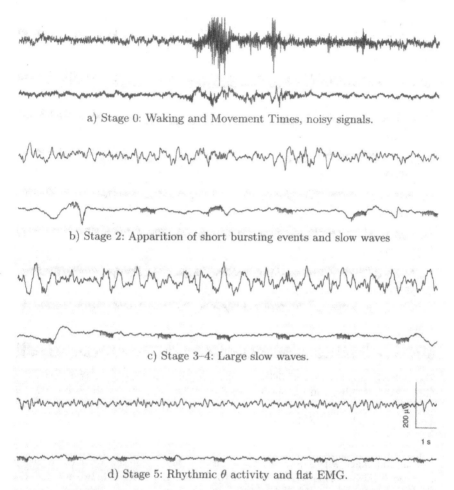

a) Stage 0: Waking and Movement Times, noisy signals.

b) Stage 2: Apparition of short bursting events and slow waves

c) Stage 3–4: Large slow waves.

d) Stage 5: Rhythmic θ activity and flat EMG.

Fig. 8.2. Electrophysiological behaviour during principal sleep stages. Each figure represents an epoch of 20 s. The same scale has been used for all figures

is used during a phase of validation (or test) for evaluating the performance with data that has not been used during the learning process. For a review of the different techniques for evaluating and preparing the data into learning set (LSet) and testing set (TSet) the reader is referred to [9, 10, 17].

Leave one out (N-1 vectors for learning and 1 vector for testing) or classical cross-validation (N/2 vectors for learning, and N/2 vectors for testing), can not be applied when working with a large database, such as ours, without huge computation times. We decided to randomly select a fixed number of data for the learning set and for the testing set, as is done in bootstrap techniques. The learning set will serve to train the classifiers, but also to calculate an optimist estimation (resubstitution techniques, empirical error) of the convergence of them. The testing set is used to obtain a pessimist estimation (cross validation techniques, real error) of them.

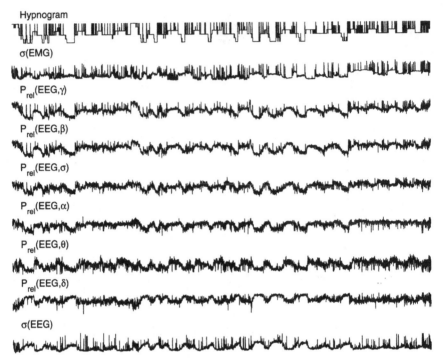

Fig. 8.3. Representation of 5000 elements of the database after transformation. Time and individuals are grouped together. When data is homogeneous, the influence of each feature is directly observable over the classification of the expert represented at the top of the figure

The difference with bootstrap techniques is that we do not reset the drawings after each drawing. This is done in order to obtain independence between estimation from the learning set and estimation from the testing set.

The choice of the number of data for the learning and testing sets can be obtained by looking over the stability of the performances of the classifiers. For a given size of the learning and testing sets, we trained a kNN classifier and a Parzen estimator. An estimation of the performance was realized over 30 subsets for both the optimistic and pessimistic error that are represented given with their standard deviation in Fig. 8.4. Classification errors reach $\approx 30\%$ and do not improve when the number of data in the sets increase over 500 samples when using both the kNN classifier or the Parzen estimator.

8.2.3 The Different Classifiers

Five common classifiers have been evaluated that can be regrouped into two distinct categories: the first category corresponds to the set of classifiers using probabilistic computations based upon the Bayes' rule to assign a class to a feature vector. The second category corresponds to classifiers delimiting regions into the representation

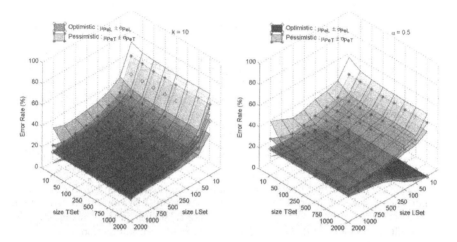

Fig. 8.4. Misclassification percentage in function of the size of the samples drawn from the database using the kNN classifier with $k = 10$ or the Parzen estimator with $\alpha = 0.5$. After 500 samples in the learning set and 500 samples in the testing set, performances are not improved

space by direct computation of frontiers. Explanation of the behavior of the different classifiers and the learning hypothesis can be found in [1, 5, 8]. Here we provide a short description of them.

Our study learning problem is to induce a classifier able to assign to a vector $\mathbf{x}$ of the representation space, a class $C \in \{\omega_i\}_{i=1}^{K}$ with respect to the knowledge database constituted of the data present in the learning set. We use the following notation: P for the probability, p the probability density, $E[.]$ the expectation operator, $|.|$ the determinant.

Bayes Rule-based Classifiers

The attribution of a vector $\mathbf{x}$ to a class is made using the Bayes' rule (8.3). The posterior conditional probability $P(\omega_i|\mathbf{x})$ is calculated for each of the K classes and the vector is given the class ω_i for which $P(\omega_i|\mathbf{x})$ is maximal (maximum a posteriori).

$$P(\omega_i|\mathbf{x}) = \frac{p(\mathbf{x}|\omega_i)P(\omega_i)}{p(\mathbf{x})} \tag{8.3}$$

$$p(\mathbf{x}) = \sum_{i=1}^{K} p(\mathbf{x}|\omega_i)P(\omega_i) \tag{8.4}$$

The learning problem consists in estimating the conditional density function $p(\mathbf{x}|\omega_i)$ from the different samples of the learning set. The different classifiers depend on the hypotheses made on this density function.

Parametric Models

The probability density function is assumed to be a multidimensional Gaussian model.

$$p(\mathbf{x}|\omega_i) = \frac{1}{(\sqrt{2\pi})^d \sqrt{|\Sigma|}} \exp\left(-\frac{1}{2}(\mathbf{x} - \boldsymbol{\mu})^t \Sigma^{-1}(\mathbf{x} - \boldsymbol{\mu})\right) \tag{8.5}$$

Its parameters (mean $\boldsymbol{\mu}$ and covariance matrices Σ) are estimated with samples drawn from the learning set:

$$\boldsymbol{\mu} = E[\mathbf{x}] = (\mu_1, \mu_2, \ldots, \mu_d)^t \tag{8.6}$$

$$\Sigma = E[(\mathbf{x} - \boldsymbol{\mu})(\mathbf{x} - \boldsymbol{\mu})^t] \tag{8.7}$$

Linear classifier: the covariance matrix $\Sigma = \Sigma_i$ is assumed to be the same for all classes. The resulting boundaries delimiting the classes are linear functions.

Quadratic classifier: the covariance matrix Σ_i is assumed to be different for each class and is estimated with representatives of each class in the learning set. The resulting boundaries delimiting the classes are quadratic functions.

Non-Parametric Models

The density function is described with

$$p(\mathbf{x}|\omega_i) = \frac{k_{n_i}}{n_i V_{n_i}} \tag{8.8}$$

with n_i number of representatives of class ω_i in the volume V_{n_i}.

k Nearest Neighbor (kNN) classifier: the probability density function is estimated by the volume occupied by a fixed number of neighbors (search of V_n with fixed k_n). It is simple to show that the decision obtained with the Bayes' rule maximization is equivalent to a voting kNN procedure. This procedure is a majority vote over the classes of the k nearest neighbors (present in the learning set) of the feature vector to classify.

Parzen estimator with Gaussian kernels: The probability density function is estimated by the sum of density kernels given a fixed volume V_n. To each sample $\mathbf{x}_{i,j}$ representative of class ω_i in the learning set, a density kernel $K(.)$ is associated. The sum over j of these n_i kernels gives the density of that class in that region and the probability density function is then

$$p(\mathbf{x}|\omega_i) = \frac{1}{n_i V_{n_i}} \sum_{j=1}^{n_i} K\left(\frac{\mathbf{x} - \mathbf{x}_{i,j}}{h_{n_i}}\right) \tag{8.9}$$

with

$$V_{n_i} = h_{n_i}{}^d = n_i{}^{-\alpha} \tag{8.10}$$

and the Gaussian kernel

$$K(u) = \frac{1}{(\sqrt{2\pi})^d} \exp\left(-\frac{1}{2}(u^2)\right) \tag{8.11}$$

These two methods require the tuning of a parameter: k, the number of neighbors for the k nearest neighbor estimator and α, for the Parzen estimator. The number of neighbors has been chosen to equal 10 and the parameter α has been set to 0.5, after evaluating the performance of the classifiers when incrementing the values of these parameters, as shown in Fig. 8.5.

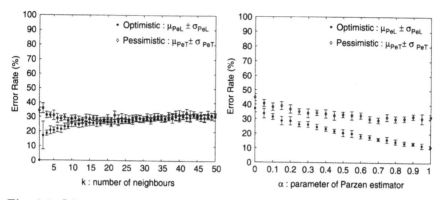

Fig. 8.5. Selection of the parameters for the k Nearest Neighbor classifier and for the Parzen estimator with Gaussian kernels. Size of the learning set and size of the test set have been set to $n_{LSet} = 500$ and $n_{TSet} = 500$. We retained $k = 10$ and $\alpha = 0.5$ for the comparison of the classifier because there is no improvement of the pessimistic error with greater values

Frontiers Based Classifiers

The frontiers of the classes in the multidimensional space are directly calculated from the data present in the learning set:

A multi layer perceptron (MLP): with 3 layers fully connected composed with 8 neurons in the input layer (hyperbolic tangent transfer function $y = 2/(1 + \exp(-2x)) - 1$), 6 neurons in the hidden layer (linear transfer function $y = x$) and 6 in the output layer (logarithmic sigmoid transfer function $y = 1/(1 + \exp(-x))$), trained by the feedforward backpropagation gradient algorithm. Weights were initiated randomly at the beginning of the learning phase. This structure is often used in discrimination [31] with an input layer connected to the representation space of $\mathbf{x}$ with $d = 8$ and the output layer connected to the desired class with $K = 6$. The choice of the number of neurons in the hidden layer has not been optimized in this study.

8.2.4 Transformations Towards Normal Distribution

Means, standard deviations, maximal and minimal values for the different retained features are given in Table 8.1. Inhomogeneity in raw data can be observed, as well as a wide spread of the data, which is typical with biological data. For example, even after doing a z-score defined by the transformation $\mathbf{z} = (\mathbf{x} - \mu)/(\sigma)$ (where μ is the mean of $\mathbf{x}$ and σ is its standard deviation), the maximal value of the sixth parameter is twenty three times the standard deviation.

In order to reduce the influence of extreme values, we applied transformations towards normal distribution on the whole set of the data, and for each parameter. These transformations are either $\log(x)$, $\log(1 + x)$, $\sqrt{x}$, $\sqrt[3]{x}$, $\log((x)/(1 - x))$, $1/(\sqrt{x})$, $\arcsin(\sqrt{x})$ depending on their effect over the different features. This was introduced by Theo Gasser [12] for normalization of EEG spectral parameters. These transformations perform more effectively, better than doing the simple z-score: the

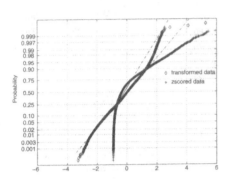

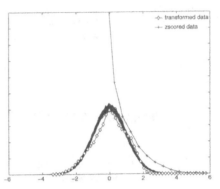

Fig. 8.6a. Effect of the transformations for $P_{\mathrm{rel}}(EEG, \beta)$ plotted in a lognormal axis system, known as Henry plot (normal probability plot). The $\log(\frac{x}{1-x})$ transformation gives a better approximation to a normal distribution represented by a line than a simple z-score

Fig. 8.6b. Effect of the transformation for $P_{\mathrm{rel}}(EEG, \beta)$ over the distribution of the data as compared to z-scored data and data obtained from Gaussian distributions simulated with the same number of realisations

inter-individual variability is reduced with the advantage to reduce tails in distributions. The effects of these transformations can be seen in Table 8.1. The maximal value of the eight parameters after transformation is no more than 6 times the standard deviation. An example of such transformations over one feature can be seen in the Henry plot shown in Fig. 8.6a or from the density plot in Fig. 8.6b.

8.3 Results

A training set and a validation set, each made up of 500 vectors randomly chosen, was built. Each classifier was trained on the first set and applied on the validation set. The performance of the classifier is given by the classification error expressed in percentage on the training set (which is optimistic) and on the validation set (which is pessimistic). This procedure was achieved ten times, which provides two times ten values for each classifier and enables the estimation of mean and variance. The results from one classifier to the other is said to be different if means are statistically different.

8.3.1 Results with Raw Data

Results from raw data are presented in Fig. 8.7a. These correspond to the mean value of the classification error on the training set and on the validation set obtained over 10 trials. For MLP, at each trial, 10 classifiers were trained with a different initialization of the weights, and the network with the minimal classification error was selected, in order to ensure the convergence of the network.

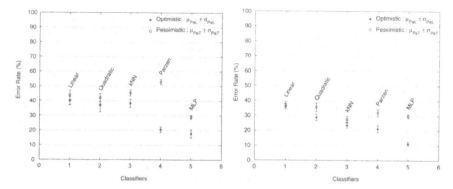

Fig. 8.7a. Means and standard deviations for the misclassification percentage obtained by 10 classifiers of each type on raw data: a) linear discrimination, b) quadratic discrimination, c) kNN with $k = 10$, d) Parzen estimator, with Gaussian kernel and e) Best MLP retained on each trial

Fig. 8.7b. same as Fig. 8.7a with Transformed Data. All classifiers have improved their results (in terms of pessimistic error), except the neural network which obtains the same results

In Fig. 8.7a, the small standard deviation of the results tends to prove that the technique we use for the evaluation is appropriate. The best result is obtained with the neural network with $29 \pm 1\%$ of misclassification error on the validation set. It is significantly different from others ($p < 0.01$, using a Wilcoxon sign rank test for paired samples). Their results vary from $53 \pm 2\%$ for the Parzen estimator to $42 \pm 3\%$ for the quadratic classifier.

The large difference between optimistic and pessimistic estimation of the percentage for Parzen estimator (classifier d) ($p < 0.01$ Wilcoxon sign rank test for paired samples) shows that the error on the training set can definitely not be used to evaluate the performance of a classifier. Indeed, using this classifier, the vector from the training set participates too much in the decision for its classification. The high percentages obtained for optimistic estimation for the classifiers a) b) and c) shows that those classifiers do not perform well on the data. This can be explained by the large tails in the distribution and by the fact that the density probability functions of the classes cannot be fitted correctly by a multidimensional Gaussian model.

8.3.2 Results with Transformed Data

Results from data with transformations are presented in Fig. 8.7b. The transformations applied to the variables are given in Table 8.1. Classifiers were trained with new coordinates obtained after these transformations.

All classifiers increased their performances ($p < 0.01$, Wilcoxon rank sum test for independent samples) except the neural network which obtains the same results. The performance of the k nearest neighbor classifier and the Parzen classifier have been significantly improved. The pessimistic error decreases from $45 \pm 2\%$ to

$28 \pm 2\%$ for the k nearest neighbor classifier and from $53 \pm 2\%$ to $32 \pm 2\%$ with the Parzen estimator. The results obtained by the k nearest neighbor classifier are then equivalent to the results obtained with the neural network. The results of the linear and quadratic classifiers are slightly better with the transformed data. The misclassification error decreased from $44 \pm 3\%$ to $37 \pm 2\%$ with the linear classifier and from $42 \pm 3\%$ to $36 \pm 3\%$ with the quadratic classifier.

These results can be explained by the fact that the neural network is not sensitive to the distributions of the data; transformations have no effect on its ability to separate space into subspaces [26]. But, the effect of the transformation leads to an improvement of the speed of convergence for the optimization of the backpropagation during the learning process explained by a better homogeneity in the distribution of the weights and features in the input layer.

On the contrary, both the Parzen estimator and the k nearest neighbor estimator use the concept of data proximity to classify a new vector. They are then penalized by extreme values. When the extreme values are moved closer by transformation, their performances equal those of the neural network. The linear and the quadratic classifier make an assumption on the shape of the classes which is still not completely verified even after data transformation.

8.4 Discussion

Disagreements between human scorers are known to vary from 10% to 20% [28]. The results obtained by these classifiers do not enter this interval, but they are not very far from them. Besides, this study enables us to compare different techniques of classification.

The advantage of the neural network is that it does not require any data transformation. The results obtained are the same with the raw data or with the transformed data. The neural network can deal with a non-Gaussian probability density function and with extreme values. However, the selection of the best neural network and the optimization of the structure of the layers are not easy tasks and can be time consuming. Though the results obtained by the nearest neighbor classifier applied to homogeneous data are the best, this method requires storing a large amount of learning vectors in memory. This can make its application difficult in practice. The main advantage of classical statistical techniques (linear and quadratic discrimination) is that the algorithms are fast.

Why do results not enter the inherent interval of disagreement between scorers? One answer is that the hypothesis of the independence of the temporal epochs is not completely true because when experts score a recording, they intrinsically know the preceding page and score the new one in consequence. A way to take into account this temporal causality is to add new columns in the database corresponding to the preceding data of the parameters (switching from a state representation to a phase one). Another way is to introduce an inference table at the end of the classification process allowing certain transitions or rejecting others.

Another answer is that the parameters retained are not as discriminative as the ones chosen visually by an expert. This is a problem which is generally encountered in automatic classification as a means to replace a human classifier. Moreover, the classifiers were trained to classify data recorded on different subjects. They had to deal not only with temporal dependence of the data as discussed before, but also with inter-individual variability. In our study, we have constructed and evaluated classifiers with no adaptation to one particular individual.

Experts are not so strict and often adapt their mind to fit the problem, but classifiers are built to fit optimized mathematical models from a learning database. The solutions proposed by these models can sometimes show the limits of the visual technique of human scoring and can be a way to refine expert knowledge.

For example, one visual interesting dilemna in the R&K manual is when a transition occured during an epoch from one stage to another: the rule is to assign the class to the predominant stage, i.e. the stage that lasts more than fifty percent of the epoch. When there are a lot of transitions, this results in many problems for the scorer and a lower productivity. This also raises doubt over the stationary hypotheses for the computing of temporal or spectral parameters. For more accurate precision, one can use recent segmentation techniques for temporal time series, where signals are segmented into non overlapping windows of variable lengths with respect to different criteria [4, 16, 18, 23]. But then, the estimation of the performance of the classifiers is not so easy.

Nowadays, R&K scoring proves its usefulness every day, but its limits are more and more admitted [13].

8.5 Conclusion

We have evaluated and compared the performance of five classifiers to automatically score polysomnographic data from various individuals into the six R&K sleep-wake stages. Though the results obtained (the misclassification percentage is about 30%) are not as good as the results obtained with the human scorers (the misclassification percentage is less than 20%), the results are interesting considering the amount of work human scoring requires. Automatic scoring may lighten the doctor's burden.

We showed that extreme values, frequently present in biological data, were a problem for all the evaluated classifiers, except for the neural network. To apply a transformation toward normal distribution appeared to be an interesting way to improve the performance of the classifiers. Both the neural network and the k nearest neighbor algorithm using transformed data gave good results. However, considering the information required to implement the two methods, we would recommend the use of the neural network.

Acknowledgments

The authors wish to thank Bénédicte Becq and Geoffrey Bramham for their unvaluable help with subtleties of the English language. They also want to thank the staff of the CRSSA/FH/PV division, who were present during this study, for their kind support.

References

1. Artioli E, Avanzolini G, Barbini P, Cevenini G, Gnudi G (1991) Classification of postoperative cardiac patients: comparative evaluation of four algorithms. Int J Biomed Comput 29:257–270
2. Aserinsky E, Kleitman N (1957) Regularly occurring periods of eye motility, and concomitant phenomena, during sleep. Science 118:273–274
3. Chapotot F, Pigeau R, Canini F, Bourdon L, Buguet A (2003) Distinctive effects of modafinil and d-amphetamine on the homeostatic and circadian modulation of the human waking EEG. Psychopharmacology 166:127–138
4. Charbonnier S, Becq G, Biot L (2004) On-line segmentation algorithm for continuously monitored data in intensive care units. IEEE T Biomed Eng 51(3):484–492
5. Cornuéjols A, Miclet L (2002) Apprentissage artificiel, Concepts et algorithmes. Eyrolles Paris
6. Dement WC, Kleitman N (1957) Cyclic variations in EEG during sleep and their relation to eye movements, body motility and dreaming. Electroencephalogr Clin Neurophysiol 9:673–690
7. Dement WC (1958) The occurence of low voltage fast electroencephalogram patterns during behavioral sleep in the cat. Electroencephalogr Clin Neurophysiol 10:291–296
8. Dubuisson B (2001) Diagnostic, intelligence artificielle et reconnaissance de formes. Hermès science Europe Paris
9. Efron B (1983) Estimating the error rate of a prediction rule: improvement on crossvalidation. J Am Stat Ass 78(382):316–330
10. Efron B, Tibshirani R (1995) Crossvalidation and the bootstrap: Estimating the error rate of a prediction rule. Technical report (477) Statistics department Stanford University
11. Frost JD (1970) An automatic sleep analyser. Electroencephalogr Clin Neurophysiol 29:88–92
12. Gasser T, Bächer P, Möchs J (1982) Transformations towards the normal distribution of broad band spectral parameters of the eeg. Electroencephalogr Clin Neurophysiol 53:119–124
13. Himanen SL, Hasan J (2000) Limitations of Rechtschaffen and Kales. Sleep Med Rev 4(2):149–167
14. Hirshkowitz M (2000) Standing on the shoulders of giants: the *Standardized Sleep Manual* after 30 years. Sleep Med Rev 4(2):169–179
15. Jouvet M, Michel F, Courjon J (1959) Sur un stade d'activité électrique cérébrale rapide au cours du sommeil physiologique. C R Soc Biol 153:1024–1028
16. Keogh E, Chu S, Hart D, Pazzani M (2001) An online algorithm for segmenting time series. In: IEEE International Conference on Data Mining, pp. 289–296
17. Kohavi R (1995) A study of cross-validation and bootstrap for accuracy estimation and model selection. In: International Joint Conference on Artificial Intelligence, pp. 1137–1145.
18. Kohlmorgen J, Müller K, Rittweger J, Pawelzik K (2000) Identification of non-stationary dynamics in physiological recordings. Biol Cyber 83(1):73–84
19. Lebart L, Morineau A, Piron M (2000) Statistique exploratoire multidimensionnelle. Dunod Paris
20. Loomis AL, Harvey EN, Hobart G (1937) Cerebral stages during sleep, as studied by human brain potentials. J Exp Psychol 21:127–144

21. Loomis AL, Harvey EN, Hobart G (1938) Distribution of disturbance patterns in the human electroencephalogram, with special reference to sleep. J Neurophysiol 1:413–418
22. Mocks J, Gasser T (1984) How to select epochs of the eeg at rest for quantitative analysis. Electroencephalogr Clin Neurophysiol 58:89–92
23. Morik K (2000) The Representation race – preprocessing for handling time phenomena. 11th European Conference on Machine Learning 1810:4–19 Springer Berlin
24. Penzel T, Conradt R (2000) Computer based sleep recording and analysis. Sleep Med Rev 4(2):131–148
25. Rechtschaffen A, Kales A (1968) A manual of standardized terminology, techniques and scoring system for sleep stages of human subjects. US Government Printing Office Washington
26. Robert C, Guilpin C, Limoge A (1997) Comparison between conventional and neural network classifiers for rat sleep-wake stage discrimination. Neuropsychobiol 35:221–225
27. Robert C, Guilpin C, Limoge A (1999) Automated sleep staging systems in rats. J Neurosci Meth 88:111–122
28. Schaltenbrand N, Lengelle R, Toussaint M, Luthringer R, Carelli R, Jacqmin A, Lainey E, Muzet A, Macher JP (1996) Sleep stage scoring using the neural network model: comparison between visual and automatic analysis in normal subjects and patients. Sleep 19(1):27–35
29. Smith JR, Negin M, Nevis AH (1969) Automatic analysis of sleep electroencephalograms by hybrid computation. IEEE T Syst Sci Cybern 5:278–284
30. Smith JR, Karacan I (1971) EEG sleep stage scoring by an automatic hybrid system. Electroencephalogr Clin Neurophysiol 31:231–237
31. Thiria S, Lechevallier Y, Gascuel O, Canu S (1997) Statistique et méthodes neuronales. Dunod, Paris

9

Prioritized Fuzzy Information Fusion for Handling Multi-Criteria Fuzzy Decision-Making Problems

Shi-Jay Chen[1] and Shyi-Ming Chen[2]

[1] Department of Computer Science and Information Engineering, National Taiwan University of Science and Technology, Taipei 106, Taiwan, R.O.C.
D8815006@mail.ntust.edu.tw
[2] Department of Computer Science and Information Engineering, National Taiwan University of Science and Technology, Taipei 106, Taiwan, R.O.C.
smchen@et.ntust.edu.tw

Abstract. The prioritized operator is very useful in dealing with multi-criteria fuzzy decision-making problems, and it can be seen as an additional selector of filter. In this article, we present a prioritized information fusion algorithm based on similarity measures of generalized fuzzy numbers for aggregating fuzzy information. The proposed information fusion algorithm can handle multi-criteria fuzzy decision-making problems in a more flexible manner due to the fact that it allows the evaluating values of the criteria to be represented by crisp values between $[0, 1]$ or generalized fuzzy numbers, where the generalized fuzzy numbers can indicate the degrees of confidence of the decision-makers' opinions.

Key words: Prioritized information fusion algorithm, generalized fuzzy numbers, similarity measures, information filtering, prioritized operator

9.1 Introduction

In [32], Yager presented a conjunction operator of fuzzy subsets called the non-monotonic/prioritized operator. In [34], Yager used the prioritized operator to deal with multi-criteria decision-making problems, and he indicated that the prioritized operator could be seen as an additional selector or filter. In [3], Bordogna et al. used the prioritized operator for fuzzy information retrieval. In [17], Hirota et al. used the prioritized operator to deal with two applications, i.e., an estimation of default fuzzy sets and a default-driven extension of fuzzy reasoning. In [35], Yager used the prioritized operator in fuzzy information fusion structures. From [3, 17, 32, 34, 35], we can see that the prioritized operator is very useful in dealing with multi-criteria fuzzy decision-making problems.

From [1, 4, 12, 13, 14, 21, 22, 23, 24, 25, 26, 27], we can see that fuzzy numbers are very useful to represent evaluating values in multi-criteria fuzzy decision-making

Shi-Jay Chen and Shyi-Ming Chen: *Prioritized Fuzzy Information Fusion for Handling Multi-Criteria Fuzzy Decision-Making Problems*, Studies in Computational Intelligence (SCI) **4**, 129–145 (2005)
www.springerlink.com

problems. Furthermore, there are some researchers [6, 10, 15] using generalized fuzzy numbers [5] to represent fuzzy information and their associated degree of uncertainty. However, the prioritized operator presented in [32] can not be used to deal with multi-criteria fuzzy decision-making problems if we use fuzzy numbers to represent evaluating values. Thus, it is obvious that to extend the prioritized operator and to develop a prioritized information fusion algorithm for aggregating the evaluating values of the criteria represented by fuzzy numbers are important research topics when dealing with multi-criteria decision-making problems.

In this article, we extend the prioritized operator presented in [32] to present a prioritized information fusion algorithm based on similarity measures of generalized fuzzy numbers for aggregating fuzzy information. The proposed prioritized information fusion algorithm can handle the multi-criteria fuzzy decision-making problems in a more flexible and more intelligent manner due to the fact that it allows the evaluating values of the criteria to be represented by generalized fuzzy numbers, where the generalized fuzzy numbers can indicate the degrees of confidence of the decision-makers' opinions.

This article is organized as follows. In Sect. 9.2, we briefly review the concepts of generalized fuzzy numbers [5], the simple center of gravity method (SCGM) [10], the similarity measure based on SCGM [11], and the operators of triangular norm (T-Norm) and triangular conorm (T-Conorm) [2, 18, 30, 36, 37]. In Sect. 9.3, we briefly review Yager's prioritized operator [32] and some properties of the prioritized operator. In Sect. 9.4, we present a prioritized information fusion algorithm for aggregating fuzzy information. In Sect. 9.5, we apply the proposed information fusion algorithm for handling multi-criteria fuzzy decision-making problems. The conclusions are discussed in Sect. 9.6.

9.2 Preliminaries

In [5], Chen represented a generalized trapezoidal fuzzy number $\tilde{A}$ in the universe of discourse X as shown in Fig. 9.1, where $\tilde{A} = (a_1, a_2, a_3, a_4; w_{\tilde{A}})$, $w_{\tilde{A}}$ denotes the maximum membership value of the generalized trapezoidal fuzzy number $\tilde{A}$, $0 < w_{\tilde{A}} \leq 1$, and a_1, a_2, a_3 and a_4 are real values. If $w_{\tilde{A}} = 1$, then the generalized trapezoidal fuzzy number $\tilde{A}$ shown in Fig. 9.1 can also be represented by $\tilde{A} = (a_1, a_2, a_3, a_4)$.

In [10], we have presented a method, called the simple center of gravity method (SCGM), to calculate the center-of-gravity (COG) point of a generalized fuzzy

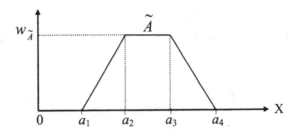

Fig. 9.1. A generalized trapezoidal fuzzy number

number based on the concept of the medium curve [29]. Let $\tilde{A}$ be a generalized trapezoidal fuzzy number, $\tilde{A} = (a_1, a_2, a_3, a_4; w_{\tilde{A}})$. The method for calculating the COG point $(x^*_{\tilde{A}}, y^*_{\tilde{A}})$ of the generalized trapezoidal fuzzy number $\tilde{A}$ is as follows:

$$
y^*_{\tilde{A}} = \begin{cases} \dfrac{w_{\tilde{A}} \times (\frac{a_3 - a_2}{a_4 - a_1} + 2)}{6}, & \text{if } a_1 \neq a_4 \text{ and } 0 < w_{\tilde{A}} \leq 1, \\[3mm] \dfrac{w_{\tilde{A}}}{2}, & \text{if } a_1 = a_4 \text{ and } 0 < w_{\tilde{A}} \leq 1, \end{cases} \tag{9.1}
$$

$$
x^*_{\tilde{A}} = \dfrac{y^*_{\tilde{A}}(a_3 + a_2) + (a_4 + a_1)(w_{\tilde{A}} - y^*_{\tilde{A}})}{2w_{\tilde{A}}}. \tag{9.2}
$$

In [11], we have presented a method to evaluate the degree of similarity between generalized trapezoidal fuzzy numbers. Assume that there are two generalized trapezoidal fuzzy numbers $\tilde{A}$ and $\tilde{B}$, where $\tilde{A} = (a_1, a_2, a_3, a_4; w_{\tilde{A}})$, $\tilde{B} = (b_1, b_2, b_3, b_4; w_{\tilde{B}})$, $0 \leq a_1 \leq a_2 \leq a_3 \leq a_4 \leq 1$, and $0 \leq b_1 \leq b_2 \leq b_3 \leq b_4 \leq 1$. Firstly, we use formulas (9.1) and (9.2) to obtain the COG points $\text{COG}(\tilde{A})$ and $\text{COG}(\tilde{B})$ of the generalized trapezoidal fuzzy numbers $\tilde{A}$ and $\tilde{B}$, respectively, where $\text{COG}(\tilde{A}) = (x^*_{\tilde{A}}, y^*_{\tilde{A}})$ and $\text{COG}(\tilde{B}) = (x^*_{\tilde{B}}, y^*_{\tilde{B}})$. Then, the degree of similarity $S(\tilde{A}, \tilde{B})$ between the generalized trapezoidal fuzzy numbers $\tilde{A}$ and $\tilde{B}$ can be calculated as follows:

$$
S(\tilde{A}, \tilde{B}) = \left(1 - \dfrac{\sum_{i=1}^4 |a_i - b_i|}{4} \right)
$$

$$
\times (1 - |x^*_{\tilde{A}} - x^*_{\tilde{B}}|)^{B(S_{\tilde{A}}, S_{\tilde{B}})} \times \dfrac{\min(y^*_{\tilde{A}}, y^*_{\tilde{B}})}{\max(y^*_{\tilde{A}}, y^*_{\tilde{B}})}, \tag{9.3}
$$

where $S(\tilde{A}, \tilde{B}) \in [0, 1]$ and $B(S_{\tilde{A}}, S_{\tilde{B}})$ is defined as follows:

$$
B(S_{\tilde{A}}, S_{\tilde{B}}) = \begin{cases} 1, & \text{if } S_{\tilde{A}} + S_{\tilde{B}} > 0, \\ 0, & \text{if } S_{\tilde{A}} + S_{\tilde{B}} = 0, \end{cases} \tag{9.4}
$$

where $S_{\tilde{A}}$ and $S_{\tilde{B}}$ are the bases of the generalized trapezoidal fuzzy numbers $\tilde{A}$ and $\tilde{B}$, respectively, defined as follows:

$$
S_{\tilde{A}} = a_4 - a_1, \tag{9.5}
$$

$$
S_{\tilde{B}} = b_4 - b_1. \tag{9.6}
$$

The larger the value of $S(\tilde{A}, \tilde{B})$, the more the similarity between the generalized fuzzy numbers $\tilde{A}$ and $\tilde{B}$. The similarity measure of generalized fuzzy numbers has the following properties and we have proven these properties in [11]:

Property 1: Two generalized fuzzy numbers $\tilde{A}$ and $\tilde{B}$ are identical if and only if $S(\tilde{A}, \tilde{B}) = 1$.
Property 2: $S(\tilde{A}, \tilde{B}) = S(\tilde{B}, \tilde{A})$.
Property 3: If $\tilde{A} = (a, a, a, a; 1.0)$ and $\tilde{B} = (b, b, b, b; 1.0)$ are two real numbers, then $S(\tilde{A}, \tilde{B}) = 1 - |a - b|$.

In [8], we modified formula (9.3) into formula (9.7) to simplify the calculation process, shown as follows:

$$S(\tilde{A}, \tilde{B}) = \left[\left(1 - \frac{\sum_{i=1}^{4} |a_i - b_i|}{4} \right) \times (1 - |x_{\tilde{A}}^* - x_{\tilde{B}}^*|) \right]^{\frac{1}{2}} \times \frac{\min(y_{\tilde{A}}^*, y_{\tilde{B}}^*)}{\max(y_{\tilde{A}}^*, y_{\tilde{B}}^*)} . \quad (9.7)$$

In [2] and [18], Alsina et al. and Höhle introduced the operators of the triangular norm (t-norm: $\wedge$) and the triangular conorm (t-conorm: $\vee$) of fuzzy sets. In [30], Sugianto summarized some t-norm operators and t-conorm operators as shown in Table 9.1.

Table 9.1. Some T-norms and T-conorms [30]

	T-norms		T-conorms
$\min(a, b)$,	Logical Product	$\max(a, b)$,	Logical Sum
$a \times b$,	Algebraic Product	$a + b - a \times b$,	Algebraic Sum
$\frac{ab}{a+b-ab}$,	Hamacher Product	$\frac{a+b-2ab}{1-ab}$,	Hamacher Sum
$\frac{ab}{1-(1-a)(1-b)}$,	Einstein Product	$\frac{a+b}{1+ab}$,	Einstein Sum
$\begin{cases} a, & \text{if } b = 1, \\ b, & \text{if } a = 1, \\ 0, & \text{otherwise.} \end{cases}$	Drastic Product	$\begin{cases} a, & \text{if } b = 0, \\ b, & \text{if } a = 0, \\ 1, & \text{otherwise.} \end{cases}$	Drastic Sum
$\max(a + b - 1, 0)$,	Bounded Product	$\min(a + b, 1)$,	Bounded Sum

In [36] and [37], we can see that t-norm and t-conorm operators can apply to the fuzzy number environment. Assume that there are two trapezoidal fuzzy numbers $\tilde{A} = (a_1, a_2, a_3, a_4)$ and $\tilde{B} = (b_1, b_2, b_3, b_4)$, where $0 \leq a_1 \leq a_2 \leq a_3 \leq a_4 \leq 1$ and $0 \leq b_1 \leq b_2 \leq b_3 \leq b_4 \leq 1$. The t-norm and t-conorm operations between the trapezoidal fuzzy numbers $\tilde{A}$ and $\tilde{B}$ are shown as follows:

$$T(\tilde{A}, \tilde{B}) = (a_1 \wedge b_1, a_2 \wedge b_2, a_3 \wedge b_3, a_4 \wedge b_4) , \quad (9.8)$$

$$S(\tilde{A}, \tilde{B}) = (a_1 \vee b_1, a_2 \vee b_2, a_3 \vee b_3, a_4 \vee b_4) , \quad (9.9)$$

where $\wedge$ and $\vee$ are any t-norms and t-conorms as shown in Table 9.1, respectively. According to the two formulas (9.8) and (9.9), we can see that the results of t-norm and t-conorm operations of fuzzy numbers are still fuzzy numbers, and the results of t-norm and t-conorm operations of fuzzy numbers are linearity. The results of t-norm and t-conorm operations of fuzzy numbers using formulas (9.8) and (9.9) are approximate results because the traditional t-norm and t-conorm operators with fuzzy numbers do not preserve the linearity [22].

9.3 Non-Monotonic/Prioritized Intersection Operator

In [31] and [32], Yager presented a non-monotonic intersection operator η. Let X be the universe of discourse and let A and B be two fuzzy subsets in X. The non-monotonic intersection operator η is defined as follows:

$$\eta(A, B) = D , \quad (9.10)$$

$$\mu_D(x) = \mu_A(x) \wedge (\mu_B(x) \vee (1 - \text{Poss}[B|A])) , \quad (9.11)$$

where $\text{Poss}[B|A] = \text{Max}_x[\mu_A(x) \wedge \mu_B(x)]$. In [33], the two operators $\wedge$ and $\vee$ can be replaced by any t-norms and t-conorms, respectively. The value of $\text{Poss}[B|A]$ measures the degree of intersection between the two fuzzy subsets A and B, and the value of $1 - \text{Poss}[B|A]$ can be regarded as a measure of conflict between the two fuzzy subsets A and B. According to [33] and [34], we can see that there are two properties of Yager's non-monotonic intersection operator shown as follows:

1. If $\text{Poss}[B|A] = 1$, then $\mu_D(x) = \mu_A(x) \wedge \mu_B(x)$. This means that the fuzzy subset A can be revised by the fuzzy subset B entirely due to the fact that the fuzzy subset B is completely compatible with the fuzzy subset A.
2. If $\text{Poss}[B|A] = 0$, then $\mu_D(x) = \mu_A(x)$. This means that the fuzzy subset A can not be revised by the fuzzy subset B due to the fact that the fuzzy subset B is completely incompatible with the fuzzy subset A.

From [33, 34] and the above two properties, we can see that the fuzzy subset B can be seen as a special rule with which we are trying to use the fuzzy subset B for selecting the best alternative based upon the fuzzy subset A due to the fact that the fuzzy subset B acts as an additional selector or filter for those alternatives which satisfy the fuzzy subset A, and the fuzzy subset B has a lower priority than the fuzzy subset A. In [34], Yager used the non-monotonic intersection operator to deal with multi-criteria decision-making problems. He pointed out that A is called the first order criterion and B is called the second order criterion, and he modified formula (9.11) into

$$\mu_D(x) = \mu_A(x) \wedge (\mu_B(x) \vee (1 - \omega(\text{Poss}[B|A]))) , \qquad (9.12)$$

where ω is a qualifier,

$$\omega(r) = \begin{cases} 0, & \text{if } r < \alpha , \\ 1, & \text{if } r \geq \alpha , \end{cases} \qquad (9.13)$$

and $\alpha \in [0, 1]$. In [35], Yager called the operator the "non-monotonic/ prioritized intersection operator" due to the fact that the operator supports a concept of priority, i.e., in $\eta(A, B)$, A has a priority over B [32].

Example 1. Assume that the algebraic product (i.e., ab) and algebraic sum (i.e., $a + b - ab$) are used to represent t-norm and t-conorm (i.e., $\wedge$ and $\vee$), respectively. Assume that there are two different criteria A and B, three alternatives x_1, x_2, x_3, and the evaluating values of the three alternatives x_1, x_2, x_3 shown as follows:

$$A = \left\{ \frac{0.9}{x_1}, \frac{0.8}{x_2}, \frac{0.5}{x_3} \right\}, \quad B = \left\{ \frac{0.1}{x_1}, \frac{0.7}{x_2}, \frac{0.3}{x_3} \right\},$$

then we can see that $\text{Poss}[B|A] = \text{Max}_x[A(x) \wedge B(x)] = \text{Max}[(0.9 \wedge 0.1), (0.8 \wedge 0.7), (0.5 \wedge 0.3)] = 0.56$. By applying formula (9.13), we can see that

$$\mu_D(x_1) = 0.9 \wedge (0.1 \vee (1.0 - 0.56)) = 0.4464 ,$$
$$\mu_D(x_2) = 0.8 \wedge (0.7 \vee (1.0 - 0.56)) = 0.6656 ,$$
$$\mu_D(x_3) = 0.5 \wedge (0.3 \vee (1.0 - 0.56)) = 0.304 .$$

Because of $\mu_D(x_2) > \mu_D(x_1) > \mu_D(x_3)$, we can see that the preferring order of the alternatives is $x_2 > x_1 > x_3$. Thus, the best alternative among the alternatives x_1, x_2 and x_3 is x_2.

9.4 Prioritized Information Fusion Algorithm

In this section, we extend the traditional prioritized operator [32] to propose a prioritized information fusion algorithm for aggregating evaluating values represented by generalized fuzzy numbers. Let U be the universe of discourse, $U = [0, k]$. Assume that there are n alternatives $x_1, x_2, \ldots,$ and x_n, and assume that there are two different prioritized criteria A and B as shown in Table 9.2, where $A = (\tilde{A}_1, \tilde{A}_2, \ldots, \tilde{A}_i, \ldots, \tilde{A}_n)$ is called the first order criterion, $B = (\tilde{B}_1, \tilde{B}_2, \ldots, \tilde{B}_i, \ldots, \tilde{B}_n)$ is called the second order criterion, $\tilde{A}_i$ and $\tilde{B}_i$ are generalized trapezoidal fuzzy numbers, $\tilde{A}_i = (a_{i1}, a_{i2}, a_{i3}, a_{i4}; w_{\tilde{A}_i})$, $\tilde{B}_i = (b_{i1}, b_{i2}, b_{i3}, b_{i4}; w_{\tilde{B}_i})$, $0 \le a_{ij} \le k, 0 \le b_{ij} \le k, 0 < w_{\tilde{A}_i} \le 1, 0 < w_{\tilde{B}_i} \le 1, 1 \le i \le n$, and $1 \le j \le 4$.

Table 9.2. Evaluating Values of the Alternatives Using Two Different Criteria A and B

	\multicolumn{5}{c}{Alternatives}				
Different Criteria	x_1	x_2	$\ldots$	x_i	$\ldots$	x_n
First order criterion A	$\tilde{A}_1$	$\tilde{A}_2$	$\ldots$	$\tilde{A}_i$	$\ldots$	$\tilde{A}_n$
Second order criterion B	$\tilde{B}_1$	$\tilde{B}_2$	$\ldots$	$\tilde{B}_i$	$\ldots$	$\tilde{B}_n$

The proposed prioritized information fusion algorithm to deal with multi-criteria fuzzy decision-making problems is now presented as follows:

Step 1: If the universe of discourse $U = [0, k]$ and $k \ne 1$, then translate the generalized trapezoidal fuzzy numbers $\tilde{A}_i = (a_{i1}, a_{i2}, a_{i3}, a_{i4}; w_{\tilde{A}_i})$ and $\tilde{B}_i = (b_{i1}, b_{i2}, b_{i3}, b_{i4}; w_{\tilde{B}_i})$, where $1 \le i \le n$, into standardized generalized trapezoidal fuzzy numbers $\tilde{A}_i^*$ and $\tilde{B}_i^*$, respectively, shown as follows:

$$\tilde{A}_i^* = \left(\frac{a_{i1}}{k}, \frac{a_{i2}}{k}, \frac{a_{i3}}{k}, \frac{a_{i4}}{k}; w_{\tilde{A}_i^*} \right)$$
$$= (a_{i1}^*, a_{i2}^*, a_{i3}^*, a_{i4}^*; w_{\tilde{A}_i^*}), \qquad (9.14)$$

$$\tilde{B}_i^* = \left(\frac{b_{i1}}{k}, \frac{b_{i2}}{k}, \frac{b_{i3}}{k}, \frac{b_{i4}}{k}; w_{\tilde{B}_i^*} \right)$$
$$= (b_{i1}^*, b_{i2}^*, b_{i3}^*, b_{i4}^*; w_{\tilde{B}_i^*}), \qquad (9.15)$$

where $0 \le a_{ij}^* \le 1, 0 \le b_{ij}^* \le 1, w_{\tilde{A}_i^*} = w_{\tilde{A}_i}, w_{\tilde{B}_i^*} = w_{\tilde{B}_i}, 1 \le i \le n$, and $1 \le j \le 4$.

Step 2: Based on formulas (9.1), (9.2) and (9.7), evaluate the degree of compatibility $C(A, B)$ between the first order criterion A and the second order criterion B shown as follows:

$$C(A, B) = \text{Max}\{\text{Min}[(\hat{x}_{\tilde{A}_1^*} \wedge \hat{x}_{\tilde{B}_1^*}), S(\tilde{A}_1^*, \tilde{B}_1^*)],$$
$$\text{Min}[(\hat{x}_{\tilde{A}_2^*} \wedge \hat{x}_{\tilde{B}_2^*}), S(\tilde{A}_2^*, \tilde{B}_2^*)],$$

$$\vdots$$

$$\text{Min}[(\hat{x}_{\tilde{A}_i^*} \wedge \hat{x}_{\tilde{B}_i^*}), S(\tilde{A}_i^*, \tilde{B}_i^*)],$$

$$\vdots$$

$$\text{Min}[(\hat{x}_{\tilde{A}_n^*} \wedge \hat{x}_{\tilde{B}_n^*}), S(\tilde{A}_n^*, \tilde{B}_n^*)], \qquad (9.16)$$

where $1 \leq i \leq n$, $C(A, B) \in [0, 1]$, and the operator $\wedge$ can be replaced by any t-conorms. The calculating value $C(A, B)$ of formula (9.16) is a real number because the values $\hat{x}_{\tilde{A}_i^*}$, $\hat{x}_{\tilde{B}_i^*}$ and $S(\tilde{A}_i^*, \tilde{B}_i^*)$ are all real numbers, where $\hat{x}_{\tilde{A}_i^*} \in [0, 1]$, $\hat{x}_{\tilde{B}_i^*} \in [0, 1]$, $S(\tilde{A}_i^*, \tilde{B}_i^*) \in [0, 1]$ and $C(A, B) \in [0, 1]$. The value $C(A, B)$ essentially measures the degree of compatibility between two sets of generalized fuzzy numbers A and B.

Step 3: Based on formulas (9.8), (9.9) and (9.16), extend formula (9.11) into

$$\tilde{D}_i = \tilde{A}_i^* \wedge (\tilde{B}_i^* \vee (1 - C(A, B))), \qquad (9.17)$$

where $\tilde{D}_i$ is a fuzzy number denoting the fusing result of the fuzzy numbers $\tilde{A}_i^*$ and $\tilde{B}_i^*$, $1 \leq i \leq n$, and the operators $\wedge$ and $\vee$ can be replaced by any t-norms and t-conorms, respectively. If $1 - C(A, B) = p$, where $p \in [0, 1]$, then we can use formulas (9.8) and (9.9) to calculate $\tilde{D}_i$ as follows:

$$\tilde{D}_i = \tilde{A}_i^* \wedge (\tilde{B}_i^* \vee (1 - C(A, B)))$$
$$= \tilde{A}_i^* \wedge (\tilde{B}_i^* \vee p)$$
$$= \Big(a_{i1}^* \wedge (b_{i1}^* \vee p),\ a_{i2}^* \wedge (b_{i2}^* \vee p),\ a_{i3}^* \wedge (b_{i3}^* \vee p),\ a_{i4}^* \wedge$$
$$(b_{i4}^* \vee p);\ {}^{(1+\lceil C(A,B) \rceil)}\!\sqrt{w_{\tilde{A}_i^*} \times w_{\tilde{B}_i^*}^{\lceil C(A,B) \rceil}}\ \Big)$$
$$= (d_{i1},\ d_{i2},\ d_{i3},\ d_{i4};\ w_{\tilde{D}_i}), \qquad (9.18)$$

where $d_{ik} = a_{ik}^* \wedge (b_{ik}^* \vee p)$, $w_{\tilde{D}_i} = {}^{(1+\lceil C(A,B) \rceil)}\!\sqrt{w_{\tilde{A}_i^*} \times w_{\tilde{B}_i^*}^{\lceil C(A,B) \rceil}}$, $1 \leq k \leq 4$, and $1 \leq i \leq n$. In formula (9.18), we use the geometric mean ${}^{(1+\lceil C(A,B) \rceil)}\!\sqrt{w_{\tilde{A}_i^*} \times w_{\tilde{B}_i^*}^{\lceil C(A,B) \rceil}}$ to represent the fusion result $w_{\tilde{A}_i^*}$ and $w_{\tilde{B}_i^*}$. If $C(A, B) = 0$, then we let $\lceil C(A, B) \rceil = 0$; if $0 < C(A, B) \leq 1$, then we let $\lceil C(A, B) \rceil = 1$.

Step 4: Because the fusing results $\tilde{D}_1, \tilde{D}_2, \ldots,$ and $\tilde{D}_n$ are generalized fuzzy numbers, apply formula (9.19) to obtain the ranking order $R(\tilde{D}_i)$ of the fusing results $\tilde{D}_i$, where $1 \leq i \leq n$, shown as follows [9]:

$$R(\tilde{D}_i) = [\hat{x}_{\tilde{D}_i} + (w_{\tilde{D}_i} - \hat{y}_{\tilde{D}_i})^{\hat{s}_{\tilde{D}_i}} \times (\hat{y}_{\tilde{D}_i} + 0.5)^{1 - w_{\tilde{D}_i}}] - 1, \qquad (9.19)$$

where $\hat{x}_{\tilde{D}_i}$ and $\hat{y}_{\tilde{D}_i}$ are calculated by formulas (9.1) and (9.2), respectively. Based on [16], the standard deviation $\hat{s}_{\tilde{D}_i}$ of the generalized fuzzy number $\tilde{D}_i$ is defined as follows:

$$\hat{s}_{\tilde{D}_i} = \sqrt{\frac{1}{3}\sum_{j=1}^{4}(d_{ij} - \bar{d}_i)^2} \ . \tag{9.20}$$

The larger the value of $R(\tilde{D}_i)$, the better the ranking of $\tilde{D}_i$ (i.e., x_i is the better alternative), where $1 \leq i \leq n$. Select the best alternative among the fusion results $\tilde{D}_i$, where $1 \leq i \leq n$.

In Step 2 of the proposed algorithm, we use the degree of compatibility $C(A, B)$ (i.e., formula (9.16)) to replace Poss[$B|A$] using in formula (9.11). The degree of compatibility $C(A, B)$ has considered two different situations shown as follows:

Situation 1: If we use the crisp values between zero and one to represent evaluating values, the degree of compatibility $C(A, B)$ is equal to Poss[$B|A$] and the degrees of similarity $S(\tilde{A}_i^*, \tilde{B}_i^*)$ play a minor role.

Situation 2: If we use the generalized fuzzy numbers to represent evaluating values, then the degrees of similarity $S(\tilde{A}_i^*, \tilde{B}_i^*)$ between the two generalized fuzzy numbers $\tilde{A}_i^*$ and $\tilde{B}_i^*$, where $1 \leq i \leq n$, play an important role to obtain the degree of compatibility $C(A, B)$.

In Step 3 of the proposed algorithm, we have extended formula (9.11) into formula (9.17), based on formulas (9.8), (9.9) and (9.16). It is obvious that formula (9.17) still satisfies the following two properties [34]:

Property 4: If $C(A, B) = 0$, then $\tilde{D}_i = \tilde{A}_i^*$, where $1 \leq i \leq n$.

Proof. If $C(A, B) = 0$, then we can see that the value $\lceil C(A, B) \rceil = 0$. From formula (9.21), we can see that

$$\tilde{D}_i = \tilde{A}_i^* \wedge (\tilde{B}_i^* \vee (1 - C(A, B)))$$
$$= \tilde{A}_i^* \wedge (\tilde{B}_i^* \vee 1)$$
$$= \left(a_{i1}^* \wedge (b_{i1}^* \vee 1), \ a_{i2}^* \wedge (b_{i2}^* \vee 1), \ a_{i3}^* \wedge (b_{i3}^* \vee 1), \ a_{i4}^* \wedge (b_{i4}^* \vee 1) \ ; \right.$$
$$\left. ^{(1+0)}\sqrt{w_{\tilde{A}_i^*} \times w_{\tilde{B}_i^*}^0} \right)$$
$$= (a_{i1}^* \wedge 1, \ a_{i2}^* \wedge 1, \ a_{i3}^* \wedge 1, \ a_{i4}^* \wedge 1; w_{\tilde{A}_i^*} \times 1)$$
$$= (a_{i1}^*, \ a_{i2}^*, \ a_{i3}^*, \ a_{i4}^*; \ w_{\tilde{A}_i^*}) \ ,$$
$$= \tilde{A}_i^* \ ,$$

where the operators $\wedge$ and $\vee$ can be replaced by any t-norm and t-conorm methods, respectively, and $1 \leq i \leq n$.

Property 5: If $C(A, B) = 1$, then $\tilde{D}_i = \tilde{A}_i^* \wedge \tilde{B}_i^*$, where $1 \leq i \leq n$.

Proof. If $C(A, B) = 1$, then the value $\lceil C(A, B) \rceil = 1$. From formula (9.21), we can see that

$$\tilde{D}_i = \tilde{A}_i^* \wedge (\tilde{B}_i^* \vee (1 - C(A, B)))$$
$$= \tilde{A}_i^* \wedge (\tilde{B}_i^* \vee 0)$$
$$= \left(a_{i1}^* \wedge (b_{i1}^* \vee 0), \ a_{i2}^* \wedge (b_{i2}^* \vee 0), \ a_{i3}^* \wedge (b_{i3}^* \vee 0), \ a_{i4}^* \wedge (b_{i4}^* \vee 0); \right.$$
$$\left. {}^{(1+1)}\!\sqrt{w_{\tilde{A}_i^*} \times w_{\tilde{B}_i^*}^1} \right)$$
$$= \left(a_{i1}^* \wedge b_{i1}^*, \ a_{i2}^* \wedge b_{i2}^*, \ a_{i3}^* \wedge b_{i3}^*, \ a_{i4}^* \wedge b_{i4}^*; \ \sqrt{w_{\tilde{A}_i^*} \times w_{\tilde{B}_i^*}} \right)$$
$$= \tilde{A}_i^* \wedge \tilde{B}_i^* \ ,$$

where the two operators $\wedge$ and $\vee$ can be replaced by any t-norm and t-conorm methods, respectively, and $1 \leq i \leq n$.

From Property 4, we can see that if the degree of compatibility $C(A, B) = 0$, then $\tilde{D}_i = \tilde{A}_i^*$, where $1 \leq i \leq n$. It implies that the first order criterion A can not be revised by the second order criterion B due to the fact that the second order criterion B is incompatible with the first order criterion A completely. Thus, we can disregard the second order criterion B in a prioritized multi-criteria fuzzy decision-making problem. From Property 5, we can see that if the degree of compatibility $C(A, B) = 1$, then $\tilde{D}_i = \tilde{A}_i^* \wedge \tilde{B}_i^*$, where $1 \leq i \leq n$. It means that the first order criterion A can be revised by the second order criterion B completely due to the fact that the second order criterion B is completely compatible with the first order criterion A. Furthermore, from formula (9.18), we can see that if the degree of compatibility $0 < C(A, B) < 1$, then the first order criterion A can be revised by the second order criterion B in different degrees of compatibility $C(A, B)$. Thus, we can see that the second order criterion B still acts as an additional selector or filter for those alternatives which satisfy the first order criterion A.

In the following, we use an example to compare the proposed information fusion method with the traditional operator η [32], where the evaluating values are represented by crisp values between zero and one. In the next section, we will use an example to illustrate that the proposed information fusion algorithm can be used in the fuzzy decision-making environment, where generalized fuzzy numbers are used to represent evaluating values.

Example 2. Assume that the algebraic product (i.e., ab) and the algebraic sum (i.e., $a + b - ab$) are used to represent t-norm and t-conorm (i.e., $\wedge$ and $\vee$). Assume that there are two different criteria A and B, and three alternatives, x_1, x_2 and x_3, where their evaluating values are as follows:

$$A = \left\{ \frac{0.9}{x_1}, \frac{0.8}{x_2}, \frac{0.5}{x_3} \right\}, B = \left\{ \frac{0.1}{x_1}, \frac{0.7}{x_2}, \frac{0.3}{x_3} \right\} .$$

The evaluating values of the two criteria A and B can be seen as:

$$A = \left\{ \frac{(0.9, 0.9, 0.9, 0.9; 1.0)}{x_1}, \frac{(0.8, 0.8, 0.8, 0.8; 1.0)}{x_2}, \frac{(0.5, 0.5, 0.5, 0.5; 1.0)}{x_3} \right\}$$
$$= \left\{ \frac{\mu_A(x_1)}{x_1}, \frac{\mu_A(x_2)}{x_2}, \frac{\mu_A(x_3)}{x_3} \right\},$$

$$B = \left\{ \frac{(0.1, 0.1, 0.1, 0.1; 1.0)}{x_1}, \frac{(0.7, 0.7, 0.7, 0.7; 1.0)}{x_2}, \frac{(0.3, 0.3, 0.3, 0.3; 1.0)}{x_3} \right\}$$
$$= \left\{ \frac{\mu_B(x_1)}{x_1}, \frac{\mu_B(x_2)}{x_2}, \frac{\mu_B(x_3)}{x_3} \right\}.$$

By applying formulas (9.7) and (9.16), we can obtain the degree of compatibility $C(A, B)$ between the two different criteria A and B as follows:

$$
\begin{aligned}
C(A, B) &= \text{Max}\{\text{Min}[(0.9 \wedge 0.1), S(\mu_A(x_1), \mu_B(x_1))], \\
&\quad \text{Min}[(0.8 \wedge 0.7), S(\mu_A(x_2), \mu_B(x_2))], \\
&\quad \text{Min}[(0.5 \wedge 0.3), S(\mu_A(x_3), \mu_B(x_3))]\} \\
&= \text{Max}\{\text{Min}[0.09, 0.2], \text{Min}[0.56, 0.9], \text{Min}[0.15, 0.8]\} \\
&= 0.56 .
\end{aligned}
$$

It is obvious that the value of $C(A, B)$ is equal to the value of $\text{Poss}[B|A]$ shown in Example 1. Based on the value of $C(A, B)$ and formula (9.18), we can get the fusion result $\tilde{D}_1$, where:

$$
\begin{aligned}
\tilde{D}_1 &= \tilde{A}_1^* \wedge (\tilde{B}_1^* \vee (1 - C(A, B))) \\
&= \tilde{A}_1^* \wedge (\tilde{B}_1^* \vee 0.44) \\
&= (0.9 \wedge (0.1 \vee 0.44), \ 0.9 \wedge (0.1 \vee 0.44), \ 0.9 \wedge (0.1 \vee 0.44), \ 0.9 \wedge \\
&\quad (0.1 \vee 0.44); \ {}^{(1+1)}\!\sqrt{1 \times 1}) \\
&= (0.9 \wedge 0.496, \ 0.9 \wedge 0.496, \ 0.9 \wedge 0.496, \ 0.9 \wedge 0.496; \ 1) \\
&= (0.4464, \ 0.4464, \ 0.4464, \ 0.4464; \ 1) \\
&= 0.4464 .
\end{aligned}
$$

In the same way, we can get the fusion results $\tilde{D}_2 = 0.6656$ and $\tilde{D}_3 = 0.304$. Because of $\tilde{D}_2 > \tilde{D}_1 > \tilde{D}_3$, we can see that the preferring order of the alternatives x_1, x_2 and x_3 is $x_2 > x_1 > x_3$. Thus, x_2 is the best alternative among the alternatives x_1, x_2, and x_3. The result coincides with the one shown in Example 1.

9.5 An Example

In this section, we use a numerical example to illustrate how to use the proposed information fusion algorithm for handling a multi-criteria decision making (MCDM) problem. In this example, we use generalized fuzzy numbers to represent the evaluating values of the two different criteria A and B. Assume that there are eight alternatives x_1, x_2, x_3, x_4, x_5, x_6, x_7, x_8 and two different priority criteria A and B, where A is the first order criterion and B is the second order criterion, and the evaluating values are represented by generalized trapezoidal fuzzy numbers defined in the universe of discourse $U = [0, 10]$ as shown in Table 9.3.

From [34], we can see that there are two multi-criteria decision-making (MCDM) frameworks, i.e., (a) without considering the second order criterion B, and (b) considering the second order criterion B as shown in Fig. 9.2, respectively.

First, we use formulas (9.14) and (9.15) to translate the generalized trapezoidal fuzzy numbers shown in Table 9.3 into standardized generalized trapezoidal fuzzy numbers as shown in Table 9.4.

Then, we consider the two different MCDM frameworks shown in Fig. 9.2 and we see the second order criterion B as an information filter. Assume that we have two different types of filters shown as follows:

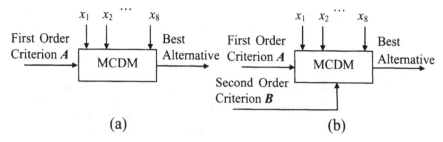

Fig. 9.2. The MCDM frameworks: (a) without considering the second order criterion B, and (b) considering the second order criterion B

(1) When the second order criterion B is seen as a "**Positive filter: B^+**": If the evaluating value $\tilde{B}_i^*$ of the second order criterion B of the alternative x_i is larger than the other evaluating values $\tilde{B}_j^*$ of the second order criterion B of alternatives x_j, where $1 \leq i \leq 8$, $1 \leq j \leq 8$, and $i \neq j$, then the alternative x_i is more fitting to the decision-makers' need than other alternatives x_j, where $1 \leq i \leq 8, 1 \leq j \leq 8$ and $i \neq j$. In this situation, the positive criterion B^+ is equal to the second order criterion B. For example, the eight alternatives x_i, where

Table 9.3. Evaluating Values Represented by Generalized Fuzzy Numbers of the Criteria A and B

Criteria/Alternatives	First Order Criterion A	Second Order Criterion B
x_1	(9, 9, 9, 9; 1.0)	(2, 3, 3, 4; 1.0)
x_2	(0, 0, 0, 0; 1.0)	(10, 10, 10, 10; 1.0)
x_3	(0, 1, 1, 2; 1.0)	(1, 2, 2, 3; 1.0)
x_4	(2, 3, 3, 4; 1.0)	(2, 2.5, 3.5, 4; 1.0)
x_5	(9, 9, 9, 9; 1.0)	(0, 0, 0, 0; 1.0)
x_6	(0, 1, 1, 2; 0.9)	(6, 7, 7, 8; 1.0)
x_7	(3, 4, 4, 5; 1.0)	(8, 8.5, 9, 9; 1.0)
x_8	(2, 3, 3, 4; 1.0)	(4, 5, 5, 6; 0.9)

Table 9.4. Evaluating Values Represented by Standardized Generalized Trapezoidal Fuzzy Numbers

Criteria/Alternatives	First Order Criterion A	Second Order Criterion B
x_1	(0.9, 0.9, 0.9, 0.9; 1.0)	(0.2, 0.3, 0.3, 0.4; 1.0)
x_2	(0.0, 0.0, 0.0, 0.0; 1.0)	(1.0, 1.0, 1.0, 1.0; 1.0)
x_3	(0.0, 0.1, 0.1, 0.2; 1.0)	(0.1, 0.2, 0.2, 0.3; 1.0)
x_4	(0.2, 0.3, 0.3, 0.4; 1.0)	(0.2, 0.25, 0.35, 0.4; 1.0)
x_5	(0.9, 0.9, 0.9, 0.9; 1.0)	(0.0, 0.0, 0.0, 0.0; 1.0)
x_6	(0.0, 0.1, 0.1, 0.2; 0.9)	(0.6, 0.7, 0.7, 0.8; 1.0)
x_7	(0.3, 0.4, 0.4, 0.5; 1.0)	(0.8, 0.85, 0.9, 0.9; 1.0)
x_8	(0.2, 0.3, 0.3, 0.4; 1.0)	(0.4, 0.5, 0.5, 0.6; 0.9)

$1 \leq i \leq 8$, can be seen as some different books and a decision maker wants to choose a book from these books. Assume that the decision maker presents a statement such as:

"I want a book that talks about **fuzzy set theory**, and *if possible I want this book to concern image processing*."

In this statement, the first order criterion A is "fuzzy set theory" and the second order criterion B is "image processing". The second order criterion B can be seen as a positive filter B^{+}. In this situation, we directly use the evaluating value $\tilde{B}_i$ of the second order criterion B in the proposed information fusion algorithm.

(2) When the second order criterion B is seen as a "**Negative filter: B^{-}**": If the evaluating value $\tilde{B}_i^{*}$ of the second order criterion B of the alternative x_i is larger than other evaluating values $\tilde{B}_j^{*}$ of the second order criterion B of alternatives x_j, where $1 \leq i \leq 8$, $1 \leq j \leq 8$, and $i \neq j$, then the alternative x_i is less fitting to the decision-makers' need than other alternatives, where $1 \leq i \leq 8$, $1 \leq j \leq 8$ and $i \neq j$. For example, the eight alternatives x_i, where $1 \leq i \leq 8$, can be seen as some different books and a decision maker wants to choose a book from these books. Assume that the decision maker presents a statement such as:

"I want a book that talks about **fuzzy set theory** and *if possible I don't want this book to concern image processing*."

In this statement, the first order criterion A is "fuzzy set theory" and the second order criterion B is "image processing". The second order criterion B can be seen as a negative filter B^{-}. In this situation, we must obtain each evaluating value $\tilde{\bar{B}}_i^{*}$, where $1 \leq i \leq 8$, of the negative criterion $B^{-} = 1 - B$ using the subtraction operator of generalized fuzzy number ψ shown as follows [5]:

$$
\begin{aligned}
1\psi\tilde{B}_i^{*} &= (1,1,1,1;1)\psi(b_{i1}^{*}, b_{i2}^{*}, b_{i3}^{*}, b_{i4}^{*}; w_{\tilde{B}_i^{*}}) \\
&= \left(1 - b_{i4}^{*}, 1 - b_{i3}^{*}, 1 - b_{i2}^{*}, 1 - b_{i1}^{*}; \min\left(1, w_{\tilde{B}_i^{*}}\right)\right), \\
&= \tilde{\bar{B}}_i^{*} \\
&= (\bar{b}_{i1}^{*}, \bar{b}_{i2}^{*}, \bar{b}_{i3}^{*}, \bar{b}_{i4}^{*}; w_{\tilde{\bar{B}}_i^{*}}),
\end{aligned}
\tag{9.21}
$$

where $0 \leq b_{i1}^{*} \leq b_{i2}^{*} \leq b_{i3}^{*} \leq b_{i4}^{*} \leq 1$, $w_{\tilde{\bar{B}}_i^{*}} = w_{\tilde{B}_i^{*}}$, and $1 \leq i \leq 8$.

In the following, we use the algebraic product (i.e., $a \times b$) and the algebraic sum (i.e., $a + b - a \times b$) to represent the t-norm and t-conorm (i.e., $\wedge$ and $\vee$), respectively.

Case 1: The MCDM framework without considering the second order criterion B and only considering the first order criterion A. Assume that the decision maker presents a statement such as:

"I want a book that talks about **fuzzy set theory**."

In this statement, the first order criterion A is "fuzzy set theory". From Table 9.4, we can see that the alternatives x_1 and x_5 very satisfy the first order criterion A; the alternatives x_3 and x_6 have the same satisfying degrees with respect to the first order criterion A; the alternative x_2 does not satisfy the first order criterion A completely. In this case, because the MCDM framework only considers the first order criterion $\boldsymbol{A}$, we can use formula (9.19) to calculate the ranking value $R(\tilde{A}_1^*)$ of the evaluating value $\tilde{A}_1^*$ shown as follows:

$$
\begin{aligned}
R(\tilde{A}_1^*) &= [\hat{x}_{\tilde{A}_1^*} + (w_{\tilde{A}_1^*} - \hat{y}_{\tilde{A}_1^*})^{\hat{s}\tilde{A}_1^*} \times (\hat{y}_{\tilde{A}_1^*} + 0.5)^{1-w\tilde{A}_1^*}] - 1 \\
&= [0.9 + (1.0 - 0.5)^0 \times (0.5 + 0.5)^{1-1}] - 1 \\
&= [0.9 + 1.0 \times 1.0] - 1 \\
&= 0.9 \ .
\end{aligned}
$$

In the same way, we can use formula (9.19) to calculate the ranking value $R(\tilde{A}_i^*)$ of each evaluating value $\tilde{A}_i^*$, where $2 \leq i \leq 8$, shown as follows: $R(\tilde{A}_2^*) = 0$, $R(\tilde{A}_3^*) = 0.0674$, $R(\tilde{A}_4^*) = 0.2674$, $R(\tilde{A}_5^*) = 0.9$, $R(\tilde{A}_6^*) = 0.038$, $R(\tilde{A}_7^*) = 0.3674$, $R(\tilde{A}_8^*) = 0.2674$. Because $R(\tilde{A}_1^*) = R(\tilde{A}_5^*) > R(\tilde{A}_7^*) > R(\tilde{A}_4^*) = R(\tilde{A}_8^*) > R(\tilde{A}_3^*) > R(\tilde{A}_6^*) > R(\tilde{A}_2^*)$, we can see that the best alternatives are x_1 and x_5, and the preferring order of these alternatives if we only consider the first order criterion $\boldsymbol{A}$ is as follows:

$$
x_1 = x_5 > x_7 > x_4 = x_8 > x_3 > x_6 > x_2 \ .
$$

Case 2: The MCDM framework both considers the first order criterion $\boldsymbol{A}$ and the second order criterion $\boldsymbol{B}$, and we see the second order criterion $\boldsymbol{B}$ as a "**Positive filter: B^+**". Assume that the decision maker presents a statement such as:

"I want a book that talks about **fuzzy set theory**, and *if possible I want this book to concern image processing*."

In this statement, the first order criterion $\boldsymbol{A}$ is "fuzzy set theory" and the second order criterion $\boldsymbol{B}$ is "image processing". In this case, we apply the proposed prioritized information fusion algorithm to deal with the MCDM framework as follows:

[Step 1]   Based on formulas (9.14) and (9.15), translate the generalized trapezoidal fuzzy numbers shown in Table 9.3 into standardized generalized trapezoidal fuzzy numbers as shown in Table 9.4.

[Step 2]   Based on formula (9.16), we can evaluate the degree of compatibility $C(\boldsymbol{A}, \boldsymbol{B}^+)$ between the first order criterion $\boldsymbol{A}$ and the positive criterion $\boldsymbol{B}^+$ as follows:

$$
\begin{aligned}
C(\boldsymbol{A}, \boldsymbol{B}^+) &= \text{Max}\{\text{Min}[(0.9 \wedge 0.3), 0.2667], \text{Min}[(0.0 \wedge 1.0), 0.0], \\
&\quad \text{Min}[(0.1 \wedge 0.2), 0.9], \text{Min}[(0.3 \wedge 0.3), 0.79], \\
&\quad \text{Min}[(0.9 \wedge 0.0), 0.1], \text{Min}[(0.1 \wedge 0.7), 0.36], \\
&\quad \text{Min}[(0.4 \wedge 0.8604), 0.4308], \text{Min}[(0.3 \wedge 0.5), 0.72]\} \\
&= \text{Max}\{0.2667, 0.0, 0.02, 0.09, 0.0, 0.07, 0.3442, 0.15\} \\
&= 0.3442 \ .
\end{aligned}
$$

[Step 3]   Based on formulas (9.8), (9.9) and (9.18), we can calculate the fusion result $\tilde{D}_1$ shown as follows:

$$\tilde{D}_1 = \tilde{A}_1^* \wedge (\tilde{B}_1^* \vee (1 - 0.3442))$$
$$= \tilde{A}_1^* \wedge (\tilde{B}_1^* \vee 0.6558)$$
$$= (a_{11}^*, a_{12}^*, a_{13}^*, a_{14}^*; w_{\tilde{A}_1^*}) \wedge ((b_{11}^*, b_{12}^*, b_{13}^*, b_{14}^*; w_{\tilde{B}_1^*}) \vee$$
$$(0.6558, 0.6558, 0.6558, 0.6558; 1.0))$$
$$= (0.9 \wedge (0.2 \vee 0.6558), \ 0.9 \wedge (0.3 \vee 0.6558), \ 0.9 \wedge$$
$$(0.3 \vee 0.6558),$$
$$0.9 \wedge (0.4 \vee 0.6558) \, ; \, \sqrt[2]{1.0 \times 1.0})$$
$$= (0.6522, 0.6832, 0.6832, 0.7141; \ 1.0) \, ,$$

where we use the t-norm operator "$a \times b$" and the t-conorm operator "$a + b - a \times b$" to denote the operators $\wedge$ and $\vee$, respectively. In the same way, we can get the fusion results $\tilde{D}_2, \tilde{D}_3, \tilde{D}_4, \tilde{D}_5, \tilde{D}_6, \tilde{D}_7$ and $\tilde{D}_8$ shown as follows:

$$\tilde{D}_2 = (0.0, 0.0, 0.0, 0.0; 1.0) \, ,$$
$$\tilde{D}_3 = (0.0, 0.0725, 0.0725, 0.1518; 1.0) \, ,$$
$$\tilde{D}_4 = (0.1449, 0.2226, 0.2329, 0.3174; 1.0) \, ,$$
$$\tilde{D}_5 = (0.5902, 0.5902, 0.5902, 0.5902; 1.0) \, ,$$
$$\tilde{D}_6 = (0.0, 0.0897, 0.0897, 0.1862; 0.9487) \, ,$$
$$\tilde{D}_7 = (0.2793, 0.3793, 0.3862, 0.4828; 1.0) \, ,$$
$$\tilde{D}_8 = (0.1587, 0.2484, 0.2484, 0.3449; 0.9487) \, .$$

[Step 4] By applying formula (9.19), we can get the ranking value $R(\tilde{D}_1)$ of the fusing result $\tilde{D}_1$ shown as follows:

$$R(\tilde{D}_1) = [\hat{x}_{\tilde{D}_1} + \left(w_{\tilde{D}_1} - \hat{y}_{\tilde{D}_1}^{\hat{s}_{\tilde{D}_1}}\right) \times (\hat{y}_{\tilde{D}_1} + 0.5)^{1 - w_{\tilde{D}_1}}] - 1$$
$$= [0.6832 + (1 - 0.3333)^{0.0253} \times (0.3333 + 0.5)^{1-1}] - 1$$
$$= 0.673 \, . \qquad (9.22)$$

In the same way, we can get the ranking values $R(\tilde{D}_2), R(\tilde{D}_3), R(\tilde{D}_4),$ $R(\tilde{D}_5), R(\tilde{D}_6), R(\tilde{D}_7)$ and $R(\tilde{D}_8)$ of the fusion results $\tilde{D}_2, \tilde{D}_3, \tilde{D}_4, \tilde{D}_5, \tilde{D}_6,$ $\tilde{D}_7$ and $\tilde{D}_8$, respectively, shown as follows: $R(\tilde{D}_2) = 0, R(\tilde{D}_3) = 0.0499,$ $R(\tilde{D}_4) = 0.2007, R(\tilde{D}_5) = 0.5902, R(\tilde{D}_6) = 0.0477, R(\tilde{D}_7) = 0.3478$ and $R(\tilde{D}_8) = 0.2064$. Because $R(\tilde{D}_1) > R(\tilde{D}_5) > R(\tilde{D}_7) > R(\tilde{D}_8) > R(\tilde{D}_4) >$ $R(\tilde{D}_3) > R(\tilde{D}_6) > R(\tilde{D}_2)$, we can see that the best alternative is x_1 and the preferring order of these alternatives if we both consider the two different criteria A and B^+ is $x_1 > x_5 > x_7 > x_8 > x_4 > x_3 > x_6 > x_2$.

From the results of Case 1 and Case 2, we can see that if we do not consider the second order criterion B as a "**Positive filter: B^+**", we can not distinguish the preferring order between the books x_1 and x_5; x_3 and x_6. However, if we use the proposed information fusion algorithm to consider the second order criterion B in the MCDM framework, we not only can distinguish the preferring order between the books x_1 and x_5; x_3 and x_6, but also can hold the preferring order of the other books.

9.6 Conclusions

In this article, we have presented a prioritized information fusion algorithm based on similarity measures of generalized fuzzy numbers for aggregating fuzzy information. We have proven that the proposed algorithm satisfies the two properties in [34]. Furthermore, from Sect. 9.5, we can see that the second order criterion B is a good filter in a multi-criteria fuzzy decision-making problem when we use generalized fuzzy numbers to represent the evaluating values. The proposed information fusion method can handle multi-criteria fuzzy decision-making problems in a more flexible manner due to the fact that it allows the evaluating values of the criteria to be represented by crisp values between [0, 1] or generalized fuzzy numbers, where the generalized fuzzy numbers also indicate the degrees of the decision makers' opinions.

References

1. Adamopoulos, G. I., Pappis, C. P. (1996) A fuzzy-linguistic approach to a multi-criteria sequencing problem. *European Journal of Operational Research*, 92(8): 628–636.
2. Alsina, C., Trillas, E., Valverde, L. (1983) On some logical connectives for fuzzy set theory. *Journal of Mathematical Analysis and Application*, 93(1): 15–26.
3. Bordogna, G., Pasi, G. (1995) Linguistic aggregation operators of selection criteria in fuzzy information retrieval. *International Journal of Intelligent Systems*, 10(2): 233–248.
4. Chan, F. T. S., Chan, M. H., Tang, N. K. H. (1997) Evaluating alternative production cycles using the extended fuzzy AHP method. *European Journal of Operational Research*, 100(2): 351–366.
5. Chen, S. H. (1985) Operations on fuzzy numbers with function principal. *Tamkang Journal of Management Sciences*, 6(1): 13–25.
6. Chen, S. H. (1999) Ranking generalized fuzzy number with graded mean integration. *Proceedings of the Eighth International Fuzzy Systems Association World Congress*, vol. 2, Taipei, Taiwan, Republic of China, pp. 899–902.
7. Chen, S. M. (1998) Aggregating fuzzy opinions in the group decision-making environment. *Cybernetics and Systems: An International Journal*, 29(4): 363–376.
8. Chen, S. J., Chen, S. M. (2002) A prioritized information fusion algorithm for handling multi-criteria fuzzy decision-making problems. *Proceedings of the 2002 International Conference on Fuzzy Systems and Knowledge Discovery*, Singapore.
9. Chen, S. J., Chen, S. M. (2003) A new method for handling multi-criteria fuzzy decision making problems using FN-IOWA operators. *Cybernetics and Systems: An International Journal*, 34(2): 109–137.
10. Chen, S. J., Chen, S. M. (2000) A new method for handling the fuzzy ranking and the defuzzification problems. *Proceedings of the Eighth National Conference on Fuzzy Theory and Its Applications*, Taipei, Taiwan, Republic of China.
11. Chen, S. J., Chen, S. M. (2003) Fuzzy risk analysis based on similarity measures of generalized fuzzy numbers. *IEEE Transactions on Fuzzy Systems*, 11(1): 45–56.

12. Chiadamrong, N. (1999) An integrated fuzzy multi-criteria decision making method for manufacturing strategies selection. *Computers and Industrial Engineering*, 37(1-2): 433–436.
13. Deng, H. (1999) Multi-criteria analysis with fuzzy pairwise comparison. *International Journal of Approximate Reasoning*, 21(3): 215–231.
14. Devedzic, G. B., Pap, E. (1999) Multi-criteria-multistages linguistic evaluation and ranking of machine tools. *Fuzzy Sets and Systems*, 102(3): 451–461.
15. Gonzalez, A., Pons, O., Vila, M. A. (1999) Dealing with uncertainty and imprecision by means of fuzzy numbers. *International Journal of Approximate Reasoning*, 21(2): 233–256.
16. Hines, W. W., Montgomery, D. C. (1990) *Probability and Statistics in Engineering and Management Science*. New York, Wiley, pp. 12–77.
17. Hirota, K., Pedrycz, W. (1997) Non-monotonic fuzzy set operations: a generalization and some applications. *International Journal of Intelligent Systems*, 12(7): 483–493.
18. Höhle, U. (1978) Probabilistic uniformization of fuzzy topologies. *Fuzzy Sets and Systems*, 1(1): 311–332.
19. Hsieh, H., Chen, S. H. (1999) Similarity of generalized fuzzy numbers with graded mean integration representation. *Proceedings of the Eighth International Fuzzy Systems Association World Congress*, vol. 2, Taipei, Taiwan, Republic of China, pp. 551–555.
20. Junghanns, A., Posthoff, C., Schlosser, M. (1995) Search with fuzzy numbers. *Proceedings of the Fourth IEEE International Conference on Fuzzy Systems*, Yokohama, Japan, 1995, pp. 979–986.
21. Karsak, E. E., Tolga, E. (2001) Fuzzy multi-criteria decision-making procedure for evaluating advanced manufacturing system investments. *International Journal of Production Economics*, 69(1): 49–64.
22. Kaufmann, A., Gupta, M. M. (1988) *Fuzzy Mathematical Models in Engineering and Management Science*, North-Holland, Amsterdam.
23. Lee, H. S. (1999) An optimal aggregation method for fuzzy opinions of group decision. *Proceedings of the 1999 IEEE International Conference on Systems, Man, and Cybernetics*, vol. 3, Tokyo, Japan, pp. 314–319.
24. Lee, S. J., Lim, S. I., Ahn, B. S. (1998) Service restoration of primary distribution systems based on fuzzy evaluation of multi-criteria. *IEEE Transactions on Power Systems*, 13(3): 1156–1163.
25. Liang, G. S. (1999) Fuzzy MCDM based on ideal and anti-ideal concepts. *European Journal of Operational Research*, 112(8): 682–691.
26. Liao, T. W., Rouge, B. (1996) A fuzzy multi-criteria decision making method for material selection. *Journal of Manufacturing Systems*, 15(1): 1–12.
27. Petrovic, R., Petrovic, D. (2001) Multi-criteria ranking of inventory replenishment policies in the presence of uncertainty in customer demand. *International Journal of Production Economics*, 71(1-3): 439–446.
28. Ramot, D., Milo, R., Friedman, M., Kandel, A. (2001) On fuzzy correlations. *IEEE Transactions on Systems, Man, and Cybernetics- Part B: Cybernetics*, 31(3): 381–390.
29. Subasic, P., Hirota, K. (1998) Similarity rules and gradual rules for analogical and interpolative reasoning with imprecise data. *Fuzzy Sets and Systems*, 96(1): 53–75.
30. Sugianto, L. F. (2001) Optimal decision making with imprecise data. *International Journal of Fuzzy Systems*, 3(4): 569–576.

31. Yager, R. R. (1987) Using approximate reasoning to represent default knowledge. *Artificial Intelligence*, 31(1): 99–112.
32. Yager, R. R. (1991) Non-monotonic set theoretic operations. *Fuzzy Sets and Systems*, 42(2): 173–190.
33. Yager, R. R. (1992) Fuzzy sets and approximate reasoning in decision and control. *Proceedings of the 1992 IEEE International Conference on Fuzzy Systems*, San Diego, California, pp. 415–428.
34. Yager, R. R. (1992) Second order structures in multi-criteria decision making. *International Journal of Man- Machine Studies*, 36(4): 553–570.
35. Yager, R. R. (1998) Structures for prioritized fusion of fuzzy information. *Information Sciences*, 108(1): 71–90.
36. Zhang, G. Q. (1992) On fuzzy number-valued fuzzy measures defined by fuzzy number-valued fuzzy integrals I. *Fuzzy Sets and Systems*, 45(2): 227–237.
37. Zhang, G. Q. (1992) On fuzzy number-valued fuzzy measures defined by fuzzy number-valued fuzzy integrals II. *Fuzzy Sets and Systems*, 48(2): 257–265.

"Published Price Information Sources"

Using Boosting Techniques to Improve the Performance of Fuzzy Classification Systems

Tomoharu Nakashima and Hisao Ishibuchi

Department of Industrial Engineering, Osaka Prefecture University
Gakuen-cho 1-1, Sakai, Osaka 599-8531, Japan
{nakashi, hisaoi}@ie.osakafu-u.ac.jp

Abstract. This paper proposes a boosting algorithm of fuzzy rule-based systems for pattern classification. In the proposed algorithm, several fuzzy rule-based classification systems are simultaneously used for classifying an input pattern. Those fuzzy rule-based classification systems are generated from different subsets of training patterns. A subset of training patterns for generating each fuzzy rule-based classification system is generated depending on the performance of previously generated fuzzy rule-based classification systems. It is expected that the proposed algorithm performs well on both training data and test data. The performance of the proposed algorithm is shown in computer simulations on two real-world pattern classification problems such as the iris data set and the appendicitis data set.

Key words: Fuzzy Systems, Pattern Classification, Ensemble Learning, Boosting

10.1 Introduction

Fuzzy Systems have been successfully developed and used in various fields [1]. Control problems have been the main application domains where fuzzy rule-based systems are used [2, 3]. Recently fuzzy rule-based systems have also been applied to pattern classification problems. There are many approaches to the automatic generation of fuzzy if-then rules from numerical data for pattern classification problems. Genetic algorithms have also been used for generating fuzzy if-then rules for pattern classification problems [4, 5, 6].

It is generally said that classification performance can be improved by combining several classification systems. For example, Ueda and Nakano [7] analytically showed that the generalization error of average output from multiple function approximators is less than that of any single function approximator. For pattern classification problems, various ensembling methods have been proposed [8, 9, 10, 11, 12, 13]. For example, Battiti and Colla [8] examined voting schemes such as a perfect unison and a majority rule for combining multiple neural networks classifiers. Cho [14] and Cho and Kim [15, 16] used fuzzy integrals for aggregating outputs from multiple neural networks. Ishibuchi et al. [17] examined the performance of two levels of voting in

Tomoharu Nakashima and Hisao Ishibuchi: *Using Boosting Techniques to Improve the Performance of Fuzzy Classification Systems*, Studies in Computational Intelligence (SCI) **4**, 147–157 (2005)
www.springerlink.com

fuzzy rule-based classification systems such as a voting by multiple fuzzy if-then rules and a voting by multiple fuzzy rule-based classification systems.

A boosting algorithm is one of the major techniques for ensembling learning systems in various research fields such as prediction, control, and pattern classification. In a typical boosting algorithm, several simple learning systems are generated using different subsets of training patterns. A subset for constructing a learning system is selected from the entire training patterns depending on the history of the performance of the previously generated learning systems. Each learning system is assigned a credit value, which is used for generating a final output. Freund and Schapire [18] proposed AdaBoost algorithm that performs well on both test patterns and training patterns. The AdaBoost successfully dealt with training patterns: Each training pattern has a weight that represents importance in the construction of a single learning system. They showed that the ensembled learning system works well even if the performance of a single learning system itself is not good.

In this paper, we propose a boosting technique for fuzzy rule-based systems for pattern classification problems. In our boosting algorithm, we also assign a weight value to each training pattern. A single fuzzy rule-based classification system is sequentially generated using a subset of the given training patterns. A subset is chosen according to the weight values of training patterns which are assigned according to the performance of the previously generated fuzzy rule-based systems. Computer simulations on two real-world pattern classification problems show that our boosting technique works well on both training patterns and test patterns.

10.2 Fuzzy Rule-Based System

10.2.1 Pattern Classification Problems

Let us assume that our pattern classification problem is an n-dimensional problem with C classes. We also assume that we have m given training patterns $\mathbf{x}_p = (x_{p1}, x_{p2}, \ldots, x_{pn})$, $p = 1, 2, \ldots, m$. Without loss of generality, each attribute of the given training patterns is normalized into a unit interval $[0, 1]$. That is, the pattern space is an n-dimensional unit hypercube $[0, 1]^n$ in our pattern classification problems.

In this paper, we use fuzzy if-then rules of the following type in our fuzzy rule-based classification systems:

$$\text{Rule } R_j : \text{If } x_1 \text{ is } A_{j1} \text{ and } \ldots \text{and } x_n \text{ is } A_{jn}$$
$$\text{then Class } C_j \text{ with } CF_j, \quad j = 1, 2, \ldots, N, \qquad (10.1)$$

where R_j is the label of the j-th fuzzy if-then rule, $A_{j1}, \ldots, A_{jn}$ are antecedent fuzzy sets on the unit interval $[0, 1]$, C_j is the consequent class (i.e., one of the given C classes), CF_j is the grade of certainty of the fuzzy if-then rule R_j, and N is the total number of fuzzy if-then rules. As antecedent fuzzy sets, we use triangular fuzzy sets as in Fig. 10.1 where we show various partitions of a unit interval into a number of fuzzy sets.

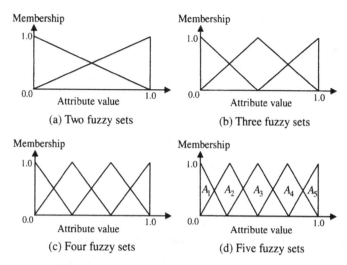

Fig. 10.1. Examples of antecedent fuzzy sets

10.2.2 Generating Fuzzy If-Then Rules

In our fuzzy rule-based classification systems, we specify the consequent class and the grade of certainty of each fuzzy if-then rule from the given training patterns [19, 20, 21].

The consequent class C_j and the grade of certainty CF_j of fuzzy if-then rule are determined in the following manner:

[Generation Procedure of Fuzzy If-Then Rules]

Step 1: Calculate $\beta_h(R_j)$ for Class h $(h = 1, \ldots, C)$ as

$$\beta_h(R_j) = \sum_{x_p \in \text{Class } h} \mu_{j1}(x_{p1}) \cdot \ldots \cdot \mu_{jn}(x_{pn}) ,$$

$$hz = 1, 2, \ldots, C . \tag{10.2}$$

Step 2: Find Class $\hat{h}$ that has the maximum value of $\beta_h(R_j)$:

$$\beta_{\hat{h}}(R_j) = \max\{\beta_1(R_j), \beta_2(R_j), \ldots, \beta_C(R_j)\} . \tag{10.3}$$

If two or more classes take the maximum value, the consequent class C_j of the rule R_j can not be determined uniquely. In this case, specify C_j as $C_j = \phi$. If a single class takes the maximum value, specify C_j as $C_j = \text{Class } \hat{h}$.

Step 3: If a single class takes the maximum value of $\beta_h(R_j)$, the grade of certainty CF_j is determined as

$$CF_j = \frac{\beta_{\hat{h}}(R_j) - \bar{\beta}}{\sum \beta_h(R_j)} , \tag{10.4}$$

where

$$\bar{\beta} = \frac{\sum_{h \neq \hat{h}} \beta_h(R_j)}{C - 1} . \tag{10.5}$$

10.2.3 Fuzzy Reasoning

By the rule generation procedure in Subsect. 10.2.2, we can generate N fuzzy if-then rules in (10.1). After both the consequent class C_j and the grade of certainty CF_j are determined for all the N fuzzy if-then rules, a new pattern $\mathbf{x}$ is classified by the following procedure [19]:

[Fuzzy Reasoning Procedure for Classification]

Step 1: Calculate $\alpha_h(\mathbf{x})$ for Class h, $h = 1, 2, \ldots, C$ as

$$\alpha_h(\mathbf{x}) = \max\{\mu_j(\mathbf{x}) \cdot CF_j | C_j = \text{Class } h, j = 1, 2, \ldots, N\},$$
$$h = 1, 2, \ldots, C, \qquad (10.6)$$

where

$$\mu_j(\mathbf{x}) = \mu_{j1}(x_1) \cdot \ldots \cdot \mu_{jn}(x_n). \qquad (10.7)$$

Step 2: Find Class h_p^* that has the maximum value of $\alpha_h(\mathbf{x})$:

$$\alpha_{h_p^*}(\mathbf{x}) = \max\{\alpha_1(\mathbf{x}), \ldots, \alpha_C(\mathbf{x})\}. \qquad (10.8)$$

If two or more classes take the maximum value, then the classification of $\mathbf{x}$ is rejected, otherwise assign Class h_p^* to $\mathbf{x}$.

10.3 Boosting Technique

In this section, we describe our boosting technique for fuzzy rule-based classification systems described in Sect. 10.2. First we describe the standard type of our boosting technique, then the modified type of our boosting technique is shown.

10.3.1 Standard Type of Boosting

Let $d_p, p = 1, \ldots, m$ be the weight value of a training pattern. A subset for generating a single fuzzy rule-based classification system is chosen according to the weight d_p. The focus of our boosting algorithm is on how to determine the weight value d_p. In our boosting technique, we first check how many times each training pattern is correctly classified during the course of the boosting process. We assign a large weight value to those training patterns which have not been correctly classified by the previously generated fuzzy rule-based systems. On the other hand, a small weight value is assigned to those training patterns which have been correctly classified many times. We also calculate the credit of each single fuzzy rule-based classification system. The credit value is used for determining the final output (i.e., the class) for the input pattern. The proposed boosting technique is described as follows:

[Boosting Algorithm]

Step 0: Set $d_p = 1$, $c_p = 0$, $p = 1, \ldots, m$, $t = 1$, and $L_t = 2$.

Step 1: Choose a subset R_t such that the weight value d_p of a training pattern $\mathbf{x}_p$ in R_t is larger than or equal to 0.5:

$$\mathbf{x}_p \begin{cases} \in R_t, & \text{if } d_p \geq 0.5, \\ \notin R_t, & \text{otherwise}, \end{cases} \qquad p = 1, \ldots, m. \qquad (10.9)$$

Step 2: Generate a fuzzy rule-based classification system S_t using a subset R_t. The number of partitions for each attribute is specified by L_t (see Fig. 10.1).

Step 3: Classify all the given training patterns $\mathbf{x}_p, p = 1, \ldots, m$, using S_t and update c_p as follows:

$$
c_p^{\text{new}} = \begin{cases} c_p^{\text{old}}, & \text{if } \mathbf{x}_p \text{ is correctly classified,} \\ c_p^{\text{old}} + 1, & \text{otherwise ,} \end{cases}
$$
$$
p = 1, \ldots, m . \tag{10.10}
$$

Step 4: Calculate the credit γ_t of the fuzzy rule-based classification system S_t as follows:

$$
\gamma_t = \sqrt{1 - \epsilon_t} , \tag{10.11}
$$

where ϵ_t is an error rate of the fuzzy rule-based classification system over the entire training patterns $\mathbf{x}_p, p = 1, \ldots, m$.

Step 5: Update d_p for the training pattern $\mathbf{x}_p$ as follows:

$$
d_p = \frac{c_p}{t}, p = 1, \ldots, m . \tag{10.12}
$$

Step 6: Evaluate the performance of the ensemble fuzzy rule-based classification systems over all the given training data $\mathbf{x}_p, p = 1, \ldots, m$. The final output (i.e., the class) for an input pattern $\mathbf{x}$ is determined as the most recommended class C_{final} by the fuzzy rule-based classification systems as follows:

$$
C_{\text{final}} = \arg \max_{c=1}^{C} \sum_{\substack{k=1 \\ C_k = c}}^{t} \gamma_t , \tag{10.13}
$$

where $C_k, k = 1, \ldots, t$ is the classification result for the input pattern in (10.8).

Step 7: If at least one of the following termination conditions is satisfied, stop the algorithm:

(a) All the given training patterns $\mathbf{x}_p, p = 1, \ldots, m$, are correctly classified by the ensemble fuzzy rule-based classification system.

(b) A prespecified maximal number of fuzzy rule-based classification systems have been generated. That is, $t = T$ where T is the prespecified maximal number of fuzzy rule-based classification systems.

If these above conditions are not satisfied, let $t = t + 1$ and $L_t = L_{t-1} + 1$, and go to Step 10.3.1.

10.3.2 Subset Selection with a Limited Number of Training Patterns

In the standard version of boosting algorithms, the distribution of classes in the subset is not considered. This may sometimes cause a biased classification system toward a major class with a largest number of training patterns. In order to avoid this problem, we introduce a subset selection scheme with a limited number of training patterns for each class. That is, we select a subset of training patterns so that the number of training patterns from each class is exactly the same. Thus, we replace Step 1 in the procedure of our boosting algorithm with the following procedure:

[Boosting Algorithm with a Limited Number of Training Patterns]

Step 1: Choose a subset R_t such that u training patterns $\mathbf{x}_p$ with the largest weight value d_p are selected for each class. The same number of selected training patterns is selected for each class.

10.4 Computer Simulations

10.4.1 Standard Version of Boosting

We evaluated the performance of the standard version of our boosting technique on two real-world data sets: the iris data set and the appendicitis data set. The iris data set is a four-dimensional three-class pattern classification problem with 50 training patterns from each class. The appendicitis data set has 106 training patterns with seven attributes and two classes.

In the computer simulations, we specified the value of T as $T = 7$ for the iris data set and $T = 5$ for the appendicitis data set since the number of generated fuzzy if-then rules by the rule generation procedure in Subsect. 10.2.2 is intractable when the number of attribute is not small. The proposed boosting algorithm terminates if all the given training data are correctly classified, or if the number of generated fuzzy rule-based classification systems reaches a prespecified value (i.e., if $t = T$). Table 10.1 and Table 10.2 show the performance of the proposed boosting algorithm.

Table 10.1. Classification rate for the iris data set

	Training Data		Test Data	
t	S_t	Ensemble	S_t	Ensemble
1	67.3%	67.3%	67.3%	67.3%
2	33.3%	67.3%	33.3%	67.3%
3	93.3%	93.3%	90.7%	90.7%
4	30.0%	92.7%	33.3%	92.7%
5	98.0%	98.7%	89.3%	92.0%
6	35.3%	96.0%	35.3%	91.3%
7	93.3%	100%	87.3%	93.3%

10.4.2 Subset Selection for Boosting

In this subsection, we examine the performance of the modified version of the boosting technique in Subsect. 10.3.2.

As Freund and Schapire [18] pointed out, a weak learner that has at least 50% classification rate is eligible as a member of the ensemble classification system in the application of boosting algorithms. In this paper, we use a simple version of fuzzy rule-based classification systems as a member of the ensemble classification system. In the computer simulations, each attribute is divided into only three fuzzy

Table 10.2. Classification rate for appendicitis data set

	Training Data		Test Data	
t	S_t	Ensemble	S_t	Ensemble
1	86.8%	86.8%	85.9%	85.9%
2	15.1%	87.8%	13.2%	85.9%
3	91.5%	92.5%	85.9%	85.9%
4	12.3%	92.4%	13.2%	86.8%
5	92.5%	88.7%	80.2%	84.0%

sets (see Fig. 10.1(b)). Furthermore, we restricted the number of antecedent fuzzy sets in fuzzy if-then rules up to two in the application to both the iris data set and the appendicitis data set. Thus, the number of generated fuzzy if-then rules in an individual fuzzy rule-based classification system is $1 +_4 C_1 \cdot 3 +_4 C_2 \cdot 3^2 = 67$ in the case of the iris data set and $1 +_7 C_1 \cdot 3 +_7 C_2 \cdot 3^2 = 211$ in the case of the appendicitis data set.

We discretized each attribute into three fuzzy subsets using an information entropy measure [22]. This method first generates candidate thresholds for discretizing an attribute at the center of the neighboring training patterns. Then the information entropy measure E for each candidate threshold is examined as follows:

$$E = \frac{|S_1|}{|S|} \sum_{c=1}^{C} \frac{|S_{1c}|}{|S_1|} \log \frac{|S_1|}{|S_{1c}|} + \frac{|S_2|}{|S|} \sum_{c=1}^{C} \frac{|S_{2c}|}{|S_2|} \log \frac{|S_2|}{|S_{2c}|}$$

$$+ \frac{|S_3|}{|S|} \sum_{c=1}^{C} \frac{|S_{3c}|}{|S_3|} \log \frac{|S_3|}{|S_{3c}|} , \qquad (10.14)$$

where S_1, S_2, and S_3 are subsets of S divided by the thresholds ($S = S_1 \cup S_2 \cup S_3$, $S_1 \neq S_2 \neq S_3$), S_{1c}, S_{2c}, and S_{3c} are the subsets of training patterns that belong to class c within the subset S_1, S_2, and S_3, respectively. Finally, the threshold with the minimum information entropy measure is determined as the best partitioning one. These procedures are iterated until all the thresholds are determined.

We evaluate the performance of the modified version of the boosting technique for the fuzzy rule-based classification systems that are described in Subsect. 10.3.2. In the computer simulations, we specified the value of u as follows: For the iris data set, we specified it as $u = 5$. That is, we select five training patterns from each class and the total number of training patterns in a subset for constructing a single fuzzy rule-based classification system is 15 for the iris data set. On the other hand, we specified the value of u as $u = 15$ for the appendicitis data set. The total number of the selected training patterns for a subset is 30. We constructed a single fuzzy rule-based classification system 14 times (i.e., $T = 14$) for the iris data set and 50 times (i.e., $T = 50$) for the appendicitis data set.

Simulation results are shown in Fig. 10.2 to Fig. 10.5. These figures show the classification rates on training data in Fig. 10.2 and Fig. 10.3 and on test data in Fig. 10.4 and Fig. 10.5. The performance on test data in Fig. 10.4 and Fig. 10.5 is measured by using a leaving-one-out method. In the leaving-one-out method, one given pattern is used as a test pattern and the fuzzy rule-based classification systems

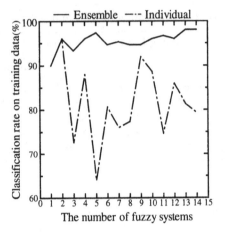

Fig. 10.2. Performance on training data (Iris data set)

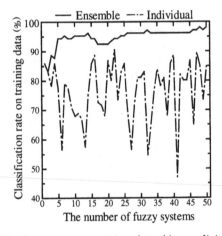

Fig. 10.3. Performance on training data (Appendicitis data set)

are generated from the other given patterns. The performance of the generated fuzzy rule-based classification systems is evaluated by classifying the test pattern. This procedure is iterated until all the given patterns are used as a test pattern. Each of these figures also shows the classification performance of the individual fuzzy rule-based classification system and the ensemble classification system. We can see that the classification rate of the ensemble classification system increases as the number of generated fuzzy rule-based classification systems increases. Its performance is better than any individual fuzzy rule-based classification system. This observation holds for the performance on both the training data and the test data.

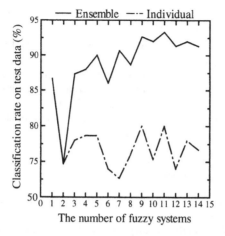

Fig. 10.4. Performance on test data (Iris data set)

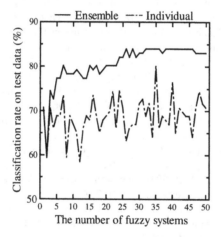

Fig. 10.5. Performance on test data (Appendicitis data set)

10.5 Conclusions

In this paper, we proposed a boosting technique of fuzzy rule-based systems for
pattern classification problems. A single fuzzy rule-based classification system is sequentially generated by using a subset of the given training patterns. The proposed
boosting technique assigns a weight to each training patterns according to the performance of the previously generated fuzzy rule-based classification systems. The final
classification is determined as the class that is most recommended in the ensembled
fuzzy rule-based classification systems. We also proposed a modified version of our
boosting technique where a prespecified number of training patterns are selected as
a member of a subset. In the computer simulations, we evaluated the performance
of our techniques on two real-world pattern classification problems such as the iris

data set and the appendicitis data set. It was shown that the proposed technique works well on both the training patterns and the test patterns.

Our future work on the boosting technique will investigate the relation between the performance and the similarity in the ensemble fuzzy classification systems. It is generally known that ensemble systems do not perform well when the members of the ensemble systems are similar to each other. Although a certain degree of diversity in ensemble systems can be attained by a boosting technique, we need a more powerful method that explicitly deal with the issue of the similarity.

References

1. Yen J (1999) Fuzzy logic – A modern perspective. IEEE Trans. on Knowledge and Data Engineering. 11(1):153–165
2. Sugeno M (1985) An introductory survey of fuzzy control. Information Science. 30(1/2):59–83
3. Lee C C (1990) Fuzzy logic in control systems: Fuzzy logic controller Part I and Part II. IEEE Trans. Systems, Man, and Cybernetics. 20:404–435
4. Ishibuchi H, Nakashima T (1999) Performance evaluation of fuzzy classifier systems for multi-dimensional pattern classification problems. IEEE Trans. on Systems, Man, and Cybernetics. Part B. 29(5):601–618
5. Ishibuchi H, Nakashima T (1999) Improving the performance of fuzzy classifier systems for pattern classification problems with continuous attributes. IEEE Trans. on Industrial Electronics. 46(6):1057–1068
6. Yuan Y, Zhang H (1996) A genetic algorithms for generating fuzzy classification rules. Fuzzy Sets and Systems. 84(1):1–19
7. Ueda N, Nakano R (1996) Generalization error of ensemble estimators. Proc. of Intl. Conf. on Neural Networks: 90–94
8. Battiti R, Colla A M (1994) Democracy in neural nets: Voting schemes for classification. Neural Networks. 7:691–707
9. Xu L, Krzyzak A, Suen C Y (1992) Methods of combining multiple classifiers and their applications to handwriting recognition. IEEE Trans. on Systems, Man, Cybernetics. 22(3):418–435
10. Benediktsson J A, Colla A M (1992) Consensus theoretic classification methods. IEEE Trans. on Systems, Man, Cybernetics. 22(4):688–704
11. Wolpert D H (1992) Stacked generalization. Neural Networks. 5:241–259
12. Jacobs R A, Jordan M I (1991) Adaptive mixtures of local experts. Neural Computation. 3:79–87
13. Hansen L K, Salamon P (1990) Neural network ensembles. IEEE Trans. on Pattern Analysis and Machine Intelligence. 12(10):993–1001
14. Cho S B (1995) Fuzzy aggregation of modular neural networks with ordered weighted averaging operators. International Journal of Approximate Reasoning. 13:359–375
15. Cho S B, Kim J H (1995) Combining multiple neural networks by fuzzy integral for robust classification. IEEE Trans. on Systems, Man, and Cybernetics. 25(2):380–384
16. Cho S B, Kim J H (1995) Multiple network fusion using fuzzy logic. IEEE Trans. on Neural Networks. 6(2):497–501

17. Ishibuchi H, Nakashima T, Morisawa T (1999) Voting in fuzzy rule-based systems for pattern classification problems. Fuzzy Sets and Systems. 103(2):223–238
18. Freund Y, Schapire R (1997) A decision-theoretic generalization of on-line learning and an application to Boosting. Journal of Computer and System Sciences. 55(1):119–139
19. Ishibuchi H, Nozaki K, Tanaka H (1992) Distributed representation of fuzzy rules and its application to pattern classification. Fuzzy Sets and Systems. 52(1):21–32
20. Ishibuchi H, Nozaki K, Yamamoto N, Tanaka H (1995) Selecting fuzzy if-then rules for classification problems using genetic algorithms. IEEE Trans. on Fuzzy Systems. 3(3):260–270
21. Ishibuchi H, Nakashima T (2001) Effect of rule weights in fuzzy rule-based classificaiton systems. IEEE Trans. on Fuzzy Systems. 9(4):506–515
22. Nakashima T, Ishibuchi H (2001) Supervised and unsupervised fuzzy discretization of continuous attributes for paattern classification problems. Proc. of Knowledge-Based Intelligent Information Engineering Systems & Allied Technologies:32–36

11

P-Expert: Implementation and Deployment of Large Scale Fuzzy Expert Advisory System

Andrew Chiou[1] and Xinghuo Yu[2]

[1]Faculty of Informatics and Communication, Central Queensland University, Rockhampton, 4702 Queensland, Australia
a.chiou@cqu.edu.au
[2]School of Electrical and Computer Engineering, RMIT University, Melbourne, 3001 Victoria, Australia
x.yu@rmit.edu.au

Abstract. This paper discusses the problem of parthenium weed infestation in Queensland and describes *P-Expert*, an expert advisory system designed to provide expert knowledge in control and management strategies of parthenium weed. P-Expert is fundamentally a hybrid fuzzy expert system incorporating technologies from fuzzy logic, relational database and multimedia systems. The primary topic of this paper will be a description of the framework of P-Expert, focussing on three main areas: (1) Layered component architecture – each component of the expert advisory system is designed as modules to facilitate maintenance, adaptability and flexibility, (2) Discourse semantics (explanatory capabilities) – provision for explanation and justifying outputs given by the expert advisory system, (3) Meta-consequent – mapping final aggregated output from a fuzzy If-Then rule onto a finite database, and (4) Deployment of large scale advisory expert system.

Key words: parthenium weed, fuzzy logic, meta-consequent, expert advisory system, implementation, deployment

11.1 Introduction

The parthenium weed (*Parthenium hysterophorus* L.) has demonstrated the ability to cause significant environmental, health and financial problems if not managed properly. At present, expert knowledge in the control and management strategies of parthenium weed is provided by government agencies, with input and participation from landholders, local government and community organisations. However, due to the limited resources and capabilities of these agencies and individuals, there is a need for an alternative means in disseminating expert knowledge. One of the alternatives is to develop a computerised expert advisory system that can be deployed to end users to provide expert knowledge and recommendations in the absence of actual human experts.

Andrew Chiou and Xinghuo Yu: *P-Expert: Implementation and Deployment of Large Scale Fuzzy Expert Advisory System*, Studies in Computational Intelligence (SCI) **4**, 159–173 (2005)
www.springerlink.com

P-Expert, the expert advisory system, is fundamentally a hybrid expert system incorporating technologies from fuzzy logic, relational databases and multimedia systems. The primary focus of this paper will be a selective description of three main areas from this framework. These are: (1) Layered component architecture, (2) Discourse semantics (explanatory capabilities), and (3) Meta-consequent function.

11.2 Parthenium Weed Problem

Parthenium has been identified as a weed of national significance, and also has the status of a declared plant under the provisions of the *Rural Lands Protection Act 1985*. It is declared under three categories in different areas in Queensland: Category P2 – where the plant must be destroyed; Category P3 – where the infestation is to be reduced; and Category P4 – where the plant should be prevented from spreading [3, 35]. Due to the often unpredictability of climatic changes which occurs in Australia, this weed has the potential to spread extensively. Economically, the spread of parthenium weed has the potential to cause losses of A\$109–129 million annually [2]. It has major impacts on both pasture and cropping industries, at an estimated loss of at least \$16 million per year for the grazing industry [7]. Parthenium weed is generally unpalatable to stock. However, cattle and sheep do consume the weed when feed is scarce. Stock, especially horses, suffer from allergic skin reactions while grazing infested paddocks. Stocking rate reductions of 25–80 percent are common, resulting in market weights of stock often being lighter [8]. Large consumption can eventually lead to taints in mutton or in some cases, kill stock [33].

In addition, parthenium weed infestation includes contamination of seed produce and total habitat change. In some cases the weed can completely dominate pastures, resulting in a monoculture of non-nutrious vegetable matter [10]. As well as losses in the grazing industry, parthenium weed also causes severe health problems. Parthenium and related genera contain sesquiterpene lactones [28]. which induce severe allergic dermatitis. Other symptoms as a result of direct contact include hay fever and asthma [20].

11.2.1 Biology and History

Parthenium hysterophorus L. is a herbaceous annual or ephemeral member of the Asteraceae, reaching a height of 2 metres in good soil and flowering within 4 to 6 weeks of germination [22]. Large plants can produce more than 15,000 seeds. Most of the seeds germinate within two years if conditions are suitable, although a portion of seed may remain viable for several years even when buried [6]. Parthenium weed grows best on alkaline to neutral clay soils, but grows less prolifically on a wide range of other soil types [9]. The water requirements of the plant are relatively high and both germination and growth can be limited by poor rainfall [36].

The infestation of parthenium weed in Australia is widely believed to have been introduced on two separate occasions. The first occurred during the 1940's due to movement of aircraft and machinery parts from the USA [23]. The second occurrence, the more serious of the two, was in 1958, where seed was brought in as a contaminant of pasture grass seed from Texas, USA [17]. The infestation originating in the Clermont area did not spread quickly until the 1970's. However, its rapid spread quickly covered 170,000 km^2 (10 percent) of Queensland by 1994 [7].

11.2.2 Control and Management Issues

Landholders are currently trying to survive alongside parthenium weed, minimising its effect through management programs which include pasture improvement, the reduction of stocking rates, spelling, correct cultivation practices and the use of herbicides [4, 8]. While this is encouraged, there are great differences between best practice recommendations and actual on-site practices. These differences may cause resource mismanagement that eventually leads to non-economical weed control and management strategies. The discrepancies arose due to difficulties in accessing expert knowledge for the region. While experts specialising in weed control and management strategies exist at the state level, there are too few located in local districts to be of constant assistance. With the unpredictable weather patterns in the Central Queensland region, best practice advice may not be consistently viable. Hence, landowners prefer to abide to guidelines handed down through tradition and sometimes, out-of-date practices.

11.3 P-Expert Project

It is estimated that a minimum of A\$4 million is saved for each percent reduction of parthenium weed infestation levels [10]. In line with the *Queensland Parthenium Strategy (1999–2000)* guidelines [35], it is envisaged that the development of P-Expert will be one way of contributing to the reduction of parthenium weed providing expert knowledge in control and management strategies in the absence of actual human experts. This will save time and human resources.

11.4 Basic Framework

P-Expert is fundamentally a hybrid expert system based on a fuzzy logic system framework incorporating technologies from fuzzy logic, relational databases and incorporating an explanation function using multimedia. P-Experts knowledge base is composed of data types originating from sources as diverse as geographical information system (GIS), human experts, anecdotal references [27], graphics, multimedia, and databases consisting of different formats. With such diversity and variation, this project poses an atypical challenge in the design and development of the P-Expert software. In addition, prior specifications for this project [37] acknowledges that no existing technology or conventional software engineering exercise is able to sufficiently provide a complete solution. The challenge for the P-Expert project is to incorporate newly introduced intelligent software methodology to compliment existing technology to meet the requirements of this project. This paper will describe the innovations introduced into P-Expert. These are:

- **Layered component architecture** – *Each component of the expert advisory system is designed as individual modules to facilitate ownership, security, accountability, maintenance, upgradeability, adaptability and flexibility,*

- **Discourse semantics (explanatory capabilities)** – *Provision for explaining and justifying outputs given by the expert advisory system, and*
- **Meta-consequent** – *Mapping of final aggregated output of fuzzy If-Then rules onto a predefined finite database.*

11.4.1 Layered Component Architecture

The physical architecture of P-Expert is designed in layers, each layer in turn consisting of functional components. Each of these components is kept separate and independent and interacts only with other components that it has been assigned to. With a clear partitioning of each function, this will help facilitate the following factors:

- **Ownership** – *As the hardware, software and data used in P-Expert originates from a complex division of government agencies located over a widespread geographical area and kept decentralised throughout the state. A layered architecture will help maintain ownership attributes.*
- **Security** – *Layering the systems architecture will prevent sensitive data used in P-Expert from public scrutiny and help prevent a breach of security.*
- **Accountability** – *Due to security and ownership issues, layering will help keep track of each component to its owner and their accountability.*
- **Maintenance** – *A large-scale system such as P-Expert should be easy to maintain. Each component, especially the rule base and knowledge base should be kept separate to ensure that editing one or the other will not affect systems integrity* [26].
- **Upgradeability** – *Layering will ensure that the function of each component is kept separate and upgradeable. Altering any of the components should not affect the rest of the system.*
- **Adaptability** – *One of the requirements of P-Expert is to have an open-ended design. That is, it should be adaptable for use in other regions or for different weed species.*
- **Flexibility** – *The system should cater to different groups of users and different user environments* [12]. *For example, via layering different user interface design can be selected from multiple instances of input/output layer to meet different user requirements.*

Conventional System Architecture

P-Experts layered component architecture concept is based on conventional expert systems [13, 19] and conventional If-Then fuzzy rule-based systems architecture [5, 14, 21, 32, 34]. In such conventional systems, although not explicitly declared as such, it is not unusual to expect to find components inherently classified into the following layers (Fig. 11.1):

- **Editor Layer** – *Var-Function Editor & If-Then Rule Editor components.*
- **Data Layer** – *Knowledge Base component.*
- **System Layer** – *Fuzzifier, Inference Engine, Defuzzier components.*
- **Input/Output Layer** – *Input User Interface, Output User Interface components.*

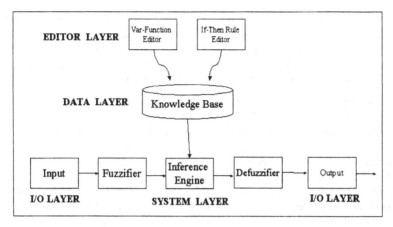

Fig. 11.1. Conventional fuzzy If-Then rule-based system architecture

P-Experts System Architecture

To meet the atypical requirements of P-Experts specifications (i.e. inclusion of multiple data types; explanation function using multimedia; and inclusion of anecdotal references as an input data type), extensive modification has been made to the conventional design. The extended framework of this revised architecture is shown in (Fig. 11.2).

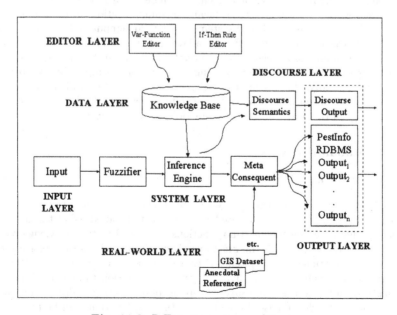

Fig. 11.2. P-Experts system architecture

The editor layer and the data layer are retained without modification, while the defuzzier component in the system layer has been replaced by the meta-consequent function. Also, the input/output layer has been separated into two distinct layers. Two new layers are added, the discourse and real-world layer, giving a total of seven different layers in the P-Expert framework. In brief, these are:

- **Editor Layer** – *Var-Function Editor & If-Then Rule Editor components.*
- **Data Layer** – *Knowledge Base component.*
- **System Layer** – *Fuzzifier, Inference Engine & Meta-Consequent function components.*
- **Input Layer** – *Input User Interface component.*
- **Output Layer** – *Discourse Output & Database Results components.*
- **Discourse Layer** – *Discourse Semantics & Discourse Output components.*
- **Real-World Layer** – *Data originating from knowledge external to the data layer.*

Note that the discourse output component has an overlapping function in both the discourse layer and output layer. The seven layers can further categorised into three major groups according to their functions. These categories are: (1) Standard layers – input, editor and data layers, (2) Discourse semantics – discourse and output layers, and (3) Meta-consequent – system, real-world and output layers.

Communication Between Layers

As physical components of P-Expert may reside over a widespread geographical network, independent governing of each component may lay in the hands of each region's agencies of policy and regulation. Policies governing security over sensitive data and systems integrity may not allow direct access by unauthorised parties to databases belonging to government agencies. Therefore, a layer does not communicate directly with another layer. To facilitate communication, layers communicate with each other via an established system of stand-alone input and output files (I/O Files). The reasons are two fold. Firstly, a layer producing a file can ensure that the output file contains only selected or read-only data. It can be further checked for security reasons and errors prior to being submitted as an input file to a recipient layer. And secondly, this would ensure that the overall system will still continue to function even if one or more layers are taken offline, for example, during maintenance or servicing. In this way, the I/O files functions as proxy data.

11.4.2 Standard Layers

Editor Layer and Data Layer

The editor layer consists of the fuzzy variable function and the if-then rule editor. Membership shapes and its associated rules are created and edited at this level. The resulting rule base and membership functions are stored in the knowledge base at the data layer. Even though the functions of the editor and data layers are similar to that of conventional fuzzy systems, P-Expert differs in content type. As P-Expert is specified as a hybrid expert advisory system, in addition to numerical data and linguistic data types, the editor layer is capable of integrating multimedia data sets as part of its knowledge base. In addition to this, explanations can be tagged onto individual rules. These features will be discussed further in the paper.

Input Layer

Reference [11] demonstrated that the usability and acceptance of expert system eventually ends with the user interface. As P-Expert is inherently a reasonably complex design, a good user interface design will ensure better user acceptance and a shorter learning curve. This is attained by designing multiple instances of different user interface components in the input layer. Hence, depending on different requirements, end users can select different user interface layouts and a choice for customisation.

11.4.3 Discourse Semantics (Explanatory Capabilities)

Since P-Expert consists of technical knowledge with laypersons as active end users, it is mandatory that its output (whether in the form of advice or recommendation) must be accompanied with an explanation to assist the user understand the rationale behind the recommendation. In many cases, an output may also offer advice that is seen by the user as not being the best "solution", therefore an explanation needs to be provided to the user. In addition, outputs can normally be overridden by factors such as:

- Government policies
- Regulations
- By-laws
- Legal issues
- Environmental issues

And hence, a seemingly "perfect" recommendation can be superseded by factors external to the predefined knowledge base. Therefore, there is a requirement for the expert advisory system to explain and justify its outcome in assisting users in making critical decisions.

P-Experts framework caters to this by introducing a discourse layer. This involves the discourse semantic functional component and the discourse output component. These components work by consolidating the inference process while the rest of the system carries out a consultation session. While the discourse semantic component actively keeps track of the processes, the discourse output component formats the explanation in the most understandable media (eg. textual, graphical etc.) before presenting it to the user. This presentation is an integral part of the output layer. However, the presentation of an explanation is independent of the actual output of the inference process. This is to prevent the discourse semantic functions from interfering or indirectly influencing the outcome of the inference process.

Due to the hybrid nature of the knowledge base's data types, an explanation may not necessarily be implemented in a primarily textual context. Hence, to handle all data types, the discourse semantic module in P-Expert will be implemented as (1) discourse using textual semantics, and (2) discourse using multimedia.

Discourse Using Textual Semantics

The standard explanation function often involves purely textual context in a conversation-interaction manner. That is, explanations are presented using printed

English. The theoretical base for the explanation function in conventional expert systems and its construction are explained in the comprehensive work by [15]. However, explanation in fuzzy systems is still in its infancy and is yet to be investigated thoroughly [16]. For the objectives of the P-Expert project, it is sufficient to extend the expression for embedding a textual explanation into a fuzzy If-Then rule described by [12]. The explanation is tagged onto each rule, where in the event this rule fires, the explanation tag is propagated into the inference process. The explanation tag in no way interferes or influences the outcome of the results. The explanation tag is simply a documentation device used by the discourse semantic component which will subsequently be passed to the discourse output component for processing as an output element.

Discourse Using Multimedia

In addition to textual semantics for its explanatory capabilities, P-Expert is also required to explain its inference process using non-textual presentation. For example, P-Expert can help explain a biological control strategy by presenting the flight and sound patterns of certain biological agents (e.g. insects) using video and audio clips. The challenge arises from the fact that P-Experts knowledge base is composed of data types ranging from GIS datasets, graphics, video clips, pictorial and audio clips of plant and biological agents. Therefore, a provision is made to include multimedia data types into its explanatory capabilities. P-Experts theoretical foundations are based on explorations in the relationship between expert systems and multimedia [1, 24, 25, 30, 31].

11.5 Meta-Consequent

While the consequent part of If-Then rules in conventional fuzzy systems lies within the constraints of the universe of discourse, the constraints of P-Expert are mapped onto a predefined large-scale database. The database is finite, that is, the contents of the data are unlikely to change in the long term. In conventional systems, the final integrated values of an inference process are obtained before the centroid values are derived, which is subsequently defuzzified for real world application. However in P-Expert, a defuzzified value would not be meaningful or useful unless it has been "interpreted" and adjusted to match current conditions (e.g. seasonal changes, weather conditions). This process is called the *meta-consequent function*. This involves the replacement of the defuzzier component in the conventional system layer with the meta-consequent component.

The meta-consequent component is complemented by the real world layer, where its components provide undefined external value known only at run-time. For example, weed control thresholds [29] are never known during the inference process until a visual inspection has been carried out. Hence, the real world layer is the deciding factor that determines how the defuzzified output is mapped onto the database, resulting in a final one-to-one or one-to-many output. This process can be (loosely) compared to functions found in search engines. The difference here is "keywords" are not provided by the user, but rather generated by the fuzzy inference process. Depending on its complexity, the meta-consequent function can be categorised as *simple consequent, multi consequent* or *complex consequent* function.

Simple Consequent Function

Reference [18] have previously proposed a post adjustment mechanism to incorporate real world data with data from the expert system's knowledge base. While the knowledge base contains pre-defined data collected from past instances, real world data represents the trends of external factors unknown at run-time (e.g. weather anomalies) and personal views (e.g. anecdotal references) not contained in the knowledge base. This post adjustment mechanism has been extensively modified for P-Experts purposes. Therefore, in the instance of a one-to-one mapping consequent, the meta-consequent function caters for the post adjustment by applying a BUT operator. This is expressed as,

$$(\textbf{IF } x \text{ is } a \textbf{ THEN } y \text{ is } b) \textbf{ BUT } (z \text{ is } c \textbf{ THEN } y \text{ is } d) \tag{11.1}$$

where the antecedent z is c is not a condition, but a factual statement.

For example,

$$(\textbf{IF } temp \text{ is } high \textbf{ THEN } outbreak \text{ is } unlikely) \textbf{ BUT}$$
$$(location \text{ is } wet \textbf{ THEN } outbreak \text{ is } very\ likely)$$

Unlike an $ELSE$ operator, the BUT part of the If-Then rule is always true. That is, the BUT operator is evaluated and executed under all conditions. It is used to supersede the consequent of original rule on the left-hand side of the BUT operator. From (11.1), this implies that b is not equal to d. However, under certain circumstances it is possible that b has the same value as d, that is, $(b = d)$.

For example,

$$(\textbf{IF } temp \text{ is } high \textbf{ THEN } outbreak \text{ is } unlikely) \textbf{ BUT}$$
$$(location \text{ is } extremely\_dry \textbf{ THEN } outbreak \text{ is } unlikely)$$

However, this case is highly unlikely and it's a mechanism provided to negate the effects of the BUT operator in rare circumstances.

Multi Consequent Function

In the instance of a one-to-many mapping consequent, the multi consequent function in P-Expert will allow branching to different membership functions within the same fuzzy variable. This will allow the same variable to have ownership over different sub-domains. We introduce a new operator, $CASE\text{-}OF$, to facilitate the operation of this function. The simplified expression is,

$$(\textbf{IF } x \text{ is } a \textbf{ THEN CASE-OF } m)$$
$$\{\textbf{CASE-OF } m_1 : y_1 \text{ is } b_1;$$
$$\vdots$$
$$\textbf{CASE-OF } m_n : y_n \text{ is } b_n\} \tag{11.2}$$

where, m is an external input from the real-world layer. An example for the software implementation for (11.2) is,

(**IF** *weather* is *wet* **THEN CASE-OF** *location*)

{**CASE-OF** *Rockhampton: infestation* is *likely;*

  CASE-OF *Mackay: infestation is unlikely;*

  CASE-OF *Gladstone: infestation is unlikely* **AND** *flowering is likely*}

Complex Consequent Function

In a one-to-one-to-many instance, the operators, *BUT* and *CASE-OF* are combined to give a mechanism to override the consequent of a rule, and yet facilitating branching under different cases. The simplified expression is a combination of (11.1) and (11.2) giving,

$$(\textbf{IF } x \text{ is } a \textbf{ THEN } y \text{ is } b) \textbf{ BUT } (z \text{ is } c \textbf{ THENCASE-OF } m)$$
$$\{\textbf{CASE-OF } m_1 : w_1 \text{ is } d_1;$$
$$\vdots$$
$$\textbf{CASE-OF } m_n : w_n \text{ is } d_n\} \qquad (11.3)$$

Where, m is an external input from the real-world layer. An example for the software implementation for (11.3) is,

(**IF** *temp* is *high* **THEN** *infestation* is *unlikely*) **BUT**

(*location* is *wet* **THEN CASE-OF** *location*)

{**CASE-OF** *Rockhampton: infestation* is *likely;*

CASE-OF *Mackay: infestation* is *unlikely;*

CASE-OF *Gladstone: infestation* is *unlikely* **AND** *flowering is likely*}

Here, we see that the *BUT* operation overrides the default consequent of the If-Then rule, and at the same time the overriding consequent is dependent on the real-world input. In this example, the location determines the final outcome.

11.6 P-Experts Deployment Infrastructure

Even though P-Expert is developed independently of the main government IT infrastructure and database, it is not practical however, for it to operate as a stand-alone software entity. The P-Expert expert advisory system has to draw its resources, such as database and telecommunication facilities, from the government agency. While this government agency have existing infrastructure that is comprehensive for its stated objectives, it does not however provide a readily available provision to facilitate the embedding of a secondary system (i.e. P-Expert). Hence, the goal of P-Expert project does not only specify the development of the software itself, but it also specifies the mechanism of how the P-Expert software can be embedded into the existing infrastructure. Even though P-Expert system is not considered an insignificant piece of software, it is nonetheless a minor addition compared to the rest of the agency's overall infrastructure. Hence, the specification for the P-Expert project

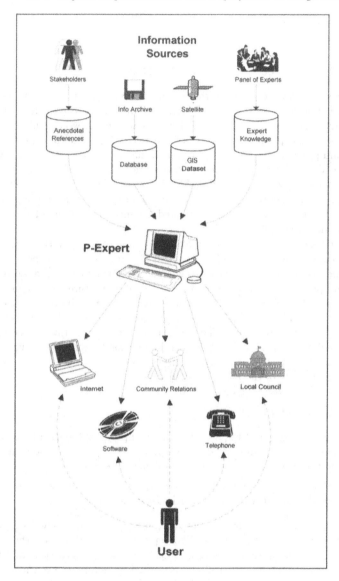

Fig. 11.3. P-Experts deployment infrastructure

also requires it to be deployed seamlessly into the existing system with minimal disruption.

The deployment infrastructure for P-Expert can be categorised into five major levels Fig. 11.3. These levels are:

1. data source,
2. knowledge base,
3. P-Expert system,

4. deployment interface, and
5. user.

11.6.1 Data Source and Knowledge Base

The data source level consists of the original sources that provide the data to the knowledge base level used by P-Expert. Satellite imagery provides current GIS dataset mapping existing infestation sites of parthenium weed. This data is used to help locate and predict the spread of the weed. The panel of experts consist of officials and practitioners of weed control and management strategies. They provide the practical knowledge in how these actions are to be carried out. This knowledge is stored at the knowledge base level. The information archive source, although not strictly an original source of data, represents data that have been pre-processed by the government agency and other affiliated organisations. Some of these data are statistics, maps and GIS datasets.

As opposed to GIS datasets produced by satellite imagery that has the latest datasets, the dataset from the information archive are past records and history of the weeds' infestation. In exception to stakeholders information source, all other data sources are pre-existing in the current agency's infrastructure. Stakeholders provide the knowledge base of the anecdotal references. These may be results of interviews and surveys in qualitative format, such as verbal and information snippets.

Due to the unconventional and unstructured format of this source of data, a fuzzy modelling component is required to be developed from ground-up to convert these references into meaningful data. This component itself is considered one of the major developments of the P-Expert project.

11.6.2 P-Expert Level

This level consists of the P-Experts system. It employs all the data from the knowledge base level. For security and data integrity reasons, P-Expert software can only read and not alter the contents of the knowledge base level.

11.6.3 Deployment Interface and User Level

As its name suggests, the deployment interface level specifies how P-Expert is deployed outwardly among end users. The deployment stage can be either on-line or off-line. The on-line option via the Internet is the most convenient and preferred method. However, taking practical and accessibility factors into considerations, off-line methods would have to be provided to end users who do not have access to the Internet. Five deployment methods are employed here to allow end users to access P-Expert, one in the on-line access and four in the off-line access. In the sole on-line access, users log on to the P-Expert server via the World Wide Web and the interaction between users and P-Expert is in real-time. The P-Expert expert advisory system can also be distributed as stand-alone software, for example as a downloadable installation or on a CD-ROM. The only difference between the stand-alone software and the on-line version is the age of the database. As for users who neither have access to the Internet nor a computer, government agencies will provide hotline services, walk-in information services and community visitations to help operate P-Expert on the behalf of the users.

11.7 Summary

Parthenium hysterophorus is a declared weed to the state of Queensland. It has the potential to cause substantial financial loss in the grazing industry. In addition, it can cause health problems to both stock and humans, and cause severe environmental problems. Managing parthenium weed occurrences and outbreaks is often handicapped by competing demands for scarce resources (including human and financial resources) at the same time. The P-Expert project was initiated to meet the requirement for increased assistance by proposing to design and implement a prototype expert advisory system that can offer expert advice to assist the control and management of the parthenium weed. P-Expert is fundamentally a hybrid expert system based on a fuzzy logic system framework incorporating technologies from fuzzy logic, relational databases and explanation functions. The three main areas P-Expert hopes to contribute to the field of fuzzy intelligent systems are: (1) Layered component architecture in the fuzzy If-Then rule based system, where components in the system are designed as separate modules to facilitate maintenance, adaptability and flexibility, (2) Discourse semantics that will provide explanatory capabilities, (3) Meta-consequent functions that will map the consequent of an If-Then rule onto a finite database, and (4) Deployment of large scale advisory expert system.

References

1. Abu-Hakima, S. (1991). Generating Hypermedia Explanations: Workshop Notes on Intelligent Multimedia Interfaces. *Ninth National Conference on Artificial Intelligence (AAAI-91)*. Anaheim: 63–68.
2. Adamson, D. C. (1996). Determining the Economic Impact of Parthenium on the Australian Beef Industry: A Comparison of Static and Dynamic Approaches. MSc Thesis. Brisbane, University of Queensland.
3. Anon (1998). "Parthenium Weed: Parthenium hysterophorus." *DNR Pest Facts*.
4. Armstrong, T. R. and L. C. Orr (1984). Parthenium hysterophorus Control with Knockdown and Residual Herbicides. *Proceedings of the Seventh Australian Weeds Conference*. Sydney, Academic Press. 2: 8–13.
5. Berkan, R. C. and S. L. Trubatch (1997). Fuzzy Systems Design Principles: Building Fuzzy IF-THEN Rule Bases, IEEE Press.
6. Butler, J. E. (1984). "Longevity of Parthenium hysterophorus L. Seed in the Soil." *Australian Weeds* 3: 6.
7. Chippendale, J. F. and F. D. Panetta (1994). "The Cost of Parthenium Weed to the Queensland Cattle Industry." *Plant Protection* 9: 73–76.
8. Condon (1992). Management Remains the Key to Parthenium Control. *Queensland Country Life*: 23.
9. Dale, I. J. (1981). "Parthenium Weed in the Americas." *Australian Weeds* 1: 8–14.
10. Dhileepan, K. and R. E. McFadyen (1997). Biological Control of Parthenium in Australia: Progress and Prospects. *Proceedings of the First International Conference on Parthenium Management*. Dharwad, University of Agricultural Sciences. 1: 40–44.

11. Dologite, D. G. and R. J. Mockler (1994). "Designing the User Interface of a Strategy Planning Advisory System: Lessons Learned." *International Journal of Applied Expert Systems* 2(1): 3–21.

12. Forslund, G. (1995). "Designing for Flexibility: A Case Study." *Expert Systems* 12(1): 27–37.

13. Gisolfi, A. and W. Balzano (1993). "Constructing and Consulting the Knowledge Base of an Expert Systems Shell." *Expert Systems* 10(1): 29–35.

14. Goel, S., V. K. Modi, et al. (1995). Design of a Fuzzy Expert System Development Shell. *Proceedings of 1995 IEEE Annual International Engineering Management Conference*: 343-346.

15. Gregor, S. and I. Benbasat (1999). "Explanations from Intelligent Systems: Theoretical Foundations and Implications for Practice." *MIS Quarterly* 23(4): 497–530.

16. Gregor, S. and X. Yu (2000). Exploring the Explanatory Capabilities of Intelligent System Technologies. *Second International Discourse With Fuzzy Logic In The New Millennium.* Mackay, Australia, Physica-Verlag: (current volume).

17. Haseler, W. H. (1976). "Parthenium hysterophorus L. in Australia." *PANS* 22: 515–517.

18. Lee, K. C. and W. C. Kim (1995). "Integration of Human Knowledge and Machine Knowledge by Using Post Adjustment: its Performance in Stock Market Timing Prediction." *Expert Systems* 12(4): 331–338.

19. Leung, K. S. and M. H. Wong (1990). "An Expert-System Shell Using Structured Knowledge: An Object-Oriented Approach." *Computer* 23(3): 38–46.

20. McFadyen, R. E. (1995). "Parthenium Weed and Human Health in Queensland." *Australian Family Physician* 24: 1455–1459.

21. Miyoshi, T., H. Koyama, et al. (1990). LIFE Fuzzy Expert System Shell. *Proceedings of the First International Symposium on Uncertainty Modelling and Analysis*: 196–201.

22. Navie, S. C., R. E. McFadyen, et al. (1996). "The Biology of Australian Weeds 27. Parthenium hysterophorus L." *Plant Protection* 11(2): 76–88.

23. Parsons, W. T. and E. G. Cuthbertson (1992). *Noxious Weeds of Australia.* Melbourne, Inkata Press.

24. Pracht, W. E. (1986). "A Graphical Interactive Structural Modelling Aid for Decision Support Systems." *IEEE Transactions on Systems, Man, and Cybernetics* 16(2): 265–270S.

25. Ragusa, J. M. and E. Turban (1994). "Integrating Expert Systems and Multimedia: A Review of the Literature." *International Journal of Applied Expert Systems* 2(1): 54–71.

26. Shiraz, G. M., P. Compton, et al. (1998). FROCH: A Fuzzy Expert System with Easy Maintenance. *IEEE International Conference on Systems, Man, and Cybernetics 1998.* 3: 2113–2118.

27. Suh, C.-K. and E.-H. Suh (1993). "Using Human Factor Guidelines for Developing Expert Systems." *Expert Systems* 10(3): 151–156.

28. Swain, T. and C. A. Williams (1977). Heliantheae – Chemical Review. *The Biology and Chemistry of the Compositae.* V. H. Heywood, J. B. Harbone and B. L. Turner. London, Academic Press. 2.

29. Swanton, C. J., S. Weaver, et al. (1999). Weed Thresholds: Theory and Applicability. *Expanding the Context of Weed Management.* D. D. Buhler. New York, Food Product Press: 9–29.

30. Swartout, W., C. Paris, et al. (1991). "Explanations in Knowledge Systems: Design for Explainable Expert Systems." *IEEE Expert* 6(3): 58–64.
31. Szuprowica, B. O. (1991). "The Multimedia Connection." Expert Systems: *Planning/Implementation/Integration* 2(4):59–63.
32. Tsutomu, M., K. Hiroshi, et al. (1993). Fuzzy Expert System Shell LIFE FEShell-Working Environment. *Proceedings of the Second International Symposium on Uncertainty Modelling and Analysis*: 153–160.
33. Tudor, G. D., A. L. Ford, et al. (1981). Taints in Meat from Sheep Grazing Parthenium Weed. *Proceedings of the Sixth Australian Weeds Conference.*
34. Umano, M., I. Hatono, et al. (1994). Fuzzy Expert System Shells. Proceedings of the Sixth International Conference on Tools with Artificial Intelligence: 219–225.
35. Walton, C. (1999). Queensland Parthenium Strategy 1999-2004, Queensland Department of Natural Resources: 2–15.
36. Williams, J. D. and R. H. Groves (1980). "The Influence of Temperature and Photoperiod on Growth and Development of Parthenium hysterophorus L." Weed Res. 20: 47–52.
37. Yu, X. and J. B. Lowry (1999). A Computerised Intelligent Planning Support System for Forest Landuse Assessment: 1999 Strategic Partnerships with Industry – Research and Training (SPIRT) Application. Rockhampton, Australian Research Council/Department of Employment, Education, Training and Youth Affairs: (unpublished report).

12

Data Mining and User Profiling
for an E-Commerce System

Ken McGarry, Andrew Martin and Dale Addison

School of Computing and Technology, University of Sunderland, St Peters
Campus, St Peters Way, Sunderland, United Kingdom SR6 ODD
ken.mcgarry@sunderland.ac.uk

Abstract. Many companies are now developing an online internet presence to sell
or promote their products and services. The data generated by e-commerce sites is
a valuable source of business knowledge but only if it is correctly analyzed. Data
mining web server logs is now an important application area for business strategy.
We describe an e-commerce system specifically developed for the purpose of demon-
strating the advantages of data mining web server logs. We also provide details of
system implementation which was developed in Java. The data generated by the
server logs is used by a rule induction algorithm to build a profile of the users, the
profile enables the web site to be personalized to each particular customer. The
ability to rapidly respond and anticipate customer behavior is vital to stay ahead
of the competition.

Key words: Data mining, personalisation, machine learning, interestingness measures

12.1 Introduction

This paper describes the development of a software tool for the exploratory analysis
of website data. The project was initiated by the School of Computing and Technology at the University of Sunderland in the UK. The need for this project arises
from the School's consultancy activities, which consists of advising small to medium
enterprizes (SME) on Information Technology and Artificial Intelligence techniques.
Many companies are interested in improving their web-based profitability by profiling and targeting customers more effectively. The system used as an exploratory
tool to develop an understanding of the issues involved in mining website data. In
addition, it is used to demonstrate these principles to SME's to encourage take-up
of the technology. The system consists of three main stages:

- The development of a small mock e-commerce system.
- The analysis of server logs, cookies, user generated forms using data mining
  techniques such as rule induction.
- The development of suitable models and rules to predict future user activity.

Ken McGarry et al.: *Data Mining and User Profiling for an E-Commerce System*, Studies in
Computational Intelligence (SCI) **4**, 175–189 (2005)
www.springerlink.com

Data mining and knowledge discovery is generally understood to be the search for interesting, novel and useful patterns in data on which organizations can base their business decisions [3]. Although statistics do have a place within the data mining paradigm, most of the techniques used are based on algorithms devised by the machine learning and artificial intelligence community. A wide variety of techniques have been successfully used such as neural networks [13, 14, 15], decision trees/rule induction [21], association rules [1] and clustering techniques [5].

Another aspect of the data mining process that needs to be addressed is the identification of what constitutes an interesting and useful pattern. There are two main methods of determining this. The first uses objective mathematical measures to assess the degree of interestingness, many such measures exist and are domain dependent [2, 7]. The second method is to incorporate the users' subjective knowledge into the assessment strategy. Each of these approaches has various characteristics, for example the subjective method requires access to a domain expert [10, 12]. The determination of interestingness is likely to remain an open research question.

Many companies are now using data mining techniques to anticipate customer demand for their products and as a result these companies are able to reduce overheads and inventory costs. Companies such as Walmart are well aware of the benefits of data mining and have used the technology to modify their business strategy. The introduction of online transactions and the nature of the data involved with this form of business has given data mining increased scope. The internet is a medium for business transactions that is available 24 hours a day, all year round and is worldwide in scope. The potential for increased business opportunities is enormous but many technical challenges need to be addressed.

The remainder of this paper is structured as follows: section two discusses the advantages and techniques used for profiling online customers, section three describes the Java based WEKA data mining package and how it is used within our system, section four presents the details of the system implementation, section five describes the development and experimental work performed on the system, section six discusses the conclusions.

12.2 Data Mining the Web

The data generated by the World Wide Web has enabled customer segmentation and targeting opportunities which were previously considered impossible or difficult. The users as they browse or purchase products from a web site create a trail of electronic data that can be used to create very specific adverts.

In business and marketing terms this concern with the customers' details for use of boosting sales is commonly referred to as Customer Relationship Marketing (CRM) and is very important to businesses. Knowing about customers through their profiles and providing them with a relevant service should create a certain amount of customer loyalty. It should also satisfy the customer and make them return to the shop or e-commerce site. Understanding the customer's preferences is important when trying to satisfy a customer and hence increase profits through further purchases or new customers.

The tasks undertaken by web mining can be grouped under several classes based upon their objectives. Referring to Fig. 12.1 web mining can be broadly divided into

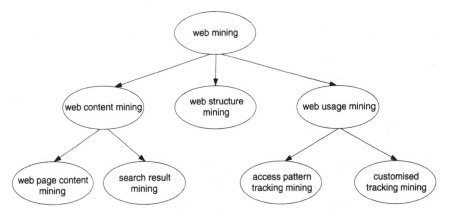

Fig. 12.1. Web mining hierarchy

content mining, structure mining and usage mining. Content mining seeks to uncover the objects and resources within a site while structure mining reveals the inter and intra connectivity of the web pages. Usage mining analyzes web server logs to track the activities of the users as they traverse a site.

A web site is often the first point of contact between a potential customer and a company. It is therefore essential that the process of browsing/using the web site is made as simple and pleasurable as possible for the customer. Carefully designed web pages play a major part here and can be enhanced through information relating to web access. The progress of the customer is monitored by the web server log which holds details of every web page visited. Over a period of time a useful set of statistics can be gathered which can be used for trouble shooting e.g. when did the customer abandon the shopping cart? or why do customers view a page but do not buy? Therefore, problems with poorly designed pages may be uncovered.

However, the main advantage of mining web server logs relate to sales and marketing. Sites like Amazon hold individual customer's previous product searches and past purchases with which to target this particular individual. In essence the data residing within the web server log can be used to learn and understand the buying preferences of customers. This information can be actively used to target similar types of customers, for example in promoting special offers or advertising new products to customers most likely to respond. For a good general introduction to web mining see the work of Mena [16, 17]. Data gathered from both the web and more conventional sources can be used to answer such questions as:

- Marketing – those likely to buy.
- Forecasts – predicting demand.
- Loyalty – those likely to defect.
- Credit – which were the profitable items.
- Fraud – when and where they occur.

Information is available from:

- Registration forms, these are very useful and the customers should be persuaded to fill out at least one. Useful information such as age, sex and location can be obtained.

- Server log, this provides details of each web page visited and timings.
- Past purchases and previous search patterns, useful for personalization of web pages.
- Cookies, these reside on the customers hard drive and enable details between sessions to be recorded.

12.3 Weka Data Mining Package

Weka was developed at the University of Waikato in New Zealand [24]. Weka is written in Java and implements many AI algorithms (classification and clustering). Weka is used primarily to mine combined user profiling and buying/browsing data patterns. The classifier used is Prism and is based on ID3 [21]. The software can be used as a stand-alone package or implemented within the users system through DLL's. For an example of Prism rules see Fig. 12.2.

If occupation = execman
and age = 18
and salary = 2 then ALC1
If occupation = compeng
and age = 19
and salary = 8 then ALC2
If occupation = ostdnt
and age = 18
and salary = 1 then ALC2
If occupation = execman
and age = 18
and salary = 2 then ALC2
If age = 21
and salary = 8
and occupation = retired then ALC3

Fig. 12.2. Prism rules

The generated rules are propositional $< IF..THEN >$ classification rules, they have several conditions in each antecedent and refer to a single consequent or class.

Weka requires all data to be in ARFF format (Attribute-Relation File Format) which in addition to the raw data contains details of each attribute and its data type. The ARFF file format is simply a list of comma separated values. Figure 12.3 gives an example of the required format.

Table 12.1 is primarily used by the Prism2Db program, where the parsed rules from the Weka data analysis output are placed. Within the shopping system, this table is used by the product recommendations servlet. The data is user profile data that has been analysed by Weka into the form of rules, with a resultant page. This table is read by the recommendation servlet and produces the rule:

```
@RELATION iris
@ATTRIBUTE sepallength NUMERIC
@ATTRIBUTE sepalwidth NUMERIC
@ATTRIBUTE petallength NUMERIC
@ATTRIBUTE petalwidth NUMERIC
@ATTRIBUTE class Iris-setosa,Iris-versicolor,Iris-virginica

@DATA
5.1,3.5,1.4,0.2,Iris-setosa
4.9,3.0,1.4,0.2,Iris-setosa
4.7,3.2,1.3,0.2,Iris-setosa
```

Fig. 12.3. ARFF File Format

Table 12.1. User profile data

Age	Salary	Occupation	Postcode	Product
18	2	execman	–	ALC1
19	8	compeng	–	ALC2
18	1	ostdnt	–	ALC2
18	2	execman	–	ALC2
18	8	retired	–	ALC3

IF age = whatever AND salary = whatever AND occupation = whatever THEN page this user is likely to visit = whatever page.

The raw data consisted of 2190 records of a log database table, which contained pre-parsed web server log information relating to which page a user had been to and which user had been to that page; it was found that Weka analysis of this data returned 289 IF-AND-THEN rules which were used as the page recommendation advice.

12.3.1 Determining the Value of Extracted Patterns

Although the technology behind the data mining revolution have been around for some time, the ability to differentiate between useful and novel information on one hand, and mundane or trivial information on the other is not implemented by many of the software packages. This paper discusses how data mining systems can be improved by the use of so-called "interestingness" measures that can be used to automate the discovery of novel and useful information. These measures include a combination of objective measures relating to the statistical strength of a discovered piece of information and subjective measures which compares the knowledge of the expert user with the discovered information. Recent work has to some extent addressed this problem but much work has still to be done [2, 7, 12].

The interested reader is encouraged to examine the following volumes related to knowledge discovery and interestingness measures for evaluating patterns. A useful collection of knowledge discovery papers edited by Piatetsky-Shapiro and Frawley covers many aspects of interestingness [20]. A further volume was produced that covers many recent advances in data mining and knowledge discovery [4]. Hilderman and Hamilton provide a detailed review of several interestingness measures in their technical report and book [8, 9].

Objective Measures

Freitas enhanced the original measures developed by Piatetsky-Shapiro by proposing that future interestingness measures could benefit by taking into account certain criteria [6, 7].

- Disjunct size. The number of attributes selected by the induction algorithms prove to be of interest. Rules with fewer antecedents are generally perceived as being easier to understand and comprehend. However, those rules consisting of a large number of antecedents may be worth examining as they may refer either to noise or a special case.
- Misclassification costs. These will vary from application to application as incorrect classifications can be more costly depending on the domain e.g. a false negative is more damaging than a false positive in cancer detection.
- Class distribution. Some classes may be under represented by the available data and may lead to poor classification accuracies etc.
- Attribute ranking. Certain attributes may be better at discriminating between the classes than others. So discovering attributes present in a rule that were thought previously not to be important is probably worth investigating further.

The statistical strength of a rule or pattern is not always a reliable indication of novelty or interestingness. Those relationships with strong correlations usually produce rules that are well known, obvious or trivial. We must therefore seek another method to detect interesting patterns.

Subjective Measures

The following subjective measures have been developed in the past and generally operate by comparing the users beliefs against the patterns discovered by the data mining algorithm. There are techniques for devising belief systems and typically involve a knowledge acquisition exercise from the domain experts. Other techniques use inductive learning learning from data and some also refine an existing set of beliefs through machine learning. Silbershatz and Tuzhilin view subjective interesting patterns as those more likely to be unexpected and actionable [22].

- Probabilistic measures. Bayesian appraoches have been used to enable conditional probabilities to be used. Silbershatz devised a system of "hard" and "soft" beliefs [22]. The soft beliefs could be revised and updated in the light of new evidence. However, any new patterns that contradict hard beliefs are would not result in modification of beliefs but would tend to indicate that these patterns are interesting.

- Syntactic distance measure. This measure is based on the degree of distance between the new patterns and a set of beliefs. More specifically, if the rule consequents are the same but the antecedents are greatly different then the system may have uncovered some interesting patterns [11].

- Logical contradiction. Although devised for association rules this technique uses the statistical strength (objective) of a rule i.e. the confidence and support to determine if a pattern can be classed as unexpected or not [18, 19]. Any conflict with between the degrees of user suggested by the users and those actually generated by the association rules.

- Actionability. Actionability refers to the organizations ability to do something useful with the discovered pattern. Therefore a pattern can be said to be interesting if it is both unexpected and actionable. This is clearly a highly subjective view of the patterns as actionability is dependent not only on the problem domain but also on the users' objectives at a given point in time [23].

A major research question is how to combine both objective and subjective measures into a unified measure. The patterns generated by our system are analyzed as a post-processing exercise. We pay particular attention to objective measures such as class imbalance.

12.4 System Development

The system is comprised of a number of interacting Java modules. Figure 12.4 shows how the modules interact. Data from the customer is gained from the initial action of registering, here details such as name, occupation, age, sex and other details are entered. Such forms are very useful but users will only fill them out given sufficient incentive such as free gifts, discounts etc. Other data becomes available when users make purchases and browse web pages, this is reported by the server log and shopping basket Java servlet. After the initial trials with a student user base we gathered enough data to train the Prism rule induction algorithm.

12.4.1 Java, Servlets and JSP

Java, Servlets, JSP and Java Beans technologies were chosen primarily to make the system completely portable across platforms making this proof-of-principle project more useful as it can be demonstrated on any platform supporting Java.

Servlets are server side objects which are run using special server technology and are accessed in several ways: They are most commonly used on websites to query databases in order to form Java Beans via a standard submission of data from a HTML form. They can also be "servlet chained" so that several servlets can pass data to another in order to compartmentalize and specialize code functions. Basically servlets are Java objects which run on a Java web server and perform workhorse tasks in the background so other code can benefit from it, namely JSPs.

Java beans are Java objects that are used in conjunction with JSPs and servlets to provide reusable components in code or to provide a place to store data. In this project the later use is the most common, the e-basket being one such example which holds a user's chosen products in memory (or more accurately, within the

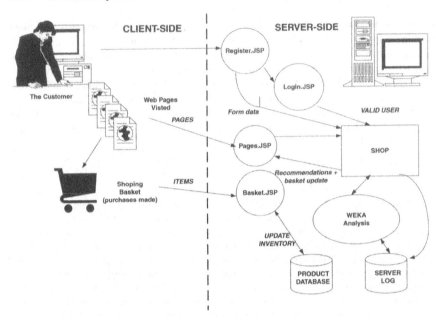

Fig. 12.4. System Overview of the Simulated Shop

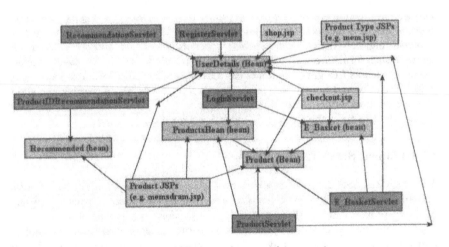

Fig. 12.5. JSP dependency and interaction

user's session, in the server's memory) whilst they decide what they are going to buy; this object is then later used to display what is in a user's basket and to buy or remove the contents of the basket.

Java Server Pages (JSP) are not client side (i.e. they are not executed in a web browser like an applet), JSPs are actually server side and provide a simpler coding interface than the servlet for output and formatting web interface code (e.g. HTML and javascript).

The shop consists of three levels of pages, starting with the main shop page which welcomes the user and offers them a choice of products. Then the subcategory pages which offer specific types of products e.g. bread.jsp and memory.jsp. Lastly, the products pages, such as memsdram.jsp and breadgranary.jsp, which display several items which may be purchased. These pages are linked into the database and update the stock count when purchases are made. The final page is the checkout.jsp page which allows the users to buy their selected products. The system gives each registered user some e-money or credit from which to make purchases. Checks are made to ensure the user is registered and logged in before a session is valid. As the users make purchases a shopping basket implemented by a Java servlet maintains track of all details e.g. cost, number of items. The servlet ensures that the user is still in credit and checks that items are in stock before purchases can occur. Valid transactions also update the users credit and the database. Any such conflicts are reported to advise the user.

Figure 12.6 shows the main page of the simulated shop.

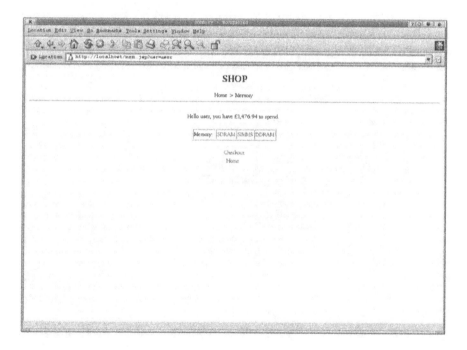

Fig. 12.6. Home Page

12.4.2 Register Servlet

The register servlet is coded to take given register details passed to it from register.jsp and perform some simple error checking (checks for blank data, illegal characters, etc.) and adds found problems in the data to a vector object. If errors are found the user is returned to register.jsp which uses the vector of errors found to tell the user what to check and re-enter correctly. If errors are not found the user's given details are inserted into the relevant user and user profile tables on the database; a hard coded amount of e-money is assigned to the user in the database. The servlet then redirects all data to the Login servlet to directly log the user in after correctly registering.

12.4.3 Register JSP

This simple JSP allows the user to enter their relevant requested details, the code has been specifically written so attributes such as age and occupation are hard coded into the system. This makes the code slightly less flexible from a maintainability point of view, but is done to allow the use of drop down boxes; this method of input avoids the necessity for extensive error checking to numbers and string lists (the list of occupations for example) as the user can only input what they can select.

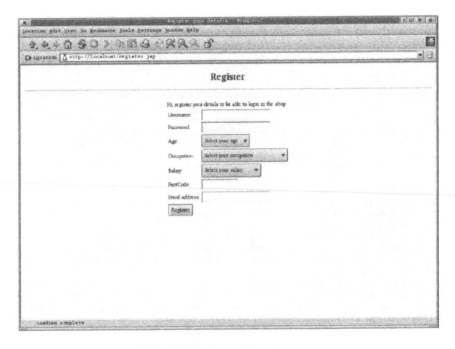

Fig. 12.7. User registration page

12.4.4 Login Servlet

To create login functionality to the shop, the best Java construct to use was the servlet as a database connection would be required to check submitted user and passwords. To create a reusable component and reduce code repetition the servlet was designed to take user and password data down from the HTTP session (or basically retrieve data sent in the standard way a HTML form would send data, for example). This gives the added advantage that the register servlet can log users in directly after registration by simply redirecting all data sent to the registration servlet to the login servlet. The process of sending this data on is called "Request dispatching", and the process of joining servlets in this way is called "Servlet chaining". The doPost() method in this servlet contains one main section of code that checks if the user exists via a database call; dependent on what the database returns the servlet will either log the user in (create an e-basket and user details object and redirect them to shop.jsp) or reject the login (send the user back to index.html).

12.4.5 The Product Servlet

This servlet is called upon every entry to a products JSP and performs the sole function of creating a Java Bean that encapsulates all product based information for that page. It obtains the page attribute from the user's session (passed by the accessing JSP) and uses this relative URL to query the database for all products that should appear on the given page. It is this information returned from the database that is placed into many Product objects, which in turn are placed into a ProductsBean, this Java Bean is placed onto the user's session for the accessing product JSP to use.

12.4.6 The Products JSP

Products JSPs are all of the JSPs which display products from the database, examples of which are memsdram.jsp, breadgranary.jsp and bookcomp.jsp. These JSPs firstly check to see if the user is logged in by performing a check on the user's session for a User Details object. If the user does not have this object then they have either clearly not logged in via the Login Servlet, or the user's session has timed out for whatever reason (i.e. closed the browser, a long pause between page accesses etc). If the user is deemed to not be logged in, all Product JSPs default to a small piece of HTML which instruct the user that they have been deemed not logged in and that they should login by browsing to the provided login link. If the user is logged in then the product JSP will get product information for that page down from the ProductsBean on the user's session, this bean should have been created prior to the page access via the Product Servlet.

All product information found within the ProductsBean is then formatted to the user's browser in HTML. During this process if products are found to be out of stock (quantity of stock equals zero) then no button is displayed for buying that product, this helps prevent any potential errors in trying to buy products that are not in stock. Products that are in stock are furnished with an add to e-basket button which allows the user to add that product to their basket, this functionality uses the E-Basket servlet to update the user's e-basket bean on their session. It is worth noting that when a user adds a product to their e-basket the quantity in stock figure on the product JSP does not immediately decrement as the product is deemed to be in the user's basket but still in stock. This means that a user can insert more products into their e-basket than are shown to be in stock, however the user can only check out as many items as they can afford, and as many items as are in stock. In addition to the products information the JSPs also display two types of recommendation information (page and product) taken from recommendation information found on the user's session.

The JSP also uses a switch on the user's session to find out whether the recommendations are based upon Weka data or are derived purely from SQL statements. A differing explanation output of the recommendation data and differing amount of recommendation appear dependant on the manner in which the data has been derived. All product JSPs have code at the top of the source that allows commonly customised aspects of the code to be changed (i.e. the URL of the page, the name of the page etc), this enables new product types to be added very quickly.

12.4.7 The Checkout JSP

The checkout.jsp in conjunction with the E-Basket Servlet, provides the user with checkout functionality for their chosen products. Using an E-Basket bean storing chosen products from product JSPs, the checkout JSP formats these details to the user's browser via HTML. As the products are formatted to the page a running total of the price is kept and dependent on whether the user has enough e-money to buy the contents of their e-basket a button is shown for buying all items, or alternatively a prompt is displayed informing the user that they do not have enough money. This acts as another barrier to the user buying items that they should not be able to. In addition to the buy items button, a remove item button is placed beside every respective item to allow a user to remove items from their e-basket.

12.4.8 The E-Basket Servlet

The E-Basket Servlet performs several main functions in relation to keeping up to date the user's e-basket Java Bean. Firstly, it allows buying of all products in the e-basket bean by iterating through the user's e-basket, checking stock quantities and user e-money and then "buying" the product by decrementing the user's e-money and product stock quantity accordingly and removing that item from the user's e-basket. Products found to be too expensive for the user's account or found not to be in stock are not "bought" and are not removed from the user's e-basket. Secondly, it allows the user to remove given items form their e-basket by iterating the e-basket for the given productID and removing a product with that ID. Thirdly, it allows the user to add products to their e-basket, this is done by querying the database for product details for the given productID and then adding a Product object to the e-basket based upon those returned details. No product is added if the product is not found in the database.

12.4.9 Web Server Platform

The Apache Tomcat server is a free program from the Apache organization and is used to serve static HTML, dynamic JSPs and Java Servlets, making it an important part of the system. Tomcat is written in Java and provides two important aspects to any Java web server. First, it can serve static pages making it highly useful and simpler for development purposes (avoiding the need to set up a second server like Apache to serve static HTML). Second, it can use Java Servlet technology to allow

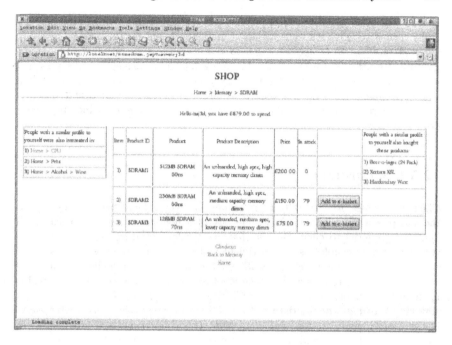

Fig. 12.8. Memory Chip page of the Simulated Shop

logic intensive code to be run native to the system as well as being able to serve JSP pages to allow dynamic content to be added to a site.

The Tomcat server was chosen to provide Servlet and JSP support; it is currently the best Java web server available and which made it the primary choice for serving the dynamic content within the shopping component of this system. The source code is available, bugs and bug fixes are well known and it is a very stable platform.

An example of the server log is shown in Fig. 12.9, which presents a snapshot of a test session where several users were logged in. The IP address of each machine is captured, along with the date and time stamp, as well as the web pages each user visited.

12.4.10 Test Results

The initial system was based on the data generated by 25 students. The site therefore had to cater for the particular preferences held by this group (mainly aged between

---

194.82.103.77 - - [19/Feb/2002:18:06:52 00] "GET /checkout.jsp HTTP/1.0" 200 793
194.82.103.78 - - [19/Feb/2002:18:07:06 00] "GET /shop.jsp HTTP/1.0" 200 1062
194.82.103.74 - - [19/Feb/2002:18:07:19 00] "GET /checkout.jsp HTTP/1.0" 200 793

Fig. 12.9. Web server extract

20–35). The rules generated after parsing the web server and items purchased generally tended to agree with the expected market segment for this particular group. The web site had to be large enough to give the necessary variability i.e. enough products to differentiate the users. Users logging in subsequently were presented with a personalized web page based on previous purchases and web pages visited. The data mining software was then used off-line to create a user profile based on the form, purchase and server data. The generated rules were later incorporated to provide a more robust user model.

12.5 Conclusions

Client companies are generally more inclined to use a technology if they can see the end product actually working. The use of a demonstration system can highlight the potential to clients better than mere sales literature or site visits alone can. The system enables us to experiment and show clients the possibilities of web server based data mining. Future work will see the addition of a client email list to contact specific customers with details of special offers, etc. Also, better use will be made of the support provided by "cookies" to track individual users more closely. The ability to perform the data mining activity on-line without user intervention is also a good candidate for implementation. This chapter has also provided an introduction to objective and subjective interestingness measures that can be used to rank the patterns produced by the data mining process.

References

1. R. Agrawal, T. Imielinski, and A. Swami. Mining association rules between sets of items in large databases. In *SIGMOD-93*, pp. 207–216, 1993.
2. C. Fabris and A. Freitas. Discovering suprising patterns by detecting occurences of simpson's paradox. In *Development in Intelligent Systems XVI (Proc. Expert Systems 99, The 19th SGES International Conference on Knowledge-Based Systems and Applied Artificial Intelligence)*, pp. 148–160, Cambridge, England, 1999.
3. U. Fayyad, G. Piatetsky-Shapiro, and P. Smyth. From data mining to knowledge discovery: an overview. In U. Fayyad, G. Piatetsky-Shapiro, P. Smyth, and R. Uthursamy, editors, *Advances in Knowledge Discovery and Data Mining*, pp. 1–34. AAAI-Press, 1996.
4. U. Fayyad, G. Piatetsky-Shapiro, P. Smyth, and R. Uthursamy, editors. *Knowledge Discovery in Databases*. AAAI/MIT Press, 1996.
5. D. Fisher. Knowledge acquistion via incremental concept clustering. *Machine Learning*, 2:139–172, 1987.
6. A. Freitas. On objective measures of rule suprisingness. In *Principles of Data Mining and Knowledge Discovery: Proceedings of the 2nd European Symposium,Lecture Notes in Artificial Intelligence*, volume 1510, pp. 1–9, Nantes, France, 1998.
7. A. Freitas. On rule interestingness measures. *Knowledge Based Systems*, 12(5-6):309–315, 1999.

8. R. Hilderman and H. Hamilton. Knowledge discovery and interestingness measures: a survey. Technical Report CS-99-04, Department of Computer Science, University of Regina, Regina, Saskatchewan, Canada S4S 0A2, 1999.
9. R. Hilderman and H. Hamilton. *Knowledge Discovery and Measures of Interest.* Kluwer Academic Publishers, 2001.
10. B. Liu, W. Hsu, and S. Chen. Using general impressions to analyze discovered classification rules. In *Proceedings of the 3rd International Conference on Knowledge Discovery and Data Mining*, pp. 31–36, 1997.
11. B. Liu, W. Hsu, L. Fun, and H. Y. Lee. Finding interesting patterns using user expectations. Technical Report TRS7/96, Department of Information Systems and Computer Science, National University of Singapore, Lower Kent Road, Singapore 119260, Republic of Singapore, 1996.
12. B. Liu, W. Hsu, L. Mun, and H. Y. Lee. Finding interesting patterns using user expectations. *IEEE Transactions on Knowledge and Data Engineering*, 11(6):817–832, 1999.
13. K. McGarry and J. MacIntyre. Data mining in a vibration analysis domain by extracting symbolic rules from RBF neural networks. In *Proceedings of 14th International Congress on Condition Monitoring and Engineering Management*, pp. 553–560, Manchester, UK, 4th-6th September 2001.
14. K. McGarry, S. Wermter, and J. MacIntyre. The extraction and comparison of knowledge from local function networks. *International Journal of Computational Intelligence and Applications*, 1(4):369–382, 2001.
15. K. McGarry, S. Wermter, and J. MacIntyre. Knowledge extraction from local function networks. In *Seventeenth International Joint Conference on Artificial Intelligence*, volume 2, pp. 765–770, Seattle, USA, August 4th-10th 2001.
16. J. Mena. *Data Mining Your Website.* Digital Press, 1999.
17. J. Mena. *Web Mining For Profit: E-Business Optimization.* Digital Press, 2001.
18. B. Padmanabhan and A. Tuzhilin. Unexpectedness as a measure of interestingness in knowledge discovery. *Decision Support Systems*, 27:303–318, 1999.
19. B. Padmanabhan and A. Tuzhilin. Knowledge refinement based on the discovery of unexpected patterns in data mining. *Decision Support Systems*, 33:309–321, 2002.
20. G. Piatetsky-Shapiro and W. J. Frawley, editors. *Knowledge Discovery in Databases.* AAAI, MIT Press, 1991.
21. J. R. Quinlan. Induction of decision trees. *Machine Learning*, 1:81–106, 1986.
22. A. Silberschatz and A. Tuzhilin. On subjective measures of interestingness in knowledge discovery. In *Proceedings of the 1st International Conference on Knowledge Discovery and Data Mining*, pp. 275–281, 1995.
23. A. Silberschatz and A. Tuzhilin. What makes patterns interesting in knowledge discovery systems. *IEEE Transactions on Knowledge and Data Engineering*, 8(6):970–974, 1996.
24. F. Witten and E. Frank. *Data Mining.* Morgan Kaufmann Publishers, 2000.

13

Soft Computing Models
for Network Intrusion Detection Systems

Ajith Abraham[1] and Ravi Jain[2]

[1]Department of Computer Science, Oklahoma State University, USA
ajith.abraham@ieee.org
[2]School of Information Science, University of South Australia, Australia
ravi.jain@unisa.edu.au

Abstract. Security of computers and the networks that connect them is increasingly becoming of great significance. Computer security is defined as the protection of computing systems against threats to confidentiality, integrity, and availability. There are two types of intruders: external intruders, who are unauthorized users of the machines they attack, and internal intruders, who have permission to access the system with some restrictions. This chapter presents a soft computing approach to detect intrusions in a network. Among the several soft computing paradigms, we investigated fuzzy rule-based classifiers, decision trees, support vector machines, linear genetic programming and an ensemble method to model fast and efficient intrusion detection systems. Empirical results clearly show that soft computing approach could play a major role for intrusion detection.

Key words: intrusion detection information security, soft computing

13.1 Introduction

The traditional prevention techniques such as user authentication, data encryption, the avoidance of programming errors and firewalls are used as the first line of defense for computer security. If a password is weak and is compromised, user authentication cannot prevent unauthorized use. Firewalls are vulnerable to errors in configuration and ambiguous or undefined security policies. They are generally unable to protect against malicious mobile code, insider attacks and unsecured modems. Programming errors cannot be avoided as the complexity of the system and application software is changing rapidly, leaving behind some exploitable weaknesses. Intrusion detection is therefore required as an additional wall for protecting systems [5, 9]. Intrusion detection is useful not only in detecting successful intrusions, but also provides important information for timely countermeasures [11, 13]. An intrusion is defined as any set of actions that attempt to compromise the integrity, confidentiality or availability of a resource. An attacker can gain access because of an error in the configuration of a system. In some cases it is possible to fool a system into giving access by misrepresenting oneself. An example is sending a TCP packet that has a

Ajith Abraham and Ravi Jain: *Soft Computing Models for Network Intrusion Detection Systems*, Studies in Computational Intelligence (SCI) **4**, 191–207 (2005)
www.springerlink.com

forged source address that makes the packet appear to come from a trusted host. Intrusions may be classified into several types [12].

- Attempted break-ins, which are detected by typical behavior profiles or violations of security constraints.
- Masquerade attacks, which are detected by atypical behavior profiles or violations of security constraints.
- Penetration of the security control system, which are detected by monitoring for specific patterns of activity.
- Leakage, which is detected by atypical use of system resources.
- Denial of service, which is detected by atypical use of system resources.
- Malicious use, which is detected by atypical behavior profiles, violations of security constraints, or use of special privileges.

The process of monitoring the events occurring in a computer system or network and analyzing them for sign of intrusions is known as intrusion detection. Intrusion detection is classified into two types: misuse intrusion detection and anomaly intrusion detection.

Misuse intrusion detection uses well-defined patterns of the attack that exploit weaknesses in system and application software to identify the intrusions. These patterns are encoded in advance and used to match against the user behavior to detect intrusion.

Anomaly intrusion detection uses the normal usage behavior patterns to identify the intrusion. The normal usage patterns are constructed from the statistical measures of the system features, for example, the CPU and I/O activities by a particular user or program. The behavior of the user is observed and any deviation from the constructed normal behavior is detected as intrusion.

We have two options to secure the system completely, either prevent the threats and vulnerabilities which come from flaws in the operating system, as well as in the application programs, or detect them and take some action to prevent them in future and also repair the damage. It is impossible in practice, and even if possible, extremely difficult and expensive, to write a completely secure system. Transition to such a system for use in the entire world would be an equally difficult task. Cryptographic methods can be compromised if passwords and keys are stolen. No matter how secure a system is, it is vulnerable to insiders who abuse their privileges. There is an inverse relationship between the level of access control and efficiency. More access controls make a system less user-friendly and more likely to not be used.

An Intrusion Detection System (IDS) is a program that analyzes what happens or has happened during an execution and tries to find indications that the computer has been misused. An intrusion detection system does not eliminate the use of preventive mechanism but it works as the last defensive mechanism in securing the system. Data mining approaches are a relatively new technique for intrusion detection.

13.2 Intrusion Detection – A Data Mining Approach

Data mining is a relatively new approach for intrusion detection. Data mining approaches for intrusion detection were first implemented in mining audit data for automated models for intrusion detection [2, 8]. The raw data is first converted into ASCII network packet information, which in turn is converted into connection level information. These connection level records contain connection features like service, duration, etc. Data mining algorithms are applied to this data to create models to detect intrusions. Data mining algorithms used in this approach are RIPPER (rule based classification algorithm), meta-classifier, frequent episode algorithm and association rules. These algorithms are applied to audit data to compute models that accurately capture the actual behavior of intrusions as well as normal activities.

The RIPPER algorithm was used to learn the classification model in order to identify normal and abnormal behavior [4]. Frequent episode algorithm and association rules together are used to construct frequent patterns from audit data records. These frequent patterns represent the statistical summaries of network and system activity by measuring the correlations among system features and the sequential co-occurrence of events. From the constructed frequent patterns the consistent patterns of normal activities and the unique intrusion patterns are identified and analyzed, and then used to construct additional features. These additional features are useful in learning the detection model more efficiently in order to detect intrusions. The RIPPER classification algorithm is then used to learn the detection model. A Meta classifier is used to learn the correlation of intrusion evidence from multiple detection models and to produce a combined detection model. The main advantage of this system is the automation of data analysis through data mining, which enables it to learn rules inductively, replacing manual encoding of intrusion patterns. However, some novel attacks may not be detected.

Audit data analysis and mining combine's association rules and classification algorithm to discover attacks in audit data [1]. Association rules are used to gather necessary knowledge about the nature of the audit data as the information about patterns within individual records can improve the classification efficiency. This system has two phases: training and detection. In the training phase a database of frequent item sets is created for the attack-free items by using only the attack-free data set. This serves as a profile against which frequent item sets found later will be compared. Next a sliding-window, on-line algorithm is used to find frequent item sets in the last D connections and compares them with those stored in the attack-free database, discarding those that are deemed normal. In this phase a classifier is also trained to learn the model to detect the attack. In the detection phase a dynamic algorithm is used to produce item sets that are considered as suspicious and used by the classification algorithm already learned to classify the item set as attack, false alarm (normal event) or as unknown. Unknown attacks are the ones which are not able to detect either as false alarms or as known attacks. This method attempts to detect only anomaly attacks.

13.3 Soft Computing Models

Soft Computing (SC) is an innovative approach to construct computationally intelligent systems consisting of artificial neural networks, fuzzy inference systems, approximate reasoning and derivative free optimization methods such as evolutionary computation etc. In contrast with conventional artificial intelligence techniques which only deal with precision, certainty and rigor the guiding principle of soft computing is to exploit the tolerance for imprecision, uncertainty, low solution cost, robustness, partial truth to achieve tractability and better rapport with reality [15].

13.3.1 Fuzzy Rule Based Systems

Fuzzy logic has proved to be a powerful tool for decision making to handle and manipulate imprecise and noisy data. The notion central to fuzzy systems is that truth values (in fuzzy logic) or membership values (in fuzzy sets) are indicated by a value on the range [0.0, 1.0], with 0.0 representing absolute falseness and 1.0 representing absolute truth. A fuzzy system is characterized by a set of linguistic statements based on expert knowledge. The expert knowledge is usually in the form of *if-then* rules.

Definition 1. *Let X be some set of objects, with elements noted as x. Thus, $X = \{x\}$.*

Definition 2. *A fuzzy set A in X is characterized by a membership function which are easily implemented by fuzzy conditional statements. In the case of fuzzy statement if the antecedent is true to some degree of membership then the consequent is also true to that same degree.*

A simple rule structure: **If** antecedent **then** consequent

A simple rule: **If** variable$_1$ is low and variable$_2$ is high **then** output is benign **else** output is malignant

In a fuzzy classification system, a case or an object can be classified by applying a set of fuzzy rules based on the linguistic values of its attributes. Every rule has a weight, which is a number between 0 and 1 and this is applied to the number given by the antecedent. It involves 2 distinct parts. First the antecedent is evaluated, which in turn involves fuzzifying the input and applying any necessary fuzzy operators and second applying that result to the consequent known as inference. To build a fuzzy classification system, the most difficult task is to find a set of fuzzy rules pertaining to the specific classification problem.

We explored three fuzzy rule generation methods for intrusion detection systems. Let us assume that we have a n-dimensional c-class pattern classification problem whose pattern space is an n-dimensional unit cube $[0, 1]^n$. We also assume that m patterns $x_p = (x_{p1}, \ldots, x_{pn}), p = 1, 2, \ldots, m$, are given for generating fuzzy *if-then* rules where $x_p \in [0, 1]$ for $p = 1, 2, \ldots, m, i = 1, 2, \ldots, n$ where $x_p \in [0, 1]$ for $p = 1, 2, \ldots, m, i = 1, 2, \ldots, n$.

Rule Generation Based on the Histogram
of Attribute Values (FR1)

In this method, use of histogram itself is an antecedent membership function. Each attribute is partitioned into 20 membership functions $f_h(.), h = 1, 2, \ldots, 20$. The smoothed histogram $m_i^k(x_i)$ of class k patterns for the ith attribute is calculated using the 20 membership functions $f_h(.)$ as follows:

$$m_i^k(x_i) = \frac{1}{m^k} \sum_{x_p \in Class\, k} f_h(x_{pi})$$

$$\text{for} \quad \beta_{h-1} \leq x_i \leq \beta_h, \; h = 1, 2, \ldots, 20$$

(13.1)

where m_k is the number of Class k patterns, $[\beta_{h-1}, \beta_h]$ is the hth crisp interval corresponding to the 0.5-level set of the membership function $f_h(.)$

$$\beta_1 = 0, \quad \beta_{20} = 1 , \tag{13.2}$$

$$\beta_h = \frac{1}{20-1}\left(h - \frac{1}{2}\right) \quad \text{for } h = 1, 2, \ldots, 19 \tag{13.3}$$

The smoothed histogram in (22.1) is normalized so that its maximum value is 1. A single fuzzy *if-then* rule is generated for each class. The fuzzy if-then rule for the kth class can be written as

$$\text{If } x_1 \text{ is } A_1^k \text{ and } \ldots \text{ and } x_n \text{ is } A_1^k \text{ then class } k , \tag{13.4}$$

where A_i^k is an antecedent fuzzy set for the ith attribute. The membership function of A_i^k is specified as

$$A_i^k(x_i) = \exp\left(-\frac{\left(x_i - \mu_i^k\right)^2}{2\left(\sigma_i^k\right)^2}\right) \tag{13.5}$$

where μ_i^k is the mean of the ith attribute values x_{pi} of class k patterns, and σ_i^k is the standard deviation. Fuzzy *if-then* rules for the two-dimensional two class pattern classification problem are written as follows:

$$\text{If } x_3 \text{ is } A_3^1 \text{ and } x_4 \text{ is } A_4^1 \text{ then class 2} \tag{13.6}$$

$$\text{If } x_3 \text{ is } A_3^2 \text{ and } x_4 \text{ is } \sqrt{a^2 + b^2} \text{ then class 3} \tag{13.7}$$

membership function of each antecedent fuzzy set is specified by the mean and the standard deviation of attribute values. For a new pattern $x_p = (x_{p3}, x_{p4})$, the winner rule is determined as follows:

$$A_3^*(x_{p3}) \cdot A_2^*(x_{p4}) = \max\left\{A_1^k(x_{p3}) \cdot A_2^k(x_{p4}) \,|k = 1, 2\right\} \tag{13.8}$$

Rule Generation Based on Partition of Overlapping Areas (FR$_2$)

Figure 13.1 demonstrates a simple fuzzy partition, where the two- dimensional pattern space is partitioned into 25 fuzzy subspaces by five fuzzy sets for each attribute (S: small, MS: medium small, M: medium, ML: medium large, L: large). A single fuzzy *if-then* rule is generated for each fuzzy subspace. Thus the number of possible fuzzy *if-then* rules in Fig. 13.1 is 25.

One disadvantage of this approach is that the number of possible fuzzy *if-then* rules exponentially increases with the dimensionality of the pattern space. Because the specification of each membership function does not depend on any information about training patterns, this approach uses fuzzy if-then rules with certainty grades. The local information about training patterns in the corresponding fuzzy subspace is used for determining the consequent class and the grade of certainty. In this approach, fuzzy if-then rules of the following type are used:

$$\text{If } x_1 \text{ is } A_{j1} \text{ and } \ldots \text{ and } x_n \text{ is } A_{jn} \text{ then class } C_j ,$$
$$\text{with } CF = CF_{j'} \; j = 1, 2, \ldots, N \tag{13.9}$$

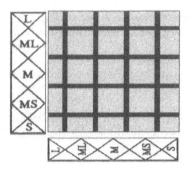

Fig. 13.1. An example of fuzzy partition

where j indexes the number of rules, N is the total number of rules, A_{ji} is the antecedent fuzzy set of the ith rule for the ith attribute, C_j; is the consequent class, and CFj is the grade of certainty. The consequent class and the grade of certainty of each rule are determined by the following simple heuristic procedure:

Step 1: Calculate the compatibility of each training pattern $x_p = (x_{p1}, x_{p2}, \ldots, x_{pn})$ with the jth fuzzy *if-then* rule by the following product operation:

$$\pi_j(x_p) = A_{j1}(x_{p1}) \times \cdots \times A_{jn}(x_{pn}), p = 1, 2, \ldots, m . \qquad (13.10)$$

Step 2: For each class, calculate the sum of the compatibility grades of the training patterns with the jth fuzzy *if-then* rule R_j:

$$\beta_{\text{class } k}(R_j) = \sum_{x_p \in class\, k}^{n} \pi(x_p), k = 1, 2, \ldots, c \qquad (13.11)$$

where $\beta_{\text{class } k}(R_j)$ the sum of the compatibility grades of the training patterns in class k with the jth fuzzy if-then rule R_j.

Step 3: Find Class A_j^* that has the maximum value $\beta_{\text{class } k}(R_j)$;

$$\beta_{\text{class} k_j^*} = \text{Max}\{\beta_{\text{class}1}(R_j), \ldots, \beta_{\text{class } c}(R_j)\} \qquad (13.12)$$

If two or more classes take the maximum value or no training pattern compatible with the jth fuzzy *if-then* rule (*i.e.,if* $\beta_{\text{Class}}k(R_j) = 0$ for $k = 1$, $2, \ldots, c$), the consequent class C_i can not be determined uniquely. In this case, let C_i be ϕ.

Step 4: If the consequent class C_i is 0, let the grade of certainty CF_j be $CF_j = 0$. Otherwise the grade of certainty CF_j *is* determined as follows:

$$CF_j = \frac{(\beta_{class\, k_j^*}(R_j) - \bar{\beta})}{\sum_{k=1}^{c} \beta_{class\, k}(R_j)} \qquad (13.13)$$

$$\text{where } \bar{\beta} = \sum_{\substack{k=1 \\ k \neq k_j^*}} \frac{\beta_{Class\, k}(R_j)}{(c-1)}$$

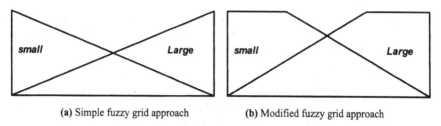

(a) Simple fuzzy grid approach   (b) Modified fuzzy grid approach

Fig. 13.2. Fuzzy partition of each attributes

The above approach could be modified by partitioning only the overlapping areas as illustrated in Fig. 13.2.

This approach generates fuzzy *if-then* rules in the same manner as the simple fuzzy grid approach except for the specification of each membership function. Because this approach utilizes the information about training patterns for specifying each membership function as mentioned in Sect. 13.2.1.1, the performance of generated fuzzy *if-then* rules is good even when we do not use the certainty grade of each rule in the classification phase. In this approach, the effect of introducing the certainty grade to each rule is not so important when compared to conventional grid partitioning.

Neural Learning of Fuzzy Rules (FR_3)

The derivation of *if-then* rules and corresponding membership functions depends heavily on the a priori knowledge about the system under consideration. However there is no systematic way to transform experiences of knowledge of human experts to the knowledge base of a Fuzzy Inference System (FIS). In a fused neuro-fuzzy architecture, neural network learning algorithms are used to determine the parameters of fuzzy inference system (membership functions and number of rules). Fused neuro-fuzzy systems share data structures and knowledge representations. A common way to apply a learning algorithm to a fuzzy system is to represent it in a special neural network-like architecture. An Evolving Fuzzy Neural Network (EFuNN) implements a Mamdani type FIS and all nodes are created during learning. The nodes representing membership functions (MF) can be modified during learning. Each input variable is represented here by a group of spatially arranged neurons to represent a fuzzy quantization of this variable. New neurons can evolve in this layer if, for a given input vector, the corresponding variable value does not belong to any of the existing MF to a degree greater than a membership threshold. Technical details of the learning algorithm are given in [16].

13.3.2 Linear Genetic Programming (LGP)

Linear genetic programming is a variant of the GP technique that acts on linear genomes [3]. Its main characteristics in comparison to tree-based GP are that the evolvable units are not expressions of a functional programming language (like LISP), but the programs of an imperative language (like c/c ++). An alternate approach is to evolve a computer program at the machine code level, using lower level representations for the individuals. This can tremendously hasten the evolution process as, no matter how an individual is initially represented, finally it always has to be represented as a piece of machine code, as fitness evaluation requires physical execution of the individuals.

The basic unit of evolution here is a native machine code instruction that runs on the floating-point processor unit (FPU). Since different instructions may have different sizes, here instructions are clubbed up together to form instruction blocks of 32 bits each. The instruction blocks hold one or more native machine code instructions, depending on the sizes of the instructions. A crossover point can occur only between instructions and is prohibited from occurring within an instruction. However the mutation operation does not have any such restriction. In this research a steady state genetic programming approach was used to manage the memory more effectively [1].

13.3.3 Decision Trees (DT)

Intrusion detection can be considered as classification problem where each connection or user is identified either as one of the attack types or normal based on some existing data. Decision trees work well with large data sets. This is important as large amounts of data flow across computer networks. The high performance of decision trees makes them useful in real-time intrusion detection. Decision trees construct easily interpretable models, which is useful for a security officer to inspect and edit. These models can also be used in the rule-based models with minimum processing [7]. Generalization accuracy of decision trees is another useful property for intrusion detection model. There will always be new attacks on the system, which are small variations of known attacks after the intrusion detection models are built. The ability to detect these new intrusions is possible due to the generalization accuracy of decision trees.

13.3.4 Support Vector Machines (SVM)

Support Vector Machines have been proposed as a novel technique for intrusion detection. SVM maps input (real-valued) feature vectors into a higher dimensional feature space through some nonlinear mapping. SVMs are powerful tools for providing solutions to classification, regression and density estimation problems. These are developed on the principle of structural risk minimization. Structural risk minimization seeks to find a hypothesis for which one can find the lowest probability of error. The structural risk minimization can be achieved by finding the hyper plane with maximum separable margin for the data [14]. Computing the hyper plane to separate the data points, i.e. training a SVM, leads to a quadratic optimization problem. SVM uses a feature called a kernel to solve this problem. A kernel transforms linear algorithms into nonlinear ones via a map into feature spaces. SVMs classify data by using these support vectors, which are members of the set of training inputs that outline a hyper plane in feature space.

13.4 Attribute Deduction in Intrusion Detection Systems

Since the amount of audit data that an IDS needs to examine is very large even for a small network, analysis is difficult even with computer assistance because extraneous features can make it harder to detect suspicious behavior patterns. Complex relationships exist between features, which are difficult for humans to discover. IDS must therefore reduce the amount of data to be processed. This is very important if real-time detection is desired. The easiest way to do this is by doing an intelligent input feature selection. Certain features may contain false correlations, which hinder the process of detecting intrusions. Further, some features may be redundant since the information they add is contained in other features. Extra features can increase computation time, and can impact the accuracy of IDS. Feature selection improves classification by searching for the subset of features, which best classifies the training data.

Feature selection is done based on the contribution the input variables made to the construction of the decision tree. Feature importance is determined by the role of each input variable either as a main splitter or as a surrogate. Surrogate splitters are defined as back-up rules that closely mimic the action of primary splitting rules. Suppose that, in a given model, the algorithm splits data according to variable "protocol_type" and if a value for "protocol_type" is not available, the algorithm might substitute "flag" as a good surrogate. Variable importance, for a particular variable is the sum across all nodes in the tree of the improvement scores that the predictor has when it acts as a primary or surrogate (but not competitor) splitter. Example, for node i, if the predictor appears as the primary splitter then its contribution towards importance could be given as $i_{importance}$. But if the variable appears as the nth surrogate instead of the primary variable, then the importance becomes $i_{importance} = (p^n)^* i_{improvement}$ in which p is the "surrogate improvement weight" which is a user controlled parameter set between (0–1) [17].

13.5 Intrusion Detection Data

In 1998, DARPA intrusion detection evaluation program created an environment to acquire raw TCP/IP dump data for a network by simulating a typical U.S. Air Force LAN [10]. The LAN was operated like a real environment, but was blasted with multiple attacks. For each TCP/IP connection, 41 various quantitative and qualitative features were extracted. Of these a subset of 494,021 data were used for our studies, of which 20% represent normal patterns [6]. Different categories of attacks are summarized in Fig. 13.4. Attack types fall into four main categories:

DoS: Denial of Service

Denial of Service (DoS) is a class of attack where an attacker makes a computing or memory resource too busy or too full to handle legitimate requests, thus denying legitimate users access to a machine. There are different ways to launch DoS attacks: by abusing a computer's legitimate features; by targeting the implementation bugs; or by exploiting a system's miss configurations. DoS attacks are classified based on the services that an attacker renders unavailable to legitimate users.

R2L: Unauthorized Access from a Remote Machine

A remote to user (R2L) attack is a class of attack where an attacker sends packets to a machine over a network, then exploits the machine's vulnerability to illegally gain local access as a user. There are different types of R2U attacks; the most common attack in this class is done using social engineering.

U2Su: Unauthorized Access to Local Super User (Root)

User to root (U2Su) exploits are a class of attacks where an attacker starts out with access to a normal user account on the system and is able to exploit vulnerability to gain root access to the system. Most common exploits in this class of attacks are regular buffer overflows, which are caused by regular programming mistakes and environment assumptions.

Probing: Surveillance and Other Probing

Probing is a class of attack where an attacker scans a network to gather information or find known vulnerabilities. An attacker with a map of machines and services that are available on a network can use the information to look for exploits. There are different types of probes: some of them abuse the computer's legitimate features; some of them use social engineering techniques. This class of attack is the most common and requires very little technical expertise.

13.6 Experiment Setup and Results

The data for our experiments was prepared by the 1998 DARPA intrusion detection evaluation program by MIT Lincoln Labs [10]. The data set contains 24 attack types that could be classified into four main categories namely *Denial of Service (DoS), Remote to User (R2L), User to Root (U2R)* and *Probing*. The original data contains 744 MB data with 4,940,000 records. The data set has 41 attributes for each connection record plus one class label. Some features are derived features, which are useful in distinguishing normal connection from attacks. These features are either continuous or discrete. Some features examine only the connections in the past two seconds that have the same destination host as the current connection, and calculate statistics related to protocol behavior, service, etc. These are called same host features. Some features examine only the connections in the past two seconds that have the same service as the current connection and are called same service features. Some other connection records were also sorted by destination host, and features were constructed using a window of 100 connections to the same host instead of a time window. These are called host-based traffic features. R2L and U2R attacks don't have any sequential patterns like DoS and Probe because the former attacks have the attacks embedded in the data packets whereas the later attacks have many connections in a short amount of time. So some features that look for suspicious behavior in the data packets like number of failed logins are constructed and these are called content features.

Table 13.1. Variables for intrusion detection data set

Variable No.	Variable Name	Variable Type	Variable Label
1	duration	continuous	A
2	protocol_type	discrete	B
3	service	discrete	C
4	flag	discrete	D
5	src_bytes	continuous	E
6	dst_bytes	continuous	F
7	land	discrete	G
8	wrong_fragment	continuous	H
9	urgent	continuous	I
10	hot	continuous	J
11	num_failed_logins	continuous	K
12	logged_in	discrete	L
13	num_compromised	continuous	M
14	root_shell	continuous	N
15	su_attempted	continuous	O
16	num_root	continuous	P
17	num_file_creations	continuous	Q
18	num_shells	continuous	R
19	num_access_files	continuous	S
20	num_outbound_cmds	continuous	T
21	is_host_login	discrete	U
22	is_guest_login	discrete	V
23	count	continuous	W
24	srv_count	continuous	X
25	serror_rate	continuous	Y
26	srv_serror_rate	continuous	X
27	rerror_rate	continuous	AA
28	srv_rerror_rate	continuous	AB
29	same_srv_rate	continuous	AC
30	diff_srv_rate	continuous	AD
31	srv_diff_host_rate	continuous	AE
32	dst_host_count	continuous	AF
33	dst_host_srv_count	continuous	AG
34	dst_host_same_srv_rate	continuous	AH
35	dst_host_diff_srv_rate	continuous	AI
36	dst_host_same_src_port_rate	continuous	AJ
37	dst_host_srv_diff_host_rate	continuous	AK
38	dst_host_serror_rate	continuous	AL
39	dst_host_srv_serror_rate	continuous	AM
40	dst_host_rerror_rate	continuous	AN
41	dst_host_srv_rerror_rate	continuous	AO

Table 13.2. Reduced variable set

C, E, F, L, W, X, Y, AB, AE, AF, AG, AI

Our experiments have three phases namely data reduction, training phase and testing phase. In the data reduction phase, important variables for real-time intrusion detection are selected by feature selection. In the training phase, the different soft computing models were constructed using the training data to give maximum generalization accuracy on the unseen data. The test data is then passed through the saved trained model to detect intrusions in the testing phase. The 41 features are labeled as shown in Table 13.1 and the class label is named as AP. This data set has five different classes namely *Normal, DoS, R2L, U2R* and *Probes*. The training and test comprises of 5,092 and 6,890 records respectively [6].

Our initial research was to reduce the number of variables. Using all 41 variables could result in a big IDS model, which could be an overhead for online detection. The experiment system consists of two stages: Network training and performance evaluation. All the training data were scaled to (0–1). The decision tree approach described in Sect. 13.4 helped us to reduce the number of variables to 12 variables. The list of reduced variables is illustrated in Table 13.2.

Using the original and reduced data sets, we performed a 5-class classification. The (training and testing) data set contains 11,982 randomly generated points from the data set representing the five classes, with the number of data from each class proportional to its size, except that the smallest class is completely included. The set of 5,092 training data and 6,890 testing data are divided in to five classes: normal, probe, denial of service attacks, user to super user and remote to local attacks. The datasets contain a total of 24 training attack types, with an additional 14 types in the test data only. Where the attack is a collection of different types of instances that belong to the four classes described earlier and the other is the normal data. The normal data belongs to class 1, probe belongs to class 2, denial of service belongs to class 3, user to super user belongs to class 4, remote to local belongs to class 5. All the IDS models are trained and tested with the same set of data.

We examined the performance of all three fuzzy rule based approaches (FR_1, FR_2 and FR_3) mentioned in Sect. 13.3.1. When an attack is correctly classified the grade of certainty is increased and when an attack is misclassified the grade of certainty is decreased. A learning procedure is used to determine the grade of certainty. Triangular membership functions were used for all the fuzzy rule based classifiers. We used 4 triangular membership functions for each input variable for

Table 13.3. Parameter settings for linear genetic programming

Parameter	Normal	Probe	DoS	U2Su	R2L
Population size	2048	2048	2048	2048	2048
Tournament size	8	8	8	8	8
Mutation frequency (%)	85	82	75	86	85
Crossover frequency (%)	75	70	65	75	70
Number of demes	10	10	10	10	10
Maximum program size	256	256	256	256	256

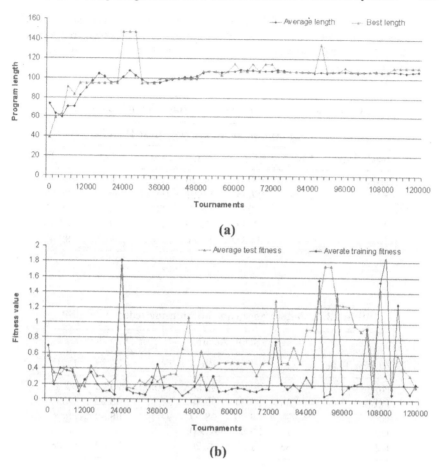

Fig. 13.3. LGP performance for the detection of normal patterns (**a**) growth in program length (**b**) average fitness

the EFuNN training (FR$_3$). A sensitivity threshold Sthr = 0.95 and error threshold Errthr = 0.05 was used for all the classes. 89 rule nodes were developed during the one pass learning [17].

The settings of various linear genetic programming system parameters are of utmost importance for successful performance of the system. The population space has been subdivided into multiple subpopulation or demes. Migration of individuals among the subpopulations causes evolution of the entire population. It helps to maintain diversity in the population, as migration is restricted among the demes. Table 13.3 depicts the parameter settings used for LGP experiments. The tournament size was set at 120,000 for all the 5 classes. Figure 13.3 demonstrates the growth in program length during 120,000 tournaments and the average fitness values for detecting normal patterns (class 1). More illustrations are available in [1].

Our trial experiments with SVM revealed that the polynomial kernel option often performs well on most of the datasets. We also constructed decision trees using the training data and then testing data was passed through the constructed classifier to classify the attacks [12].

Table 13.4. Performance comparison using full data set

Attack Type	Classification Accuracy on Test Data Set (%)					
	FR_1	FR_2	FR_3	DT	SVM	LGP
Normal	40.44	100.00	98.26	99.64	99.64	99.73
Probe	53.06	100.00	99.21	99.86	98.57	99.89
DOS	60.99	100.00	98.18	96.83	99.92	99.95
U2R	66.75	100.00	61.58	68.00	40.00	64.00
R2L	61.10	100.00	95.46	84.19	33.92	99.47

Table 13.5. Performance comparison using reduced data set

Attack Type	Classification Accuracy on Test Data Set (%)					
	FR_1	FR_2	FR_3	DT	SVM	LGP
Normal	74.82	79.68	99.56	100.00	99.75	99.97
Probe	45.36	89.84	99.88	97.71	98.20	99.93
DOS	60.99	60.99	98.99	85.34	98.89	99.96
U2R	94.11	99.64	65.00	64.00	59.00	68.26
R2L	91.83	91.83	97.26	95.56	56.00	99.98

A number of observations and conclusions are drawn from the results illustrated in Tables 13.4 and 13.5. Using 41 attributes, the FR2 method gave 100% accuracy for all the 5 classes, showing the importance of fuzzy inference systems. For the full data set, LGP outperformed decision trees and support vector machines in terms of detection accuracies (except for one class).

The reduced dataset seems to work very well for most of the classifiers except the fuzzy classifier (FR2). For detecting U2R attacks FR2 gave the best accuracy. Due to the tremendous reduction in the number of attributes (about 70% less), we are able to design a computational efficient intrusion detection system. Since a particular classifier could not provide accurate results for all the classes, we propose to use an ensemble approach as demonstrated in Fig. 13.4. The proposed ensemble model could detect all the attacks with high accuracy (lowest accuracy being 99.64%) with only 12 input variables. Ensemble performance is summarized in Table 13.6.

In some classes the accuracy figures tend to be very small and may not be statistically significant, especially in view of the fact that the 5 classes of patterns differ in their sizes tremendously. For example only 27 data sets were available for

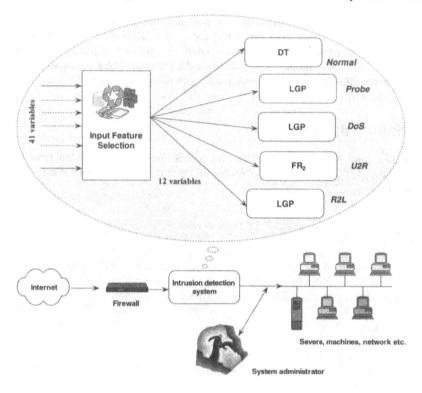

Fig. 13.4. IDS architecture using an ensemble of intelligent paradigms

training the U2R class. More definitive conclusions can only be made after analyzing more comprehensive sets of network traffic.

13.7 Conclusions

In this chapter, we have illustrated the importance of soft computing paradigms for modeling intrusion detection systems. For real time intrusion detection systems,

Table 13.6. Performance of the ensemble method

Attack Type	Ensemble Classification Accuracy on Test Data (%)
Normal	100.00
Probe	99.93
DOS	99.96
U2R	99.64
R2L	99.98

LGP would be the ideal candidate as it can be manipulated at the machine code level. Overall, the fuzzy classifier (FR2) gave 100% accuracy for all attack types using all the 41 attributes. The proposed ensemble approach requires only 12 input variables. More data mining techniques are to be investigated for attribute reduction and enhance the performance of other soft computing paradigms.

Acknowledgements

Authors would like to thank S. Chebrulu and S. Peddabachigari (Oklahoma State University, USA) for the various contributions during the different stages of this research.

References

1. Abraham A., Evolutionary Computation in Intelligent Web Management, Evolutionary Computing in Data Mining, Ghosh A. and Jain L.C. (Eds.), Studies in Fuzziness and Soft Computing, Springer Verlag Germany, 2004.
2. Barbara D., Couto J., Jajodia S. and Wu N., ADAM: A Testbed for Exploring the Use of Data Mining in Intrusion Detection. SIGMOD Record, 30(4), pp. 15–24, 2001.
3. Brameier. M. and Banzhaf. W., A comparison of linear genetic programming and neural networks in medical data mining, Evolutionary Computation, IEEE Transactions on, Volume: 5(2), pp. 17–26, 2001.
4. Cohen W., Learning Trees and Rules with Set-Valued Features, American Association for Artificial Intelligence (AAAI), 1996.
5. Denning D., An Intrusion-Detection Model, IEEE Transactions on Software Engineering, Vol. SE-13, No. 2, pp. 222–232, 1987.
6. KDD Cup 1999 Intrusion detection data set: <http://kdd.ics.uci.edu/databases/kddcup99/kddcup.data_10_percent.gz>
7. Brieman L., Friedman J., Olshen R., and Stone C., Classification of Regression Trees. Wadsworth Inc., 1984.
8. Lee W. and Stolfo S. and Mok K., A Data Mining Framework for Building Intrusion Detection Models. In Proceedings of the IEEE Symposium on Security and Privacy, 1999.
9. Luo J. and Bridges S. M., Mining Fuzzy Association Rules and Fuzzy Frequency Episodes for Intrusion Detection, International Journal of Intelligent Systems, John Wiley & Sons, Vol. 15, No. 8, pp. 687–704, 2000.
10. MIT Lincoln Laboratory. <http://www.ll.mit.edu/IST/ideval/>
11. Mukkamala S., Sung A.H. and Abraham A., Intrusion Detection Using Ensemble of Soft Computing Paradigms, Third International Conference on Intelligent Systems Design and Applications, Intelligent Systems Design and Applications, Advances in Soft Computing, Springer Verlag, Germany, pp. 239–248, 2003.
12. Peddabachigari S., Abraham A., Thomas J., Intrusion Detection Systems Using Decision Trees and Support Vector Machines, International Journal of Applied Science and Computations, USA, 2004.
13. Summers R.C., Secure Computing: Threats and Safeguards. McGraw Hill, New York, 1997.

14. Vapnik V.N., The Nature of Statistical Learning Theory. Springer, 1995.
15. Zadeh L.A., Roles of Soft Computing and Fuzzy Logic in the Conception, Design and Deployment of Information/Intelligent Systems, Computational Intelligence: Soft Computing and Fuzzy-Neuro Integration with Applications, O. Kaynak, L.A. Zadeh, B. Turksen, I.J. Rudas (Eds.), pp 1–9, 1998.
16. Kasabov N., Evolving Fuzzy Neural Networks – Algorithms, Applications and Biological Motivation, in Yamakawa T and Matsumoto G (Eds), Methodologies for the Conception, Design and Application of Soft Computing, World Scientific, pp. 271–274, 1998
17. Shah K., Dave N., Chavan S., Mukherjee S., Abraham A. and Sanyal S., Adaptive Neuro-Fuzzy Intrusion Detection System, IEEE International Conference on Information Technology: Coding and Computing (ITCC'04), USA, IEEE Computer Society, Volume 1, pp. 70–74, 2004.

14

Use of Fuzzy Feature Descriptions to Recognize Handwritten Alphanumeric Characters

G.E.M.D.C. Bandara[1], R.M. Ranawana[2], and S.D. Pathirana[3]

[1,3]Faculty of Engineering, University of Peradeniya, Peradeniya 20400, Sri Lanka
[2]University of Oxford, Computing Laboratory, Oxford, OX1 2HY, United Kingdom
dcb@pdn.ac.lk[1], romesh.ranawana@comlab.ox.ac.uk[2], susp@pdn.ac.lk[3]

Abstract. The main challenge in handwritten character recognition involves the development of a method that can generate descriptions of the handwritten objects in a short period of time. Due to its low computational requirement, fuzzy logic is probably the most efficient method available for on-line character recognition. The most tedious task associated with using fuzzy logic for online character recognition is the building of the rule-base that would describe the characters to be recognized. The problem is complicated as different people write the same character in complete different ways. This work describes a method that can be used to generate a fuzzy value database that describes the characters written by different individuals.

14.1 Introduction

The objective of Handwritten Character Recognition is the recognition of data that describes handwritten objects. On-line handwriting recognition deals with a time ordered sequence of data. Many methodologies have been developed for the recognition of handwritten characters. Some of the more recent methodologies include the use of Bayesian inference [3, 5, 6, 8, 9] Neural Networks, Fuzzy Logic and Genetic Algorithms. When compared with other description methods, fuzzy description is probably the most efficient in terms of computational usage. Due to this reason, fuzzy logic is an appropriate method for on-line character recognition.

The algorithms presented here are concerned with the recognition of singular characters written in sequence, one after another on a writing tablet, in a similar way as the characters are introduced into a hand-held PDA (Personal Diary Assistant) or into documents created using recent Microsoft Word. The most tedious task associated with the recognition of alphanumeric characters using fuzzy logic is the design and creation of the fuzzy rule base. Another problem associated with handwritten alphanumeric character recognition is the difficulty in associating exact fuzzy parameters to certain characters, due to differences associated with the way they are written by different people. Figure 14.1 shows an example of how different people write the letter "b". If the fuzzy rule base is automatically generated according to a certain individual's handwriting instead of a rule base designed using

G.E.M.D.C. Bandara et al.: *Use of Fuzzy Feature Descriptions to Recognize Handwritten Alphanumeric Characters*, Studies in Computational Intelligence (SCI) **4**, 209–232 (2005)
www.springerlink.com
© Springer-Verlag Berlin Heidelberg 2005

Fig. 14.1. Character "b" written by different individuals

parameters inserted by the designer of the system, then the system can be more effective in recognizing the handwritten characters of that particular individual.

The challenge with on-line character recognition is to develop a process that can generate these descriptions within a short period. This is because anybody who uses an on-line character recognition system will need the system to recognize the character immediately, with a minimum time delay. This problem is not present with off-line character recognition. Thus, the algorithms used should be short as well as efficient.

When compared with other description methods, fuzzy description is probably the fastest method available. This is due to the following reasons:

1. Fuzzy feature descriptions model human perception of features.
2. Training of the system is not required. Other methods (e.g. Neural Networks) require an extensive training set, as well as a considerable period for training of the system to be able to perceive a wide variety of characters.
3. The mathematics that fuzzy systems require is very fundamental (i.e. min, max, addition and multiplication). Thus, the calculations involved are minor and simple.

The methods described in this paper can be applied to any on-line alphanumeric character recognition system. They describe the processing sequence after the time ordered sequence of coordinates has been sent to the computer.

The proposed system operates in two modes, namely, the training mode and the recognition mode. This chapter describes the proposed system by first presenting the fuzzy characteristic values that are calculated for each written character, and then shows how these values are stored into the system. Sections 14.4 and 14.5 describe how the system can be used to identify a written character. The paper ends with some conclusions and future work outlines.

14.1.1 Algorithm Used

Each character is divided into a number of segments. The segmentation method is presented in one of our previous papers [1], and is based on the observation that all segments in any handwritten character in English begin with the direction of the movement of the pen changing by more than 90°. Thus, the method is based on the angle the line connecting the coordinate points forms with the x-axis of the O-x-y grid. Whenever a pen-up takes place and the pen is put down again, to write a part of the same character, a new segment is recognized. Once the separate segments of a given character have been properly identified (Fig. 14.2), the individual characteristics of each given segment can be calculated.

The segmentation for the letter "b" is shown in (Fig. 14.2) and the proposed algorithm for character recognition is given in (Fig. 14.3).

Fig. 14.2. The segmentation of character "b"

The input into the Fuzzy Feature Recognition System (Fig. 14.3) is a time-ordered sequence of coordinates, and it is up to the system to decode this data into a set of characters. This incoming information will also include information about the number of times the writer lifted his/her pen and the number of times that the pen was put back down on the piece of paper or writing tablet.

The algorithm used within the fuzzy system is shown in (Fig. 14.3).

Each step of the algorithm is described in detail below.

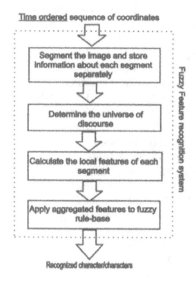

Fig. 14.3. Algorithm of the online handwritten character recognition

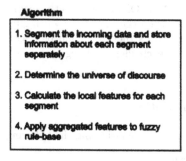

Fig. 14.4. Fuzzy algorithm used for online handwritten character recognition

14.2 Initial Processing

14.2.1 Data Structures Used

The following data structures are used for storing information about each identified segment in the incoming data set.

A data type called Coordinates is used to represent each point.

TYPE Coordinate
>   *x as integer* (The x coordinate)
>   *y as integer* (The y coordinate)

End TYPE

Information about each segment is stored in a separate variable

TYPE Segment
>   *points(0 To MAX_POINTS) As Coordinate*
>   *number of points As Integer*
>   *title As Double* (straight line between the end points)
>   *MAXX As Integer*
>   *MAXY As Integer*
>   *MINY As Integer*
>   *MHP As Double* (Mu Horizontal position)
>   *MVP As Double* (Mu Vertical position)
>   *MLENH As Double* (Mu Length Horizontal)
>   *MLENV As Double (Mu Length Vertical)*
>   *MLENS As Double* (Mu length Slant)
>   *MARC As Double* (Mu Arc-ness)
>   *MSTR As Double* (Mu Straightness)
>   *MVL As Double* (Mu Vertical line)
>   *MHL As Double* (Mu Horizontal line)
>   *MPS As Double* (Mu positive slant (/))
>   *MNS As Double* (Mu Negative slant (\))
>   *MCL As Double* (Mu C-like)
>   *MDL As Double* (Mu D-like)
>   *MAL As Double* (Mu A-like)
>   *MUL As Double* (Mu U-like)
>   *MOL As Double* (Mu O-like)

End Type

Where the value for MAX_POINTS depicts the maximum number of points per segment. This value depends on the size of the writing pad.

The above-mentioned data structure should be created at the beginning of the program. The data about each identified segment is stored separately in such a structure.

14.2.2 Image Segmentation

The incoming data is segmented considering the following features:

1. Pen-ups and pen-downs
2. Abrupt changes in pen movement direction

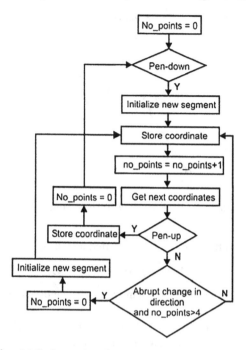

Fig. 14.5. Incoming data segmentation algorithm

Image Segmentation Based on Pen-ups and Pen-downs

Each time a position in the data where the pen has been put on the paper or writing tablet is recognized, a variable to store data about that segment is created and initialized. All data following that event are then recorded as belonging to the new segment until one of the following events takes place:

1. The pen is taken up again
2. The direction of writing changes by more than 90°.

The coordinate points belonging to a recognized segment are then stored in the created segment variable.

Image Segmentation According to Changes in Writing Direction

The image segmentation method used is based on the observation that all segments in any handwritten character in English begin with the direction of movement changing more than 90°. The image segmentation method used here is based on the angle of the coordinate points with the x-axis of the O-x-y grid. The algorithm used for the calculation of this angle is described in Sect. 14.2.2.

The coordinates of four points that are about four pen-point distances away from each other are taken and the difference in angle between the two lines is calculated. The reason for taking points a certain distance away from each other is that, if consecutive points are taken, the angle calculated will be one of the following: 0°, 45°, 90°, 135°, 180°, 225°, 270° or 315°.

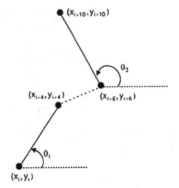

Fig. 14.6. Image segmentation according to the changes in the direction of writing

Then the angle difference can be calculated as follows:

$$Angle\_difference = |\theta_1 - \theta_2| \qquad (14.1)$$

If the angle difference is larger than 180°, a value of 180 is subtracted from it, i.e.

$$Angle\_difference = Angle\_difference - 180° \qquad (14.2)$$

Finally, a fuzzy value is calculated to represent the magnitude change of direction.

$$\mu_{ANG\_DIFF} = \frac{Angle\_difference}{180°} \qquad (14.3)$$

where μ_{ANG_DIFF}, is confined to the universe of discourse $[0,1]$.

Membership Function 1 (Fig. 14.7) is then used for the purpose of evaluation of this value. It can be used to determine if the intersection of the two lines constitutes a new segment or not. The corresponding linguistic values are "*very small*", "*small*", "*small medium*", "*medium*' "*large medium*", "*large*" and "*very large*".

If the value μ_{ANG_DIFF}, is "*medium*" or larger, it can be concluded that a new segment has now been started and the present segment can be concluded. A new variable is then initialized to accommodate the new segment.

Once it is determined that the whole character has been written, the recognized segments can then be sent into the fuzzy feature recognition system.

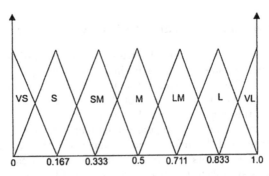

Fig. 14.7. Membership functions for evaluating the angle difference

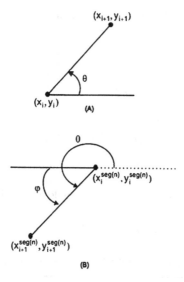

Fig. 14.8. Representation of the angle θ (**A**), and φ (**B**)

This method is based on the assumption that all segments are at least four coordinate points long.

14.2.3 Calculation of the Angle that a Given Line Makes with the x-Axis of the o-x-y Grid of the Image

A method for calculating the angle of a line that joins two given points with the x-axis of the image is described in this section.

This method proposes a scheme for calculating the value of θ when the values of (x_i, y_i) and (x_{i+1}, y_{i+1}) are given. To calculate this, an intermediate value, φ, is used. (Fig. 14.4) shows both angles.

The calculation of this angle requires the following steps:

1. Calculation of φ
2. Use of φ to calculate θ

Calculation of φ

$$\varphi = \tan^{-1}\left(\frac{|y_{i+1}^{seg(n)} - y_i^{seg(n)}|}{|x_{i+1}^{seg(n)} - x_i^{seg(n)}|}\right) \tag{14.4}$$

Table 14.1. Calculation of the angle θ using φ

Condition	$\theta^{\mathrm{seg}(n)}$
$\left(x_i^{\mathrm{seg}(n)} < x_{i+1}^{\mathrm{seg}(n)}\right)$ and $\left(y_i^{\mathrm{seg}(n)} < y_{i+1}^{\mathrm{seg}(n)}\right)$	φ
$\left(x_i^{\mathrm{seg}(n)} > x_{i+1}^{\mathrm{seg}(n)}\right)$ and $\left(y_i^{\mathrm{seg}(n)} < y_{i+1}^{\mathrm{seg}(n)}\right)$	$180 - \varphi$
$\left(x_i^{\mathrm{seg}(n)} > x_{i+1}^{\mathrm{seg}(n)}\right)$ and $\left(y_i^{\mathrm{seg}(n)} > y_{i+1}^{\mathrm{seg}(n)}\right)$	$180 + \varphi$
$\left(x_i^{\mathrm{seg}(n)} < x_{i+1}^{\mathrm{seg}(n)}\right)$ and $\left(y_i^{\mathrm{seg}(n)} > y_{i+1}^{\mathrm{seg}(n)}\right)$	$360 - \varphi$

(Note: For each segment, the angle of the straight line between the first and last points of that segment is also calculated by the use of the above method. This angle will be denoted by θ.)

Table 14.2. Coordinates of the universe of discourse

$x_{\min} = \underset{i=1}{\overset{N}{\mathrm{Min}}}(x_i)$	$y_{\min} = \underset{i=1}{\overset{N}{\mathrm{Min}}}(y_i)$	$x_{\max} = \underset{i=1}{\overset{N}{\mathrm{Max}}}(x_i)$	$y_{\max} = \underset{i=1}{\overset{N}{\mathrm{Max}}}(y_i)$

Use of φ to Calculate θ

The following table (Table 14.1) can be used for this step of the calculation:

14.3 Determination of the Universe of Discourse

To identify certain shape features and their positions, the first step is the determination of the universe of discourse. i.e. the smallest coordinate space or rectangle that the character fits into. This rectangle is defined by the parameters:

All computations following this step use these values.

14.4 Calculation of the Local Features for Each Segment

14.4.1 Determination of the Relative Position of Given Segment

The first step in this process is the determination of the center of the given segment. The calculation of this is required for the determination of the position of the

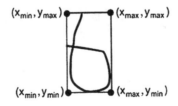

Fig. 14.9. Coordinates of the universe of discourse of a character

Table 14.3. Coordinate of the universe of discourse of a given segment

$$x_{min}^{seg(n)} = \underset{i=1}{\overset{N}{Min}} \left(x_i^{seg(n)} \right) \qquad\qquad y_{min}^{seg(n)} = \underset{i=1}{\overset{N}{Min}} \left(y_i^{seg(n)} \right)$$

$$x_{max}^{seg(n)} = \underset{i=1}{\overset{N}{Max}} \left(x_i^{seg(n)} \right) \qquad\qquad y_{max}^{seg(n)} = \underset{i=1}{\overset{N}{Max}} \left(y_i^{seg(n)} \right)$$

Note. These coordinates are graphically represented in (Fig. 10.10).

given segment with respect to the universe of discourse determined earlier. Table 14.3 provides the coordinates of the universe of discourse for a given segment. The maximum and minimum values for all coordinates that depict the given segment are calculated.

The coordinates of the center-points are then calculated as follows:

$$x_{CENTER}^{seg(n)} = \frac{x_{min}^{seg(n)} + x_{max}^{seg(n)}}{2} \tag{14.5}$$

$$y_{CENTER}^{seg(n)} = \frac{y_{min}^{seg(n)} + y_{max}^{seg(n)}}{2} \tag{14.6}$$

The relative position of the given segment is then expressed as follows:

$$\mu_{HP}^{seg(n)} = \frac{x_{CENTER}^{seg(n)} - x_{min}^{seg(n)}}{x_{max}^{seg(n)} - x_{min}^{seg(n)}} \tag{14.7}$$

$$\mu_{VP}^{seg(n)} = \frac{y_{CENTER}^{seg(n)} - y_{min}^{seg(n)}}{y_{max}^{seg(n)} - y_{min}^{seg(n)}} \tag{14.8}$$

Where the terms HP and VP stand for "Horizontal Position" and "Vertical Position" respectively.

The relative position of the given segment can then be described by using the membership functions given in (Fig. 14.11)

Where the linguistic terms L, LC, NL, C, NR, RC and R stand for "*left*", "*left center*", "*nearly left*", "*center*", "*nearly right*", "*right center*" and "*right*" respectively.

Where the linguistic terms B, BC, NB, C, NT, TC and T stand for "*bottom*", "*bottom center*", "*nearly bottom*", "*center*", "*nearly top*", "*top center*" and "*top*" respectively.

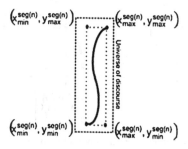

Fig. 14.10. Coordinates of the universe of discourse for a given segment

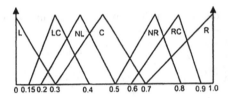

Fig. 14.11. Membership functions to determine the relative horizontal position of a given segment

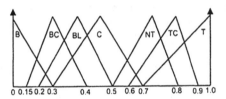

Fig. 14.12. Membership functions to determine the relative vertical position of a given segment

Membership function 1 (Fig. 14.11) is used to determine the relative horizontal position of the segment and Membership function 2 (Fig. 14.12) is used to determine the relative vertical position of the segment.

The segment can then be compared against these values to see if it is positioned correctly.

14.4.2 Geometrical Feature Detection

The next step in this process is to determine whether the given segment is a straight line or an arc. All subsequent calculations are based on this determination.

The method studied by Pal and Majumder [3] is used for this purpose. It shows that the straightness of a segment is determined by fitting a straight line with the minimum least squares error. Similarly, in a given segment, the ratio of the distance between end-points to its total arc length shows its arc-ness.

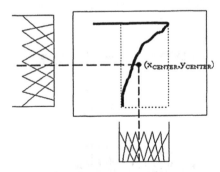

Fig. 14.13. Location of the center and geometrical feature detection using already created membership functions

We measure the arc-ness and the measure of straightness by the use of (4.5) and (4.6) given below:

$$\mu_{\text{STRAIGHTNESS}}^{\text{seg}(n)} = \left[\frac{d_{P_0 P_N}^{\text{seg}(n)}}{\sum_{k=0}^{N-1} d_{P_k P_{k+1}}^{\text{seg}(n)}} \right] \tag{14.9}$$

$$\mu_{\text{ARC-NESS}}^{\text{seg}(n)} = \left[1 - \frac{d_{P_0 P_N}^{\text{seg}(n)}}{\sum_{k=0}^{N-1} d_{P_k P_{k+1}}^{\text{seg}(n)}} \right] \tag{14.10}$$

Where $d_{P_k P_{k+1}}^{\text{seg}(n)}$ stands for the straight-line distance between point k and point $(k+1)$ on the nth segment. The number of elements in the segment is depicted by N.

Then, a decision is made on whether the segment is a straight line or an arc. This decision is based on a threshold value (i.e. an α-cut). For example, if $\mu_{\text{STRAIGHTNESS}}^{\text{seg}(n)}$ is larger than 0.6, then the segment can be determined as a straight line. If the segment has a $\mu_{\text{ARC-NESS}}^{\text{seg}(n)}$ value in excess of 0.6, the segment can be determined as an arc. Both these values cannot be higher than the selected threshold values due to the following statement (14.7) holding for all segments:

$$\mu_{\text{ARC-NESS}}^{\text{seg}(n)} + \mu_{\text{STRAIGHTNESS}}^{\text{seg}(n)} = 1 \tag{14.11}$$

i.e. the segment cannot have a high value for both these fuzzy values.

The next sequence of calculations is based on the values of $\mu_{\text{ARC-NESS}}^{\text{seg}(n)}$ and $\mu_{\text{STRAIGHTNESS}}^{\text{seg}(n)}$. The program flow then diverges into two sequences based on these values, as shown in (Fig. 14.14)

14.4.3 Calculations When the Given Segment is Determined to be a Straight Line

Determination of the Line Type

If a segment is identified as a straight line, then the orientation or the angle of inclination of this segment distinguishes it further into one of the following features:

- Vertical line
- Horizontal line

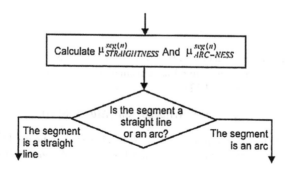

Fig. 14.14. Program flow according to the type of the segment

- Positive slant
- Negative slant

The corresponding membership function is $\Lambda(\theta; b, c)$ as given in (14.8), where,

- θ represents the angle of orientation of the straight line.
- b is the bandwidth, and its value is $90°$
- c at the maximum membership value, which is unity

$$\Lambda(x; b, c) = \begin{cases} 1 - 2 \bullet |(\frac{x-c}{b})| ; & (c - \frac{b}{2}) \leq x \, (c + \frac{b}{2}) \\ 0, & \text{otherwise} \end{cases} \qquad (14.12)$$

Vertical Line (|)

A vertical line has an ideal orientation of $90°$ or $270°$. Therefore, the fuzzy linguistic term "vertical line" is defined as a triangular membership function $\Lambda(\theta; b, c)$, by the following expression (4.9):

$$\mu_{VL} = \text{MAX}(\Lambda(\theta, 90, 90), \Lambda(\theta, 90, 270)) \qquad (14.13)$$

Horizontal Line (−)

A horizontal line has an ideal orientation of $0°$, $180°$ or $360°$. Therefore, the fuzzy linguistic term "horizontal line" is defined as a triangular membership function $\Lambda(\theta; b, c)$, by the following expression (14.10):

$$\mu_{HL} = \text{MAX}(\Lambda(\theta, 90, 0), \Lambda(\theta, 90, 180), \Lambda(\theta, 90, 360)) \qquad (14.14)$$

Positive Slant (/)

A positive slant has an ideal orientation of $45°$ or $225°$. Therefore, the fuzzy linguistic term "horizontal line" is defined as a triangular membership function $\Lambda(\theta; b, c)$, by the following expression (4.11):

$$\mu_{PS} = \text{MAX}(\Lambda(\theta, 90, 45), \Lambda(\theta, 90, 225)) \qquad (14.15)$$

Negative Slant (\)

A negative slant has an ideal orientation of $135°$ or $315°$. Therefore, the fuzzy linguistic term "horizontal line" is defined as a triangular membership function $\Lambda(\theta; b, c)$, by the following expression (4.12):

$$\mu_{NS} = \text{MAX}(\Lambda(\theta, 90, 135), \Lambda(\theta, 90, 315)) \qquad (14.16)$$

The following membership function (Fig. 14.15) can then be used to evaluate the already calculated fuzzy values to ascertain the degree of membership to a certain class of straight line.

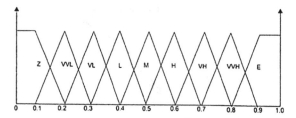

Fig. 14.15. Degree of membership of a certain straight line

where the linguistic terms Z, VVL, VL, L, M, H, VH, VVH and E stand for *"zero"*, *"Very Very Low"*, *"Very low"*, *"Low"*, *"Medium"*, *"High"*, *"Very High"*, *"Very Very High"* and *"Extreme"* respectively.

For example, the number "1" is represented by a straight line that has a vertical line degree of membership of either "E" or "VVH". If μ_{HL} is any other, it can be determined that the given segment does not depict the number "1".

After the orientation of the line has been determined, the next important aspect is the determination of the length of the given line with respect to the universe of discourse.

When calculating the relative line length, the type of straight line should also be considered. Thus, the calculations depend on whether the line is a horizontal line, a vertical line or a slanted line as shown in (Fig. 14.16).

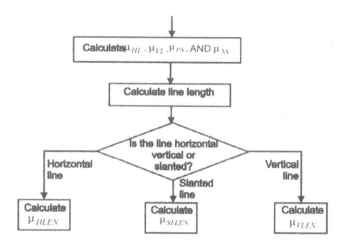

Fig. 14.16. Algorithm to decide type of the straight line

Determination of the Relative Length of the Given Straight Line

Regardless of what type of line is being considered, the length of the line should first be determined. The straight-line distance between the first and the last elements of the segment are considered for this. i.e. $d_{P_0 P_N}^{\text{seg}(n)}$

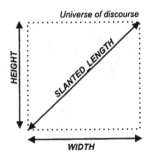

Fig. 14.17. Measurements required deciding the length of a straight line

The following measurements should also be made (Fig. 14.17):

1. Width of the universe of discourse ($WIDTH$)
2. Height of the universe of discourse ($LENGTH$)
3. Diagonal length of the universe of discourse ($SLANT\_LENGTH$)

Then, depending on the type of line, the following measurements can be made:

For a Horizontal Line:

$$\mu_{\text{HLEN}}^{\text{seg}(n)} = \frac{d_{\text{POPN}}^{\text{seg}(n)}}{\text{WIDTH}} \tag{14.17}$$

For a Vertical Line:

$$\mu_{\text{VLEN}}^{\text{seg}(n)} = \frac{d_{\text{POPN}}^{\text{seg}(n)}}{\text{HEIGHT}} \tag{14.18}$$

For a Slanted Line:

$$\mu_{\text{SLLEN}}^{\text{seg}(n)} = \frac{d_{\text{POPN}}^{\text{seg}(n)}}{\text{SLANT\_LENGTH}} \tag{14.19}$$

where the terms HLEN, VLEN and SLLEN stand for "*Horizontal length*", "*Vertical length*" and "*Slant length*" respectively.

The possibility is there for these values to exceed the universe of discourse for fuzzy functions, [0,1]. Thus, after calculating them, a check is made to see if it has exceeded the value "1", and if so, the value is reduced to 1. This step is required to confine the values to the universe of discourse [0,1].

The following membership function (Fig. 14.18) can then be used to evaluate the relative length of the given straight line:

For example, a straight line that has a vertical line of length "E" with respect to the universe of discourse could represent the number "1".

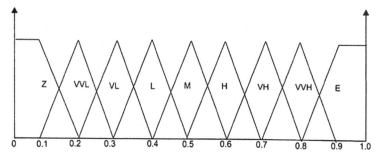

Fig. 14.18. Evaluation of the relative length of a given straight line

14.4.4 Calculations When the Given Segment is Determined to be an Arc

Once a segment has been determined to be a curve, it is necessary to determine the type of curve. For handwritten alphanumeric characters, the types of curves contain in characters usually fall into one of the following categories (Table 14.4):

Table 14.4. Types of curves present in handwritten alphaneumeric characters

Type	Example
A-like (AL)	∩ ⌒ ⌒ ⌒
U-like (UL)	⊔ ⊔ ⊔ ⊔
C-like (CL)	C (C C
D-like (DL)	⊃ ⊃ ⊃ ⊃
O-like (OL)	○ ○ ○ ○
Hockey Stick (HSL, HSR)	J J L L
Walking Stick (WSL, WSR)	⌐ Γ ⌐ ⌐

The distinction of these shape categories is accomplished by using the

- Angle of rotation
- Angle of slope joining the end points of the segment
- Measure of arc-ness
- The relative length and area covered by the segment.

All these measures are relative and normalized in the fuzzy domain of discourse [0,1], and can be combined with fuzzy aggregation operators. The output of this evaluation can classify a segment to several features and by this associate several possible meanings to its shape.

C-like Curve (CL)

To distinguish a C-like curve from a D-like curve, we use the quantitative statement of the left or right convexity direction. In a curve which is vertical (i.e. either C-like or D-like), the global minimum of horizontal projections x_{min} is relatively much lower than the weighted average of x projections, w_S, w_E, of its end points, x_S, x_E, then it is more likely to be a C-like curve.

$$\mu_{CL} = \min \left(1, \frac{\sum_{i=0}^{n} l_{x_i}}{n} \right) \qquad (14.20)$$

where

$$l_{x_i} = \begin{cases} 1; & x_i \langle \frac{(x_S + x_E)}{2} \\ 0; & \text{otherwise} \end{cases} \qquad (14.21)$$

In the above equation l_{x_i} is the binary function, which possesses the truth values over the whole segment regarding the point position. i.e. segment point is on the left hand side of the median of the end point x-projections, and the binary function is equal to 1, else 0.

This summation function is then normalized to the universe of discourse. [0,1]

D-like Curve (DL)

In a curve which is vertical, if the global maximum of horizontal projections x_{max} is relatively much higher than a weighted average of x projections of its end points x_S, x_E, then it is very likely to be a D-like curve.

This statement is represented by the summation of a binary function r_{x_i}, which posses truth values over the whole segment regarding the point position. i.e. if a segment point is on the right hand side of the median of the end point x-projections, the binary function is equal to 1, else it is 0.

This summation function is then normalized to the universe of discourse [0,1]

$$\mu_{DL} = \min \left(1, \frac{\sum_{i=0}^{n} r_{x_i}}{n} \right) \qquad (14.22)$$

where,

$$r_{x_i} = \begin{cases} 1; & x_i \rangle \frac{(x_S + x_E)}{2} \\ 0; & \text{otherwise} \end{cases} \qquad (14.23)$$

A-like Curve (AL)

In a curve which is horizontal (i.e. an A-like curve or a U-like curve), if the global maximum of horizontal projections y_{max} is relatively much higher than a weighted average (w_S, w_E) of y projections of its end points (y_S, y_E), then it is very likely to be an A-like curve.

This statement is represented by the summation of a binary function a_{x_i}, which posses truth values over the whole segment regarding the point position. i.e. if a segment point is above the median of the end point y-projections, the binary function is equal to 1, otherwise it is 0.

This summation function is then normalized to the universe of discourse. $[0,1]$

$$\mu_{AL} = \min\left(1, \frac{\sum_{i=0}^{n} a_{y_i}}{n}\right) \tag{14.24}$$

where,

$$a_{y_i} = \begin{cases} 1; & y_i \rangle \frac{(y_S + y_E)}{2} \\ 0; & \text{otherwise} \end{cases} \tag{14.25}$$

U-like Curve (UL)

In a curve which is horizontal, if the global maximum of horizontal projections $y_{\min}$ is relatively much lower than a weighted average (w_S, w_E) of y projections of its end points (y_S, y_E), then it is very likely to be a U-like curve.

This statement is represented by the summation of a binary function b_{x_i}, which possess truth values over the whole segment regarding the point position. i.e. if a segment point is below the median of the end point y-projections, the binary function is equal to 1, otherwise it is 0.

This summation function is then normalized to the universe of discourse. $[0,1]$

$$\mu_{UL} = \min\left(1, \frac{\sum_{i=0}^{n} b_{y_i}}{n}\right) \tag{14.26}$$

where,

$$b_{y_i} = \begin{cases} 1; & y_i \rangle \frac{y_S + y_E}{2} \\ 0; & \text{otherwise} \end{cases} \tag{14.27}$$

O-like Curve

The following method was used to determine a fuzzy measure for the O-like-ness of the given segment.

Let $(x_{\text{CENTER}}^{\text{seg}(n)}, y_{\text{CENTER}}^{\text{seg}(n)})$ denote the center of the curve. This is the same as the value calculated in determining the relative position of the curve.

The expected radius of the curve is denoted by (4.24).

$$rad_{\text{EXPECTED}}^{\text{seg}(n)} = \frac{(x_{\max}^{\text{seg}(n)} - x_{\min}^{\text{seg}(n)}) + (y_{\max}^{\text{seg}(n)} - y_{\min}^{\text{seg}(n)})}{4} \tag{14.28}$$

The actual radius of the curve can be determined by summing up the straight line distances from the center of the segment to each element belonging to the segment and dividing it by the number of elements in the segment (14.25).

$$rad_{\text{ACTUAL}}^{\text{seg}(n)} = \frac{\sum_{k=0}^{N} d_{p_k(x_{\text{CENTER}}^{\text{seg}(n)}, y_{\text{CENTER}}^{\text{seg}(n)})}}{N} \tag{14.29}$$

The expected diameter for a curve with the radius is then calculated using the expression 4.26.

$$diameter_{\text{EXPECTED}}^{\text{seg}(n)} = 2\Pi rad_{\text{EXPECTED}}^{\text{seg}(n)} \tag{14.30}$$

The actual diameter of the given segment can be calculated by summing the straight line distance between consecutive elements in the segment (4.27).

$$diameter_{\text{ACTUAL}}^{\text{seg}(n)} = \sum_{k=0}^{N-1} d_{p_k p_{k+1}} \tag{14.31}$$

where N stands for the number of elements in the segment.

The following calculations are then carried out to determine a fuzzy value for the O-like ness of the given segment (4.28, 4.29, 4.30, and 4.31).

$$\mu_{\text{OL2}}^{\text{seg}(n)} = \begin{cases} f(x); & f(x) < 1 \\ \frac{1}{f(x)}; & f(x) > 1 \end{cases} \tag{14.32}$$

where,

$$f(x) = \frac{diameter_{\text{ACTUAL}}^{\text{seg}(n)}}{diameter_{\text{EXPECTED}}^{\text{seg}(n)}} \tag{14.33}$$

$$\mu_{\text{OL2}}^{\text{seg}(n)} = \begin{cases} g(x); & g(x) < 1 \\ \frac{1}{g(x)}; & g(x) > 1 \end{cases} \tag{14.34}$$

where,

$$g(x) = \frac{rad_{\text{ACTUAL}}^{\text{seg}(n)}}{rad_{\text{EXPECTED}}^{\text{seg}(n)}} \tag{14.35}$$

and,

$$\mu_{\text{OL}}^{\text{seg}(n)} = \min\left(\mu_{\text{OL1}}^{\text{seg}(n)}, \mu_{\text{OL2}}^{\text{seg}(n)}\right) \tag{14.36}$$

Other curve features are calculated as follows (4.33, 4.34, 4.35, and 4.36):

$$\mu_{\text{HOL}} = WGM(\max(\mu_{\text{UL}}, \mu_{\text{DL}}), \mu_{\text{PS}}) \tag{14.37}$$

$$\mu_{\text{HOR}} = WGM(\max(\mu_{\text{UL}}, \mu_{\text{CL}}), \mu_{\text{NS}}) \tag{14.38}$$

$$\mu_{\text{STL}} = WGM(\max(\mu_{\text{AL}}, \mu_{\text{DL}}), \mu_{\text{NS}}) \tag{14.39}$$

$$\mu_{\text{STR}} = WGM(\max(\mu_{\text{AL}}, \mu_{\text{CL}}), \mu_{\text{PS}}) \tag{14.40}$$

where the terms AL, UL, CL, DL, OL, HOL, HOR, STL and STR stand for "A-like", "U-like", "C-like", "D-like", "O-like", "Hockey stick left", "Hockey stick right", "stick left" and "stick right" respectively. The function WGM (Weighted Generalization Means) is defined by (4.37)

$$WGM(a_1, a_2, \ldots a_m; w_1, w_2, \ldots w_m) = \left[\sum_{i=1}^{m} (w_i a_i)^\beta\right]^{\frac{1}{\beta}} \tag{14.41}$$

where $\sum_{i=1}^{m} w_i = 1$.

These methods were proposed by Malaviya et al. [7] of the German National Research Center for Information Technology.

Table 14.5. Categorization of alphanumeric characters according to the number of segments

Number of Segments	Example Character
1	0,1,C
2	2,3,6,7,8,9,D,G,J,L,P,Q,L,S,T,U,V,X
3	1,3,4,5,A,B,F,H,I,K,N,R,Y,Z
4	E,M,W

14.5 Application of the Calculated Features to the Fuzzy Rule Base

14.5.1 Creation of the Fuzzy Rule Base

The rule base is divided into different sections based on the number of segments contained in the character. Since all segments are checked individually, the program flow enters only the relevant section to process the extracted segment eliminating the chances to cause confusion over similar segments. The division of the rule-base further simplifies the search for the solution and speeds up the recognition process. For example:

Some characters like **1** and **3** are represented twice in the table as they can be written in different ways.

The fuzzy rules for each of these characters are written in the database as follows:

If (seg(1).<Fuzzy Value1> Is <Linguistic value1>) AND (seg(1).<Fuzzy Value1> Is <Linguistic value1>) AND ...AND (seg(NO_SEG).<Fuzzy Value n> Is <Linguistic value n>)

THEN output=<Character>

If the character has only one segment, it has only one rule. If the character has two segments, it has two rules. This is to give the user the ability to write the character segments in any order that he/she wants. Similarly, if the character has 3 segments, it is represented by six rules in the rule-base.

i.e. If the number of segments is n, the character is represented by $(n!)$ number of rules.

Example 1. Following is a set of rules used in a preliminary test of the proposed system.

Rule to Recognize the Character "1"

$$IF\ (\mu^{\text{seg}(1)}_{\text{STRAIGHTNESS}} = VVH)\ AND\ (\mu^{\text{seg}(1)}_{\text{VL}} = VVH)\ THEN\ Output = \text{``1''};$$

Rule to Recognize Character "7"

Rule 1

$$IF \ (\mu_{\text{STRAIGHTNESS}}^{\text{seg}(1)} = VVH) \ AND \ (\mu_{\text{HL}}^{\text{seg}(1)} = VVH) \ AND$$
$$(\mu_{\text{VP}}^{\text{seg}(1)} = TOP) \ AND \ (\mu_{\text{HLEN}}^{\text{seg}(1)} = VVH) \ AND \ ((\mu_{\text{STRAIGHTNESS}}^{\text{seg}(2)} = E)$$
$$OR \ (\mu_{\text{STRAIGHTNESS}}^{\text{seg}(2)} = VVH)) \ AND \ ((\mu_{\text{PS}}^{\text{seg}(2)} = E)$$
$$OR \ (\mu_{\text{PS}}^{\text{seg}(2)} = VVH)) \ AND \ (\mu_{\text{SLLEN}}^{\text{seg}(2)} = E) \ THEN \ Output = \text{"7"};$$

Rule 2

$$IF \ (\mu_{\text{STRAIGHTNESS}}^{\text{seg}(2)} = VVH) \ AND \ (\mu_{\text{HL}}^{\text{seg}(2)} = VVH) \ AND$$
$$(\mu_{\text{VP}}^{\text{seg}(2)} = TOP) \ AND \ (\mu_{\text{HLEN}}^{\text{seg}(2)} = VVH) \ AND \ ((\mu_{\text{STRAIGHTNESS}}^{\text{seg}(1)} = E)$$
$$OR \ (\mu_{\text{STRAIGHTNESS}}^{\text{seg}(1)} = VVH)) \ AND \ ((\mu_{\text{PS}}^{\text{seg}(1)} = E) \ OR \ (\mu_{\text{PS}}^{\text{seg}(1)} = VVH))$$
$$AND \ (\mu_{\text{SLLEN}}^{\text{seg}(1)} = E) \ THEN \ Output = \text{"7"};$$

The time consuming step of this method is the derivation of this set of rules. Once the set of rules has been clearly identified, the program then functions with a high degree of efficiency. The program developed to test this method provides a recognition rate in excess of 90%.

14.5.2 Application of Each Character to the Fuzzy Rule-Base

Each character is then applied to the rule base to ascertain whether its descriptions match with any of the rules on the fuzzy rule base. If there is a match, the rules consequence can be used to determine the character.

As the number of segments contain in the character can be gauged, the descriptions are only matched with that part of the fuzzy rule base that describes characters with determined number of segments. Thus, the number of rules that have to be matched on the rule base is drastically reduced.

14.5.3 The Database

The entire system is centralized around a main database that contains information about each character, the number of segments and the individual characteristics of the segments. The database consists of two main tables called *Characters* and *Segments*. The relationship between the two tables is a 1:m relationship. Here, a character can contain many segments, but a segment can belong to one and only one character. The table *Characters* is described in Table 14.6.

Table 14.6. A description of the fields in the table *Characters*

Field Name	Description
Character_code	This is a string value. This field holds a unique identifier for each character stored in the database.
Character	This is a single char value. This field contains the actual name of the character stored, e.g. "A", "a" etc.
Number_of_segments	This is an integer field. This field contains the number of segments in the character

i.e. character (character_ code, character, number_of_segments)

This table is sorted according to the number of segments. The table *Segments* was implemented in the following manner:

```
Character code as char
Segment_No as integer
//The following is the list of fuzzy descriptional values for the
given segment
MHP As Char      (μ Horizontal position)
MVP As Char      (μ Vertical position)
MLENH As Char    (μ Length Horizontal)
MLENV As Char    (μ Length Vertical)
MLENS As char    (μ length Slant)
MARC As Char     (μ Arc-ness)
MSTR As Char     (μ Straightness)
MVL As Char      (μ Vertical line)
MHL As Char      (μ Horizontal line)
MPS As Char      (μ positive slant (/))
MNS As Char      (μ Negative slant (\))
MCL As Char      (μ C-like)
MDL As Char      (μ D-like)
MAL As Char      (μ A like)
MUL As Char      (μ U-like)
MOL As Char      (μ O-like)
```

14.5.4 The Training Mode

A writing tablet was provided for the user and the user was expected to draw the character on the tablet for training. Once the character has been drawn, the system performs the following steps:

1. Determines the number of segments in the character
2. Identifies the universe of discourse for the entire character
3. Isolates the individual segments
4. Calculates the fuzzy values for each segment as described in Sect. 14.2.
5. These numerical values are then mapped [1] to corresponding fuzzy linguistic values using the membership function illustrated in (Fig. 14.19).

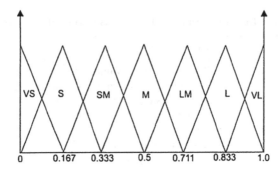

Fig. 14.19. The membership functions used to describe the characteristics of the character

Here, the linguistic terms VS, S, SM, M, LM, L and VL correspond to *"very small"*, *"small"*, *"small medium"*, *"medium"*, *"large medium"*, *"large"* and *"very large"*, respectively. Subsequently, the system queried the user to identify the alphanumeric character corresponding to that character. Once all the above-mentioned calculations have been made, a new entry is created in the table *Characters* and the corresponding values for the given character are inserted. An entry for each segment in the character is created in the table *Segments* and the corresponding values are entered. The order in which the segments were recognized was also preserved on the database. This is maintained using the attribute "Segment_No".

14.5.5 The Recognition Mode

In recognition mode, the user is required to write the character to be recognized on the writing tablet. The first four steps performed under the training mode are then performed as well on the written character. The calculated numerical values (as presented in Sect. 14.2), which describe the characteristics of the character are then stored in a variable. This is referred to as *ch*. This variable is an array containing information about each segment of the character. For example, the straightness of the second segment can be obtained by using the variable *ch* [2]. Once these steps have been performed, focus shifts onto the table *Characters*. As the table is sorted according to the number of segments, a query is generated for the characters that have an identical number of segments to the character just drawn. Then the character is recognized using a min-max inference over the selected characters produced by the previous query.

For each characteristic of the characters, (e.g. MSTR (straightness), HLEN, AL etc.) numerical values are calculated for the character to be recognized against the membership function which stored in the database for that characteristic, in the training phase. The minimum of fuzzy membership values over all the characteristics of the given character, are calculated and then calculate the maximum of the resulted values over all characters generated by the previous query is calculated (characters with the same number of segments as the character to be recognized). The character with the maximum value will be the recognized character. As min-max composition is associative, the order in which the characters are evaluated is not significant.

Table 14.7. Sample data from the database for the fuzzy descriptors of the character
"b"

Fuzzy Value	Segment 1	Segment 2
HP	SM	M
VP	M	LM
HLEN	VH	L
VLEN	VH	SM
SLLEN	L	VS
SX	M	S
SY	VS	M
EX	VS	S
EY	VL	L
MSTR	VL	VS
MARC	VS	VL
VL	VH	VS
HL	VS	VS
PS	VS	VS
NS	VS	VS
VC	VS	VS
HC	VS	L
AL	VS	VS
DL	VS	VL
CL	VS	VS
UL	VS	LM
OL	VS	VS

14.5.6 Results

The following is a sample database entry for character "b". Using the method described in [2], this character was recognized to contain two segments. Table 14.7 presents what is stored in the database for character "b", following the method described in Sect. 14.4. Similar values are stored in the training phase for all alphanumeric characters.

The system was tested by training the system on the handwriting of five different individuals. It demonstrated a recognition rate of 86% on average.

14.6 Conclusion

In preliminary tests, this method was found to be an extremely reliable and relatively simple method for generating the fuzzy description of handwritten alphanumeric characters. The generated database was found to be extremely efficient in terms of computational usage for the recognition of the handwritten characters.

This solution is not deemed to be the ultimate solution for the recognition of handwritten characters using an automatically generated database, but it is a solution which is extremely simple to implement and use, and should suffice for most small and medium scale systems. Such a system would be most useful for a system like a PDA, which emphasizes the usage of small and efficient programs. The next step in the development of this system will be to use neuro-fuzzy networks for implementing the system.

References

1. Bandara GEMDC, Pathirana SD, Ranawana RM (2002) "Use of Fuzzy Feature Descriptions to recognize Handwritten Alphanumeric Characters", 1st Conference on Fuzzy Systems and Knowledge Discovery, held at Singapore
2. Bandara GEMDC, Pathirana SD, Ranawana RM (2002) "A Short Method for On-line Alphanumeric Character Recognition", NAFIPS – FLINT 2002, held at New Orleans, USA
3. Gyeonghwan Kim, Sekwang Kim (2002) "Feature Selection Using Genetic Algorithms for Handwritten Character Recognition", Proceeding of the Sixth ACM SIGKDD International Conference on Knowledge Discovery and Data Mining
4. Kartalopoulos SV (1996) "Understanding Neural Networks and Fuzzy Logic", IEEE Press, pp. 167–168
5. Koerich Alessandro L, Leydier Yann (2002) "A Hybrid Large Vocabulary Word Recognition System using Neural Networks with Hidden Markov Models", IWFHR, http://citeseer.nj.nec.com/koerich02hybrid.html
6. Kwok-Wai Cheung, Dit-Yan Yeung (2002) "Bidirectional Deformable Matching with Application to Handwritten Character Extraction", IEEE Transactions on Pattern Analysis and Machine Intelligence, Vol. 24, No. 8
7. Malaviya A and Peters L (1995) "Extracting meaningful handwriting features with fuzzy aggregation method", 3rd International conference on document analysis and recognition, ICDAR'95, pp. 841–844, Montreal, Canada
8. Pal SK, Majumder DKD (1986) "Fuzzy mathematical approach to pattern recognition", A Halsted Press Book, Wiley & Sons, New Delhi
9. Seong-Whan Lee, Young-Joon Kim (1995) "A new type of recurrent neural network for handwritten character recognition", Third International Conference on Document Analysis and Recognition (Volume 1)

15

Soft Computing Paradigms for Web Access Pattern Analysis

Xiaozhe Wang[1], Ajith Abraham[2] and Kate A. Smith[1]

[1] School of Business Systems, Faculty of Information Technology, Monash University, Clayton, Victoria 3800, Australia
{catherine.wang,kate.smith}@infotech.monash.edu.au
[2] Department of Computer Science, Oklahoma State University, 700 N Greenwood Avenue, Tulsa, OK 741060700, USA
ajith.abraham@ieee.org

Abstract. Web servers play a crucial role to convey knowledge and information to the end users. With the popularity of the WWW, discovering the hidden information about the users and usage or access pattern is critical to determine effective marketing strategies and to optimize the server usage or to accommodate future growth. Many of the currently available or conventional server analysis tools could provide only explicit statistical data without much useful knowledge and hidden information. Therefore, mining useful information becomes a challenging task when the Web traffic volume is enormous and keeps on growing. In this paper, we propose Soft Computing Paradigms (SCPs) to discover Web access or usage patterns from the available statistical data obtained from the Web server log files. Self Organising Map (SOM) is used to cluster the data before the data is fed to three popular SCPs including Takagi Sugeno Fuzzy Inference System (TSFIS), Artificial Neural Networks (ANNs) and Linear Genetic Programming (LGP) to develop accurate access pattern forecast models. The analysis was performed using the Web access log data obtained from the Monash University's central Web server, which receives over 7 million hits in a week. Empirical results clearly demonstrate that the proposed SCPs could predict the hourly and daily Web traffic volume and the developed TSFIS gave the overall best performance compares with other proposed paradigms.

Key words: Web Mining, Clustering, Self-Organising Map, Hybrid Systems, Fuzzy Logic, Neural Networks, Genetic Programming

15.1 Introduction and Motivation for Research

The World Wide Web is continuously growing with a rapid increase of in-formation transaction volume and number of requests from Web users around the world. To provide Web administrators with more meaningful information for improving the quality of Web information service performances, the discovering of hidden knowledge and information about Web users' access or usage patterns has become a

Xiaozhe Wang et al.: *Soft Computing Paradigms for Web Access Pattern Analysis*, Studies in Computational Intelligence (SCI) **4**, 233–250 (2005)
www.springerlink.com

necessity and a critical task. As such, this knowledge could be applied directly for marketing and management of e-business, e-services, e-searching, e-education and so on.

However, the statistical data available from normal Web log files or even the information provided by commercial Web trackers can only pro-vide explicit information due to the limitations of statistical methodology. Generally, Web information analysis relies on three general sets of information given a current focus of attention: (i) past usage patterns, (ii) degree of shared content and (iii) inter-memory associative link structures [15], which are associated with three subsets of Web mining: (i) Web usage mining, (ii) Web content mining and (iii) Web structure mining. The pattern discovery of Web usage mining consists of several steps including statistical analysis, clustering, classification and so on [16]. Most of the existing research is focused on finding patterns but relatively little effort has been made to discover more useful or hidden knowledge for detailed pattern analysis and predicting. Computational Web Intelligence (CWI) [20], a recently coined paradigm, is aimed at improving the quality of intelligence in Web technology [14].

In order to achieve accurate pattern discovery for further analysis and predicting tasks, we proposed a Self Organizing Map (SOM) [9] to cluster and discover patterns from the large data set obtained from Web log files. The clustered data were further used for different statistical analysis and discovering hidden relationships. To make the analysis more intelligent we also used the clustered data for predicting Web server daily and hourly traffic including request volume and page volume. By using three Soft Computing Paradigms (SCPs) [19] including the Takagi Sugeno Fuzzy Inference System (TSFIS) [17], Artificial Neural Networks (ANNs) [21] and Linear Genetic Programming (LGP) [4], we explored the prediction of average daily request volume in a week (1 to 5 days ahead) and the hourly page volume in a day (1, 12 and 24 hours ahead). Empirical results (analysis and prediction) clearly demonstrate that the proposed SCPs could predict the hourly and daily Web traffic based on the limited available usage transaction data and clustering analysis results.

We explored the Web user access patterns of Monash University's Web server located at http://www.monash.edu.au. We made use of the statistical data provided by "Analog" [3], a popular Web log analyzer that can generate statistical and textual data and information of different aspects of Web users' access log records, as well as weekly based reports include page requests traffic, types of files accessed, domain summary, operating system used, navigation summary and so on. We illustrate the typical Web traffic patterns of Monash University in Fig. 15.1 showing the daily and hourly traffic volume (number of requests and the volume of pages requested) for the week starting 14-Jul-2002, 00:13 to 20-Jul-2002, 12:22. For more user access data logs please refer Monash Server Usage Statistics [13].

In a week, the university's Website receives over 7 million hits. The log files cover different aspects of visitor details like domains, files accessed, requests of daily and hourly received, page requests, etc. Since the log data are collected and reported separately based on different features without interconnections/links, it is a real challenge to find hidden information or to extract usage patterns. Due to the enormous traffic volume and chaotic access behavior, the prediction of the user access patterns becomes more difficult and complex. The complexity of the data volume highlights the need for hybrid soft computing systems for information analysis and trend prediction.

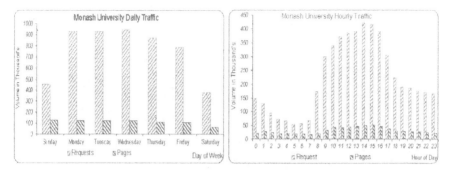

Fig. 15.1. Daily and hourly web traffic patterns of the Monash University main server

In Sects. 15.2 and 15.3, we present a hybrid framework comprising of a SOM to cluster and discover user access patterns. We also employ statistical analysis to mine more useful information using the Web Usage Data Analyzer (WUDA) [18]. In order to make the analysis more intelligent we also used the clustered data to predict the daily request volume and the hourly page volume, as explained in Sect. 15.4. We also pre-sent some basic theoretical background about the three SCPs including TSFIS, ANNs and LGP to predict the usage pattern trends comprising of daily request volume in a week (1 to 5 days ahead) and the hourly page volume in a day (1, 12 and 24 hours ahead). In Sect. 15.5 the experimentation results and comparative performance of the three different soft computing techniques are demonstrated to show the advantages and limitations of each model. Finally, in Sect. 15.6 some conclusions and directions future work are given.

15.2 Web Mining Framework using Hybrid Model

The hybrid framework combines SOM and SCPs operating in a concurrent environment, as illustrated in (Fig. 15.2). In a concurrent model, SOM assists the SCPs continuously to determine the required parameters, especially when certain input variables cannot be measured directly. Such combinations do not optimize the SCPs but only aids to improve the performance of the overall system. Learning takes place only in the SOM and the SCPs remain unchanged during this phase. The pre-processed data (after cleaning and scaling) is fed to the SOM to identify the data clusters. The clustering phase is based on SOM (an unsupervised learning algorithm), which can accept input objects described by their features and place them on a two-dimensional (2D) map in such a way that similar objects are placed close together. The clustered data is then used by WUDA for discovering different patterns and knowledge.

As shown in Fig. 15.3, data X, Y and Z may be segregated into three clusters according to the SOM algorithm. Data X is associated with Cluster 3 strongly only, but data Y and Z have weak associations with the other clusters. For example, data Y is associated with Cluster 2 but can also be considered to have a weak association with Cluster 3. And data Z is associated with both Clusters 2 and 3 even though

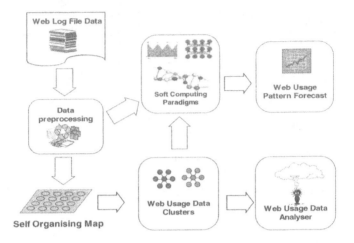

Fig. 15.2. Architecture of the hybrid Web mining model

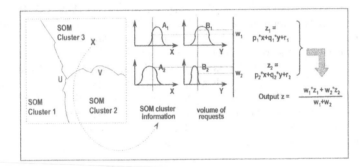

Fig. 15.3. Fuzzy association of data with SOM generated clusters

it is itself within Cluster 1. The degree of association of the data with a particular cluster is modeled as an additional input variable to predict the Web traffic patterns in our pro-posed hybrid system.

Soft computing was first proposed by Zadeh [19] to construct a new generation of computationally intelligent hybrid systems consisting of Neural Networks (NNs), Fuzzy Inference System (FIS), approximate reasoning and derivative free optimization techniques. In this paper, we performed a comparative study of TSFIS, ANNs (trained using backpropagation algorithms) and LGP to predict the hourly and daily Web traffic patterns.

15.3 Web Log Data Clustering and Experimental Analysis using WUDA

Web usage mining normally contains four processing stages including data collection, data preprocessing, pattern discovery and pattern analysis [6]. The data source selected for our case study is the Web traffic data generated by the "Analog" Web access log analyzer. It is common practice to embed Web trackers or Web log analysis tools to analyze log files for providing useful information to Web administrators. After browsing through some of the features of the best trackers available on the market, it is easy to conclude that other than generating basic statistical data they really cannot provide much meaningful information. In order to overcome the drawbacks of available Web log analyzers, the hybrid approach is proposed to discover hidden information and usage pattern trends which could aid Web managers to improve the management, performance and control of Web servers. In our approach, after selecting the required data from the large data source, all the log data were cleaned, formatted and scaled to feed into the SOM clustering algorithm. The SOM data clusters could be presented as 2D maps for each Web traffic feature, and WUDA was used for detailed user access and usage patterns analysis.

15.3.1 Data Pre-processing

We used the data from 01 January 2002 to 07 July 2002. Selecting useful data is an important task in the data pre-processing stage. After some preliminary analysis, we selected the statistical data comprising the traffic data on the domain on an hourly and daily basis, including request volume and page volume in each data type, to generate the cluster models for finding user access and server usage patterns. To build up a precise model and to obtain more accurate analysis, it is also important to remove irrelevant and noisy data as an initial step in the pre-processing task. Since SOM cannot process text data, any data in text format has to be encoded, according to a specific coding scheme, into numerical format. Further, the datasets were scaled 0–1. Besides the two inputs, "volume of requests (bytes)" and "volume of pages" directly from the original data set, we also included an additional input "time index" to distinguish the access time sequence of the data. The most recently accessed data were indexed higher while the least recently accessed data were placed at the bottom [2]. This is critical because Web usage data has time dependent characteristics.

15.3.2 Data Clustering using SOM

With the increasing popularity of the Internet, millions of requests (with different interests from different countries) are received by Web servers of large organizations. Monash University is a truly international university with its main campus located in Australia and campuses in South Africa and Malaysia. The university has plans to extend its educational services around the globe. Therefore, the huge traffic volume and the dynamic nature of the data require an efficient and intelligent Web mining framework.

In Web usage mining research, the method of clustering is broadly used in different projects by researchers for finding usage patterns or user pro-files. Among all the popular clustering algorithms, SOM has been success-fully used in Web mining projects [10, 12]. In our approach, using high dimensional input data, a 2D map of Web usage pat-terns with different clusters could be formed after the SOM training process. The related transaction entries are grouped into the same cluster and the relationship between different clusters is explicitly shown on the map. We used the Viscovery SOMine to simulate the SOM. All the records after the pre-processing stage were used by the SOM algorithm and the clustering results were obtained after the unsupervised learning. We adopted a trial and error approach by comparing the normalized distortion error and quantization error to decide the various parameter settings of the SOM algorithm. From all the experiments with different parameter settings, the best solution was selected when minimum errors were obtained.

15.3.3 WUDA to Find Domain Patterns

From the SOM clustering process, five clusters were mapped according to the user access from country of origin or domain. To analyze the difference among the clusters, we have illustrated the comparison of the unique country/domain and the averaged request volume for each cluster in Table 15.1.

Table 15.1. Analysis of request volume for domain cluster map

Cluster Number	Unique Country (or Domain)	Request Volume (Averaged)
1	160	2009.91
2	157	2325.18
3	162	3355.73
4	1 (.au)	199258.93
5	2 (.com & .net)	1995725.00

As evident from Table 15.1, Clusters 4 and 5 are distinguished from the rest of the clusters. Cluster 4 has almost 6 times of the volume of requests compared with the average request volume of the other 3 clusters (Clusters 1, 2 and 3), and Cluster 5 has the maximum number of requests which is nearly 10 times that of Cluster 4. However, by comparing the number of domain countries, Clusters 1, 2 and 3 all have around 150 different domain sources, Cluster 4 contains only Australian domains and Cluster 5 accounts only for *.com and *.net users. The majority of the requests originated from Australian domains followed by *.com and *.net users. This shows that even though Monash University's main server is accessed by users around the globe, the majority of the traffic originates within Australia.

To identify the interesting patterns from the data clusters of Clusters 1, 2 and 3, we have used a logarithmic scale in the Y-axis to plot the time index value in Fig. 15.4 for clarity. For Cluster 1 (marked with "◇"), Cluster 2 (marked with "□") and Cluster 3 (marked with "Δ"), the number of requests of each cluster are very similar and also shared by similar numbers of users from different countries which made it difficult to identify their difference depending on the volume of requests and pages. However, Clusters 1, 2 and 3 can be distinguished with reference to the time of access. So, Clusters 1, 2 and 3 have very similar patterns of access of requests, but their time of access is separated very clearly. Cluster 2 accounts for the most recent visitors and Cluster 3 represents the least recent visitors. Cluster 1 accounts for the users that were not covered by Clusters 2 and 3. Therefore, the different users were clustered based on the time of accessing the server and the volume of requests.

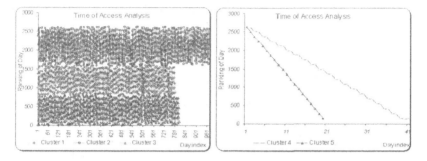

Fig. 15.4. WUDA for time of access in domain cluster map

15.3.4 WUDA to Analyze Hourly Web Traffic Request Patterns

The training process from SOM generated four clusters for the Web traffic requests on the hourly basis. The developed cluster map, indicating the hour of the day when the request was made, is illustrated in Fig. 15.5.

From the developed cluster map depicted in Fig. 15.5, it is very difficult to tell the difference between each cluster, as the requests according to the different hours in a day are scattered. But from Fig. 15.6, it may be concluded that Cluster 2 (marked with "◊") and Cluster 3 (marked with "□") have much higher requests of pages (nearly double) than Cluster 1 (marked with "Δ") and Cluster 4 (marked with "x"). This shows that 2 groups of clusters are separated based on the volume of requests for different hours of the day.

By analyzing the feature inputs of the SOM clusters Fig. 15.6, it is difficult to find more useful information. However, by looking at each hour, as shown in (Fig. 15.7), more meaningful information can be obtained. Clusters 2 and 3 are mainly responsible for the traffic during office hours (09:00–18:00), and Clusters 1 and 4 account for the traffic during the university off peak hours. It is interesting to note that the access patterns for each hour could be analyzed from the cluster results

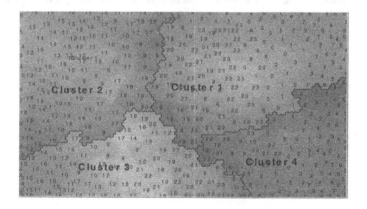

Fig. 15.5. Hourly Web traffic requests cluster map

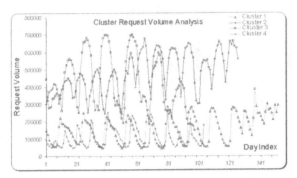

Fig. 15.6. WUDA for request volume in hourly cluster map

with reasonable classification features. By combining the information from Figs. 15.6 and 15.7, the hourly access patterns could be understood very clearly.

15.3.5 WUDA to Discover Daily Requests Clusters

Due to the dynamic nature of the Internet, it is difficult to understand the daily traffic pattern using conventional Web log analyzers. We attempted to cluster the data depending on the total activity for each day of the week using "request volume", "page volume" and "time index" as input features. The training process using SOM produced seven clusters and the developed 2D cluster map is shown in Fig. 15.8.

First, each cluster represents the traffic for only a certain access period by checking the "time index" inputs in each cluster records, and the ranking of the clusters are ordered as 2, 6, 1, 4, 3, 7 and 5 according to the descending order of the access time. In Table 15.2, WUDA reveals that the clusters are further separated according to the day of the week with interesting patterns. Clusters 3 and 6 account for access records which happened during the weekend (Saturday and Sunday). The big group consists of Clusters 1, 2, 4, 5 and 7, which account for the transactions with heavy traffic volume during normal working weekdays (Monday to Friday). With further

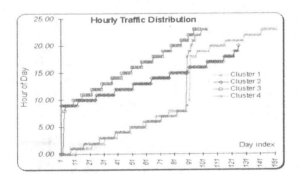

Fig. 15.7. WUDA for hour of day access in hourly cluster map

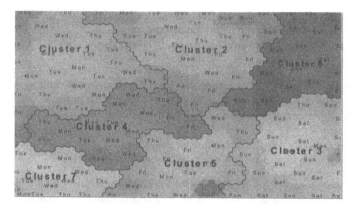

Fig. 15.8. Daily Web traffic requests cluster map

Table 15.2. WUDA for time of access (date) in daily traffic cluster map

	Cluster 1	Cluster 2	Cluster 3	Cluster 4	Cluster 5	Cluster 6	Cluster 7
Monday	19.23%	11.54%	4.17%	18.75%	12.50%	11.11%	28.57%
Tuesday	26.92%	15.38%	4.17%	18.75%	12.50%	0.00%	21.43%
Wednesday	23.08%	11.54%	0.00%	31.25%	18.75%	0.00%	21.43%
Thursday	19.23%	15.38%	4.17%	18.75%	18.75%	0.00%	28.57%
Friday	11.54%	30.77%	8.33%	12.50%	31.25%	0.00%	0.00%
Saturday	0.00%	3.85%	37.50%	0.00%	6.25%	50.00%	0.00%
Sunday	0.00%	11.54%	41.67%	0.00%	0.00%	38.89%	0.00%

detailed checking, Clusters 2 and 7 are different from other clusters because Cluster 2 covered heavier traffic on Friday but Cluster 7 missed Friday.

15.4 Soft Computing Paradigms

In contrast with conventional Artificial Intelligence techniques, which only deal with precision, certainty and rigor, the guiding principle of Soft Computing is to exploit the tolerance for imprecision, uncertainty, low solution cost, robustness and partial truth to achieve tractability and better rapport with reality [19]. In this research, we proposed 3 SCPs including TSFIS, ANNs and LGP for developing accurate predicting models for usage pattern trends based on the Website access log data.

15.4.1 Takagi Sugeno Fuzzy Inference Systems (TSFIS)

The world of information is surrounded by uncertainty and imprecision. The human reasoning process can handle inexact, uncertain and vague concepts in an appropriate manner. Usually, the human thinking, reasoning and perception process cannot be expressed precisely. These types of experiences can rarely be expressed or measured using statistical or probability theory. Fuzzy logic provides a framework to model uncertainty, human way of thinking, reasoning and the perception process. Fuzzy if-then rules and fuzzy reasoning are the backbone of FIS, which are the most important modeling tools based on fuzzy set theory. Fuzzy modeling can be pursued using the following steps:

- Select relevant input and output variables. Determine the number of linguistic terms associated with each input/output variables. Also choose the appropriate family of parameterized membership functions, fuzzy operators, reasoning mechanism, etc.
- Choose a specific type of fuzzy inference system.
- Design a collection of fuzzy if-then rules (knowledge base).

We made use of the Takagi Sugeno Fuzzy Inference Systems in which the conclusion of a fuzzy rule is constituted by a weighted linear combination of the crisp inputs rather than a fuzzy set (Sugeno 1985). A basic TSFIS if-then rule has the following structure, where p_1, q_1 and r_1 are linear parameters:

$$\text{if } x \text{ is } A_1 \text{ and } y \text{ is } B_1, \text{ then } z_1 = p_1 x + q_1 y + r_1 \tag{15.1}$$

A conventional FIS makes use of a model of the expert who is in a position to specify the most important properties of the process. Expert knowledge is often the main source to design FIS. According to the performance measure of the problem environment, the membership functions, rule bases and the inference mechanism are to be adapted. Evolutionary computation [1] and neural network learning techniques are used to adapt the various fuzzy parameters. Recently, a combination of evolutionary computation and neural network learning has also been investigated.

In this research, we used the Adaptive Neuro-Fuzzy Inference System (ANFIS) [11] framework based on neural network learning to fine tune the rule antecedent parameters and a least mean square estimation to adapt the rule consequent parameters of the TSFIS. A step in the learning procedure has two parts. In the first part the input patterns are propagated, and the optimal conclusion parameters are estimated by an iterative least mean square procedure, while the antecedent parameters (membership functions) are assumed to be fixed for the current cycle through the training set. In the second part the patterns are propagated again, and in this epoch, back propagation is used to modify the antecedent parameters, while the conclusion parameters remain fixed. Please refer to [11] for more details.

15.4.2 Artificial Neural Networks (ANNs)

ANNs were designed to mimic the characteristics of the biological neurons in the human brain and nervous system [21]. Learning typically occurs by example through training, where the training algorithm iteratively adjusts the connection weights (synapses). Back propagation (BP) is one of the most famous training algorithms for multilayer perceptrons. BP is a gradient descent technique to minimize the error E for a particular training pattern. For adjusting the weight (w_{ij}) from the ith input unit to the jth output, in the batched mode variant the descent is based on the gradient ∇E ($\frac{\delta E}{\delta w_{ij}}$) for the total training set:

$$\Delta w_{ij}(n) = -\varepsilon * \frac{\delta E}{\delta w_{ij}} + \alpha * \Delta w_{ij}(n-1) \tag{15.2}$$

The gradient gives the direction of error E. The parameters ε and α are the learning rate and momentum respectively.

15.4.3 Linear Genetic Programming (LGP)

LGP is a variant of the Genetic Programming (GP) technique that acts on linear genomes [5]. The LGP technique used for our current experiment is based on machine code level manipulation and evaluation of programs. Its main characteristic in comparison to tree-based GP is that the evolvable units are not the expressions of a functional programming language (like LISP), but the programs of an imperative language (like C). In the automatic induction of machine code by GP [5], individuals are manipulated directly as binary code in memory and executed directly without passing an interpreter during fitness calculation. The LGP tournament selection procedure puts the lowest selection pressure on the individuals by allowing only two individuals to participate in a tournament. A copy of the winner replaces the loser of each tournament. The crossover points only occur between instructions. Inside instructions the mutation operation randomly replaces either the instruction identifier, a variable or the constant from valid ranges. In LGP the maximum size of the program is usually restricted to prevent programs without bounds.

15.5 Daily and Hourly Traffic Patterns Prediction using SCPs

Besides the inputs "volume of requests" and "volume of pages" and "time index", we also used the "cluster location information" provided by the SOM output as an additional input variable. The data was re-indexed based on the cluster information. We attempted to develop SCPs based models to predict (a few time steps ahead) the Web traffic volume on an hourly and daily basis. We used the data from 17 February 2002 to 30 June 2002 for training and the data from 01 July 2002 to 06 July 2002 for testing and validation purposes. We also investigated the daily web traffic prediction performance without the "cluster information" input variable.

15.5.1 Takagi Sugeno Fuzzy Inference System (TSFIS)

We used the popular grid partitioning method (clustering) to generate the initial rule base. This partition strategy requires only a small number of membership functions for each input.

Daily Traffic Prediction

We used the MATLAB environment to simulate the various experiments. Given the daily traffic volume of a particular day the developed model could predict the traffic volume up to five days ahead. Three membership functions were assigned to each input variable. Eighty-one fuzzy if-then rules were generated using the grid based partitioning method and the rule antecedent/consequent parameters were learned after fifty epochs. We also investigated the daily web traffic prediction performance

Table 15.3. Training and test performance for web traffic volume prediction

	Root Mean Squared Error (RMSE)			
	FIS (With Cluster Input)		FIS (Without Cluster Input)	
Predicting Period	Training	Test	Training	Test
1 day	0.01766	0.04021	0.06548	0.09565
2 days	0.05374	0.07082	0.10465	0.13745
3 days	0.05264	0.06100	0.12941	0.14352
4 days	0.05740	0.06980	0.11768	0.13978
5 days	0.06950	0.07988	0.13453	0.14658

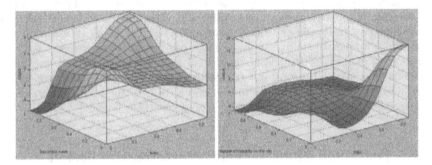

Fig. 15.9. Surface showing day of the week/index and number of requests/index

without the "cluster information" input variable. Table 15.3 summarizes the performance of the fuzzy inference system for training and test data, both with and without cluster information. Figure 15.9 illustrates the learned surface between the output and different input variable (the day of the week/index and number of requests/index).

Figure 15.10 depicts the test results for the prediction of daily Web traffic volume one day, two days, three days, four days and five days ahead.

Hourly Traffic Prediction

Three membership functions were assigned to each input variable. Eighty-one fuzzy if-then rules were generated using the grid based partitioning method and the rule antecedent/consequent parameters were learned after 40 epochs. We also investigated the volume of hourly page requested volume prediction performance without the "cluster information" input variable. Table 15.4 summarizes the performance of the FIS for training and test data.

Figure 15.11 illustrates the test results for 1 hour, 12 hours and 24 hours ahead prediction of the volume of hourly page volume.

From Tables 15.3 and 15.4 it is evident that the developed TSFIS could predict the patterns for several time steps ahead even though the models gave the most accurate results for 1 time step ahead. Hence our further study is focused on

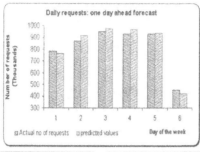

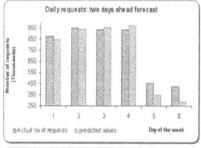

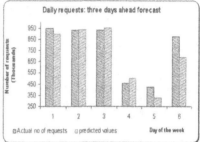

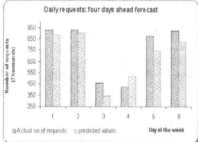

Fig. 15.10. Test results of daily prediction of Web traffic volume

Table 15.4. Training and test performance for hourly page volume prediction

| | Root Mean Squared Error (RMSE) | | | |
| | FIS (With Cluster Input) | | FIS (Without Cluster Input) | |
Predicting Period	Training	Test	Training	Test
1 hour	0.04334	0.04433	0.09678	0.10011
12 hours	0.06615	0.07662	0.11051	0.12212
24 hours	0.05743	0.06761	0.10891	0.11352

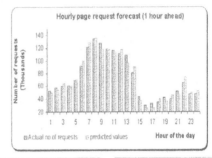

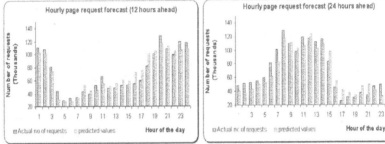

Fig. 15.11. Test results of hourly prediction of volume of page requested

developing ANN and LGP models to predict 1 time step ahead (hourly and daily access patterns).

15.5.2 Artificial Neural Networks (ANNs)

We used a feedforward NNs with 14 and 17 hidden neurons (single hidden layer) respectively for predicting the daily and hourly requests. The learning rate and momentum were set at 0.05 and 0.2 respectively and the network was trained for 30,000 epochs. The network parameters were decided after a trial and error approach. Meantime, the inputs without cluster information were also explored in the experiments. The obtained training and test results are depicted in the Table 15.5.

15.5.3 Linear Genetic Programming (LGP)

We used the "Discipulus" simulating workbench to develop the model using LGP. The settings of the various parameters are of the utmost importance for successful performance of the developed model. We used a population size of 500, 200000 tournaments, a crossover and mutation rate of 0.9 and a maximum program size of 256. As with the other paradigms, both with and without cluster information for the inputs were experimented on and the training and test errors are depicted in Table 15.5.

Table 15.5. Training and test performance of the 3 SCPs

		Predicting Period					
		Daily (1 day ahead)			Hourly (1 hour ahead)		
		RMSE			RMSE		
SCPs	Cluster Input	Train	Test	CC	Train	Test	CC
FIS	Without	0.0654	0.0956	0.9876	0.0654	0.0956	0.9768
FIS	With	0.0176	0.0402	0.9953	0.0433	0.0443	0.9841
ANN	Without	0.0541	0.0666	0.9678	0.0985	0.1030	0.8764
ANN	With	0.0345	0.0481	0.9292	0.0546	0.0639	0.9493
LGP	Without	0.0657	0.0778	0.0943	0.0698	0.0890	0.9561
LGP	With	0.0543	0.0749	0.9315	0.0654	0.0516	0.9446

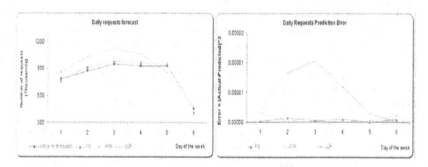

Fig. 15.12. One day ahead request prediction and error using SCPs

15.5.4 Experimentation Results Comparison with Three SCPs

Figure 15.12 illustrates the comparison of the performance of the different SCPs for 1 day ahead daily prediction of Web traffic volume and the error obtained using the test dataset. Figure 15.13 depicts the comparative performance of the SCPs' 1 hour ahead prediction of Web traffic volume and the error obtained using the test dataset.

15.6 Conclusions and Discussions

The discovery of useful knowledge, user information and access patterns allows Web based organizations to predict user access patterns and helps in future developments, maintenance planning and also to target more rigorous advertising campaigns aimed at groups of users [7]. Our analysis on Monash University's Web access patterns reveals the necessity to incorporate computational intelligence techniques for mining useful information. WUDA of the SOM data clusters provided useful information related to user access patterns. As illustrated in Tables 15.3, 15.4 and 15.5 all the three considered SCPs could easily approximate the daily and hourly Web access

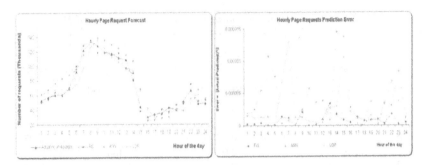

Fig. 15.13. One hour ahead request prediction and error using SCPs

trend patterns. Among the three SCPs, the developed FIS predicted the daily Web traffic and hourly page requests with the lowest RMSE on test set and with the best correlation coefficient. As discussed in [8], neuro-fuzzy hybrid systems for time series analysis and prediction can take advantages of fuzzy systems and neural networks to give excellent prediction accuracy. When compared to LGP, the developed NNs performed better (in terms of RMSE) for daily prediction but for hourly prediction LGP gave better results. Our experiment results also reveal the importance of the cluster information to improve the predict accuracy of the FIS. These techniques might be useful not only to Web administrators but also to Website tracker software vendors.

We relied on the numerical/text data provided by the Web log analyzer that generates statistical data by analyzing Web logs. Due to incomplete details, we had to analyze the usage patterns for different aspects separately, preventing us from linking some common information between the different aspects, trends, patterns etc. For example, the domain requests and the daily or hourly requests are all stand-alone information and are not interconnected. Therefore, a direct analysis from comprehensive Web logs that covers different interlinked features might be more helpful.

In this research, we considered only the Web traffic data during the university's peak working time. Our future research will also incorporate off-peak months (summer semesters) and other special situations such as unexpected events and server log technical failures. We also plan to incorporate more data mining techniques to improve the functional aspects of the concurrent neuro-fuzzy approach.

References

1. Abraham A, Nath B (2000) Evolutionary Design of Fuzzy Control Systems – An Hybrid Approach. In: Wang JL (ed) Proceedings of the Sixth International Conference on Control, Automation, Robotics and Vision, (CD ROM Proceeding), Singapore
2. Aggarwal C, Wolf JL, Yu PS (1999) Caching on the World Wide Web. IEEE Transaction on Knowledge and Data Engineering 11(1): 94–107
3. Analog (2002) Website Log Analyser. at URL: http://www.analog.cx

4. Banzhaf W, Nordin P, Keller RE, Francone FD (1998) Genetic Programming: An Introduction On The Automatic Evolution of Computer Programs and Its Applications. Morgan Kaufmann Publishers, Inc.
5. Brameier M, Banzhaf W (2001) A comparison of linear genetic programming and neural networks in medical data mining, Evolutionary Computation. IEEE Transactions on Evolutionary Computation, 5(1): 17–26
6. Chang G, Healey MJ, McHugh JAM, Wang JTL (2001) Web Mining. In: Mining the World Wide Web – An Information Search Approach. Kluwer Academic Publishers, pp. 93–104
7. Cooley R, Mobasher B, Srivastava J (1997) Web Mining: Information and Pattern Discovery on the World Wide Web. In: Proceedings of the 9th IEEE International Conference on Tools with Artificial Intelligence (ICTAI'97). Newport Beach, CA, pp. 558–567
8. Dote Y (2003) Intelligent Hybrid Systems for Nonlinear Time Series Analysis and Prediction Using Soft Computing. Invited Talk in the 3rd International Conference on Intelligent Systems, Design and Applications. Tulsa, USA.
9. Kohonen T (1990) The Self-Organizing Maps. In: Proceedings of the IEEE. Vol. 78, pp. 1464–1480
10. Honkela T, Kaski S, Lagus K, Kohonen T (1997) WEBSOM – Self Organizing Maps of Documents Collections. In: Proceedings of Workshop on Self-Organizing Maps (WSOM'97). Espoo, Finland, pp. 310–315
11. Jang R (1992) Neuro-Fuzzy Modeling: Architectures, Analyses and Applications, PhD Thesis, University of California, Berkeley, USA
12. Kohonen T, Kaski S, Lagus K, Salojrvi J, Honkela J, Paatero V, Saarela A (2000) Self Organization of a Massive Documents Collection. IEEE Transaction on Neural Networks 11(3): 574–585
13. Monash (2002) Server Usage Statistics, Monash University. Australia, at URL:http://www.monash.edu.au
14. Pal SK, Talwar V, Mitra P (2002) Web Mining in Soft Computing Framework: Relevance, State of the Art and Future Directions. IEEE Transaction on Neural Networks 13(5): 1163–1177
15. Pirolli P, Pitkow J, Rao R (1996) Silk from a Sow's Ear: Extracting Usable Structures from the Web. In: Proceedings of Conference on Human Factors in Computing Systems. Vancouver, British Columbia, Canada, pp. 118–125
16. Srivastava J, Cooley R, Deshpande M, Tan PN (2000) Web Usage Mining: Discovery and Applications of Usage Patterns from Web Data. SIGKDD Explorations 1(2): 12–23
17. Sugeno M (1985) Industrial Applications of Fuzzy Control. Elsevier Science Pub Co.
18. Wang X, Abraham A, Smith KA (2002) Web Traffic Mining Using a Concurrent Neuro-Fuzzy Approach. In: Computing Systems: Design, Management and Applications. Santiago, Chile, pp. 853–862
19. Zadeh LA (1998) Roles of Soft Computing and Fuzzy Logic in the Conception, Design and Deployment of Information/Intelligent Systems. In: Kaynak O, Zadeh LA, Turksen B, Rudas IJ (eds) Computational Intelligence: Soft Computing and Fuzzy-Neuro Integration with Applications. pp. 1–9

20. Zhang YQ, Lin TY (2002) Computational Web Intelligence: Synergy of Computational Intelligence and Web Technology. In: Proceedings of 2002 World Congress on Computational Intelligence, IEEE International Conference on Fuzzy Systems (FUZZ-IEEE'02). Honolulu, Hawaii, USA
21. Zurada JM (1992) Introduction to Artificial Neural Systems. West Publishing Company

Discovery of Fuzzy Multiple-Level Web Browsing Patterns

Shyue-Liang Wang[1], Wei-Shuo Lo[2], and Tzung-Pei Hong[3]

[1] Department of Computer Science, New York Institute of Technology, New York, USA
[2] Institute of Information Engineering, I-Shou University, Kaohsiung
[2] Department of Business Administration, Mei-Ho Institute of Technology, Pingtun
[3] Department of Electrical Engineering, National University of Kaohsiung Taiwan, R.O.C

Abstract. Web usage mining is the application of data mining techniques to discover usage patterns from web data. It can be used to better understand web usage and better serve the needs of rapidly growing web-based applications. Discovery of browsing patterns, page clusters, user clusters, association rules and usage statistics are some usage patterns in the web domain. Web mining of browsing patterns including simple sequential patterns and sequential patterns with browsing times has been studied recently. However, most of these works focus on mining browsing patterns of web pages directly. In this work, we introduce the problem of mining browsing patterns on multiple levels of a taxonomy comprised of web pages. The browsing time on each web page is used to analyze the retrieval behavior. Since the data collected are numeric, fuzzy concepts are used to process them and to form linguistic terms. A web usage-mining algorithm to discover multiple-level browsing patterns from linguistic data is thus proposed. Each page uses only the linguistic term with maximum cardinality in later mining processes, thus making the number of fuzzy regions to be processed the same as the number of pages. Computation time can thus be greatly reduced. In addition, the inclusion of concept hierarchy (taxonomy) of web pages produces browsing patterns of different granularity. This allows the views of users' browsing behavior from various levels of perspectives.

Key words: Web browsing patterns, Sequential patterns, Web mining, Concept hierarchy, Fuzzy concepts, Browsing behavior

16.1 Introduction

Web mining can be viewed as the use of data mining techniques to automatically retrieve, extract and evaluate information for knowledge discovery from web documents and services [28]. It has been studied extensively in recent years due to practical applications of extracting useful knowledge from inhomogeneous data sources on the World Wide Web. Web mining can be divided into three classes: web content

Shyue-Liang Wang et al.: *Discovery of Fuzzy Multiple-Level Web Browsing Patterns*, Studies in Computational Intelligence (SCI) 4, 251–266 (2005)
www.springerlink.com

mining, web structure mining and web usage mining [13]. Web content mining focuses on the discovery of useful information from web contents, data and documents. Web structure mining deals with mining the structure of hyperlinks within the web itself. Web usage mining emphasizes on the discovery of user access patterns from secondary data generated by users' interaction with the web.

In the past, several web mining approaches for finding user access patterns and user interesting information from the World Wide Web were proposed [7, 8, 11, 12, 33]. Chen and Sycara proposed the WebMate system to keep track of user interests from the contents of the web pages browsed. This can thus help users to easily search data from the Web [8]. Chen et al. mined path-traversal patterns by first finding the maximal forward references from log data and then obtaining large reference sequences according to the occurring numbers of the maximal forward references [7]. Cohen et al. sampled only portions of the server logs to extract user access patterns, which were then grouped as volumes [11]. Files in a volume could then be fetched together to increase the efficiency of a web server. Spliliopoulou et al. [33] proposed the Web Utilization Miner to discover interesting navigation patterns. The human expert dynamically specifies the interestingness criteria for navigation patterns, which could be statistical, structural and textual. To discover the navigation patterns satisfying the criteria, it exploits an innovative aggregated storage representation for the information in the web server log. The Web Site Information Filter System (WebSIFT) [35] is a web usage mining framework, that uses the content and structure information from a Web site, and finally identifies the interesting results from mining usage data. WebSIFT divides the web usage mining process into three principal parts that correspond to the three phases of usage mining: preprocessing, pattern discovery, and pattern analysis. The input of the mining process includes server logs (access, referrer, and agent), HTML files, and optional data. The preprocessing process constructs a user session file with the input data to derive a site topology and to classify the pages of a site. The user session file will be converted to the transaction file and output to the next phase. The pattern discovery process uses techniques such as statistics, association rules, clustering, sequential patterns to generate rules and patterns. In the pattern analysis process, the site topology and page classification are fed into the information filter, and the output of pattern discovery is analyzed by using procedural SQL and OLAP.

A Web browsing pattern is a kind of user access pattern that considers users' browsing sequences of web pages. In fact, it is similar to the discovery of sequential patterns from transaction databases. The problem of mining sequential patterns was introduced in [1]. Let $I = \{i_1, i_2, \ldots, i_m\}$ be a set of literals, called *items*. An *itemset* is a non-empty unordered set of items. A *sequence* is an ordered list of itemsets. An itemset i is denoted as $(i_1, i_2, \ldots, i_m)$, where i_j is an item. A sequence s is denoted as $(s_1 \rightarrow s_2 \rightarrow \ldots s_q)$, where s_j is an itemset. Given a database D of customer transactions, each transaction T consists of fields: customer-id, transaction-time, and the items purchased in the transaction. All the transactions of a customer can together be viewed as a sequence. This sequence is called a *customer-sequence*. A customer *supports* a sequence s if s is contained in the customer-sequence for this customer. The *support* for a sequence is the fraction of total customers who support this sequence. A sequence with support greater than a user-specified minimum support is called a large sequence. In a set of sequences, a sequence is maximal if it is not contained in any other sequences. The problem of finding sequential patterns is to find the maximal sequences among all sequences that have supports greater than a certain user-specified minimum support.

Many efficient algorithms for discovering maximal sequential patterns have been proposed [1, 6, 29, 31, 34, 37, 38, 41, 42]. In application to web browsing patterns, techniques for mining simple sequential browsing patterns and sequential patterns with browsing times have been proposed [8, 11, 12, 13, 20, 28, 33]. However, most of these works focus on mining browsing patterns of web pages directly. In this work, we introduce the problem of mining browsing patterns on multiple levels of a taxonomy comprised of web pages. In addition, browsing time is considered and processed using fuzzy set concepts to form linguistic terms. The proposed algorithm thus discovers multiple-level relevant browsing behavior from linguistic data and promotes the discovery of coarse granularity of web browsing patterns.

The rest of our paper is organized as follows. Related works are introduced in Sect. 16.2. Notation used in this paper is given in Sect. 16.3. Section 16.4 presents the mining algorithm of fuzzy multiple-level browsing patterns. Section 16.5 gives an example to illustrate the feasibility of the proposed algorithm. A conclusion is given at the end of the paper.

16.2 Related Works

It is useful to extract knowledge via data from the real world and to represent it in practical usage form. By using fuzzy sets, linguistic representation makes it easy to draw knowledge into efficient fuzzy rules, and easy to explain the knowledge to human beings. Several fuzzy learning algorithms for inducing rules from a given set of data have been designed and used for specific domains [2, 3, 5, 14, 15, 16, 17, 19, 21, 23]. Strategies based on decision trees were proposed in [9, 10, 32]. Wang et al. proposed a fuzzy version space learning strategy for managing vague information [36]. Hong et al. also proposed a fuzzy data mining approach, which integrated fuzzy-set concepts and conventional data mining algorithm to find association rules from quantitative data [18]. The approach consisted of three main steps.

Step 1: Transform each quantitative value in the transaction data into a fuzzy set using the given membership functions.

Step 2: Generate large itemsets by calculating the fuzzy cardinality of each candidate itemset.

Step 3: Induce fuzzy association rules from the large itemsets found in Step 2.

In addition, Chan and his co-workers used the fuzzy set theory and data mining technology to solve a classification problem [4]. Maddouri et al. [25] proposed a fuzzy incremental production rule induction method that could generate imprecise and uncertain IF-THEN rules from data records. Several fuzzy clustering methods [22], such as Fuzzy C-Means (FCM), have already been exploited in the context of fuzzy modeling. A new linguistic model, based on a characterization of both the structure of linguistic concepts and the uncertainty distribution in knowledge acquisition, was also presented in [24]. It used the concept of compatibility clouds, which integrated randomness and fuzziness, to capture the qualitative knowledge through a quantitative way.

In this paper, fuzzy set concepts will be used in our proposed algorithm to mine clients' multiple-level browsing behavior on a web site.

16.3 Notation

The following notation is used in our proposed algorithm:

n:	the total number of log data;
m:	the total number of files in the log data;
c:	the total number of clients in the log data;
n_i:	the number of log data from the i-th client, $1 \leq i \leq c$;
D_i:	the browsing sequence of the i-th client, $1 \leq i \leq c$;
D_{id}:	the d-th log transaction in D_i, $1 \leq d \leq n_i$;
I^g:	the g-th file, $1 \leq g \leq m$;
R^{gk}:	the k-th fuzzy region of I^g, $1 \leq k \leq |I^g|$, where $|I^g|$ is the number of fuzzy regions for I^g;
v_{id}^g:	the browsing duration of file I^g in D_{id};
f_{id}^g:	the fuzzy set converted from v_{id}^g;
f_{id}^{gk}	the *membership* value of v_{id}^g in region R^{gk};
f_i^{gk}:	the *membership* value of region R^{gk} in the i-th client sequence D_i;
$count^{gk}$:	the scalar cardinality of region R^{gk};
$max\text{-}count^g$:	the maximum count value among $count^{gk}$ values;
$max\text{-}R^g$:	the fuzzy region of file I^g with $max\text{-}count^g$;
α:	the predefined minimum support value;
λ:	the predefined minimum confidence value;
C_r:	the set of candidate sequences with r files;
L_r:	the set of large sequences with r files.

16.4 Fuzzy Multiple-Level Web Browsing Patterns

This section describes the proposed data mining algorithm of fuzzy multiple-level web browsing patterns. The log data are first extracted, sorted, and reorganized into users' browsing sequences. The browsing time of each web page is then transformed into linguistic terms using fuzzy sets. Based on the Apriori All sequential mining approach [1], browsing patterns are discovered in a top-down level-wise searching manner. Browsing patterns with no ancestor rules or patterns with interest support greater than the predefined threshold will be output as interesting web browsing patterns. The detail of the proposed web mining algorithm is described as follows.

INPUT: A server log, a predefined taxonomy of web pages, a set of membership functions, a predefined minimum support value α, and a predefined interest support threshold R.

OUTPUT: A set of interesting fuzzy multiple-level browsing patterns.

Step 1:   Select the web pages with file names including .asp, .htm, .html, .jva .cgi and closing connection from the log data; keep only the fields *date, time, client-ip* and *file-name*. Denote the resulting log data as D.

Step 2:   Encode each web page file name using a sequence of number and the symbol "*", with the t-th number representing the branch number of a certain web page on level t.

Step 3: Form a browsing sequence D_j for each client c_j by sequentially listing his/her n_j tuples (web page, duration), where n_j is the number of web page browsed by client c_j. Denote the d-th tuple in D_j as D_{jd}.

Step 4: Set $k = 1$, where k is used to store the level number being processed.

Step 5: Re-encode the web page file names by retaining the first k digits and replacing the rest of digits by "*" in each browsing sequence.

Step 6: Transform the time duration v_{id}^g of the file name I^g appearing in D_{id} into a fuzzy set f_{id}^g represented as

$$\left(\frac{f_{id}^{g_1}}{R^{g_1}} + \frac{f_{id}^{g_2}}{R^{g_2}} + \cdots + \frac{f_{id}^{g_l}}{R^{g_l}} \right)$$

using the given membership functions, where I^g is the g-th file name, R^{g_k} is the k-th fuzzy region of item I^g, $f_{id}^{g_k}$ is v_{id}^g's fuzzy membership value in region R^{g_k}, and l is the number of fuzzy regions for I^g.

Step 7: Find the membership value $f_i^{g_k}$ of each region R^{g_k} in each browsing sequence D_i as

$$f_i^{g_k} = \underset{d=1}{\overset{|D_i|}{\text{MAX}}} \, f_{id}^{g_k},$$

where $|D_i|$ is the number of tuples in D_i.

Step 8: Calculate the scalar cardinality of each region R^{g_k} as:

$$count^{g_k} = \sum_{i=1}^{c} f_i^{g_k},$$

where c is the number of browsing sequences.

Step 9: Find max-$count^g = \underset{k=1}{\overset{l}{\text{MAX}}} \left(count^{g_k} \right)$, where $1 \leq g \leq m, m$ is the number of files in the log data, and l is the number of regions for file I^g. Let max-R^g be the region with max-$count^g$ for file I^g. $max - R^g$ will be used to represent the fuzzy characteristic of file I^g in later mining processes.

Step 10: Check whether the value max-$count^g$ of a region $max - R^g$, $g = 1$ to m, is larger than or equal to the predefined minimum support value α. If a region $max - R^g$ is equal to or greater than α, put it in the set of large 1-sequences (L_1). That is,

$$L_1 = \{max - R^g \,|max - count^g \geq \alpha, \; 1 \leq g \leq m\}$$

Step 11: If L_1 is null, then exit the algorithm; otherwise, do the next step.

Step 12: Set $r = 1$ where r is used to represent the length of sequential patterns currently kept.

Step 13: Generate the candidate sequence C_{r+1} from L_r in a way similar to that in the AprioriAll algorithm [1]. Restated, the algorithm first joins L_r and L_r, under the condition that $r - 1$ items in the two itemsets are the same and with the same orders. Different permutations represent different candidates. The algorithm then keeps in C_{r+1} the sequences which have all their sub-sequences of length r existing in L_r.

Step 14: Do the following substeps for each newly formed $(r + 1)$-sequence s with contents $(s_1, s_2, \ldots, s_{r+1})$ in C_{r+1}:

(a) Calculate the fuzzy value f_i^s of s in each browsing sequence D_i as:

$$f_i^s = \underset{k=1}{\overset{r+1}{\text{Min}}} f_i^{s_k},$$

where region s_k must appear after region s_{k-1} in D_i. If two or more same subsequences exist in D_i, then f_i^s is the maximum fuzzy value among those of these subsequences.

(b) Calculate the scalar cardinality of s as:

$$count^s = \sum_{i=1}^{c} f_i^s,$$

where c is number of browsing sequences.

(c) If $count^s$ is larger than or equal to the predefined minimum support value α, put s in L_{r+1}.

Step 15: If L_{r+1} is null, then do the next step; otherwise, set $r = r + 1$ and repeat Step 13 to 15.

Step 16: Set $k = k + 1$. If the new level k exists, repeat Step 5 through Step 15 for browsing patterns in each level.

Step 17: Output the browsing patterns without ancestor patterns (by replacing the web pages in a pattern with their ancestors in the taxonomy) to users as interesting patterns.

Step 18: For each remaining pattern s (representing $s_1 \rightarrow s_2 \rightarrow \dots \rightarrow s_r \rightarrow s_{(r+1)}$), find the closest ancestor t (representing $t_1 \rightarrow t_2 \rightarrow \dots \rightarrow t_r \rightarrow t_{(r+1)}$), and calculate the support interest measure $I_{\text{support}}(s)$ of s as:

$$I_{\text{support}}(s) = \frac{count_s}{\frac{\prod_{k=1}^{r+1} count_{Sk}}{\prod_{k=1}^{r+1} count_{lk}} \times count_t}$$

Output the patterns with their support interest measure larger than or equal to the predefined interest threshold R to users as interesting patterns.

16.5 An Example

In this section, we describe an example to demonstrate the proposed mining algorithm of fuzzy multiple-level web browsing patterns. The example shows how the proposed algorithm can be used to discover fuzzy sequential patterns from the web browsing log data shown in Table 16.1. The membership functions for the browsing duration on a web page are shown in Fig. 16.1. The browsing duration is divided into three fuzzy regions: Short, Middle, and Long. In addition, the predefined taxonomy for web pages are shown in Fig. 16.2. The predefined minimum support α and interest support threshold R are set at 2 and 1.5 respectively. The proposed data-mining algorithm proceeds as follows.

Step 1: Select the web pages with file names including .asp, .htm, .html, .jva .cgi and closing connection from Table 16.1. Keep only the fields *date*, *time*, *client-ip* and *file-name*. Denote the resulting log data as Table 16.2.

Table 16.1. A part of log data used in the example

Date	Client-IP	Server-IP	Server-port	File-name
2002-06-01	140.117.72.1	140.117.72.88	11	News.htm
2002-06-01	140.117.72.1	140.117.72.88	11	Exective.htm
2002-06-01	140.117.72.1	140.117.72.88	11	Stock.htm
....	....	....	....	....
2002-06-01	140.117.72.1	140.117.72.88	11	Univesity.htm
....	....	....	....	....
2002-06-01	140.117.72.2	140.117.72.88	11	Golf.htm
2002-06-01	140.117.72.2	140.117.72.88	11	Stock.htm
....	....	....	....	....
2002-06-01	140.117.72.3	140.117.72.88	11	Golf.htm
....	....	....	....	....
2002-06-01	140.117.72.4	140.117.72.88	11	Closing connection
....	....	....	....	....
2002-06-01	140.117.72.5	140.117.72.88	11	Golf.htm
....	....	....	....	....
2002-06-01	140.117.72.6	140.117.72.88	11	Stock.htm
....	....	....	....	....
2002-06-01	140.117.72.6	140.117.72.88	11	Closing connection

Step 2: Each file name is encoded using the predefined taxonomy shown in (Fig.
16.2). Results are show in Table 16.3.

Step 3: The web pages browsed by each client are listed as a browsing sequence.
Each tuple is represented as (web page, duration), as shown in Table 16.4.
The resulting browsing sequences from Table 16.4 are shown in Table 16.5.

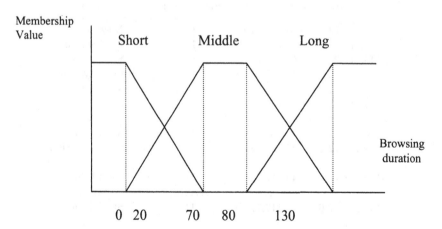

Fig. 16.1. The membership functions used in this example

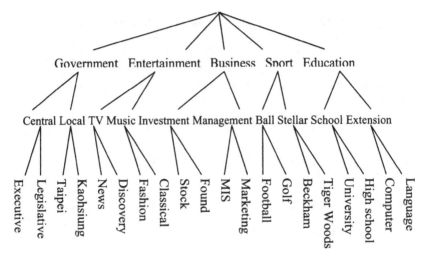

Fig. 16.2. The predefined taxonomy used in this example

Step 4: k is initially set at 1, where k is used to store the level number being processed.

Step 5: The re-encoded browsing sequences for level k are shown in Table 16.6.

Step 6: The time duration of each file in each browsing sequence is represented as fuzzy set. Take the web page 2^{**} in the first browsing sequence as an example. The time duration "30" of file 2^{**} is converted into the fuzzy set (0.8/Short + 0.2/Middle + 0/High) by the given membership functions

Table 16.2. The resulting log data for web mining

Client-IP	Server-IP	Server-port	File-name
140.117.72.1	140.117.72.88	11	News.htm
140.117.72.1	140.117.72.88	11	Exective.htm
140.117.72.1	140.117.72.88	11	Stock.htm
140.117.72.1	140.117.72.88	11	University.htm
140.117.72.2	140.117.72.88	11	Golf.htm
140.117.72.2	140.117.72.88	11	Stock.htm
....	....	....	....
140.117.72.3	140.117.72.88	11	Golf.htm
140.117.72.4	140.117.72.88	11	Closing connection
....	....	....	....
140.117.72.5	140.117.72.88	11	Golf.htm
....	....	....	....
140.117.72.6	140.117.72.88	11	Stock.htm
....	....	....	....
140.117.72.6	140.117.72.88	11	Closing connection

(Fig. 16.1). This transformation is repeated for the other files and browsing sequences. The results are shown in Table 16.7.

Step 7: The membership value of each region in each browsing sequence is found. Take the region 4^{**}.Middle for client 2 as an example. Its membership vale is max $(0.8, 0.0, 0.6) = 0.8$. The membership values of the other regions can be calculated similarly.

Step 8: The scalar cardinality of each fuzzy region in all the browsing sequences is calculated as the count value. Take the fuzzy region 4^{**}.Middle as an example. Its scalar cardinality $= (0.6 + 0.8 + 0.8 + 0.0 + 1 + 1) = 4.2$. This step is repeated for the other regions, and the results are shown in Table 16.8.

Step 9: The fuzzy region with the highest count among the three possible regions for each file is selected. Take file 1^{**} as an example. Its count is 0.0 for "Short", 0.8 for "Middle", and 0.2 for "Long", since the count for Middle is the highest among the three counts, the region Middle is thus used to represent the file 1^{**} in later mining process. This step is repeated for the other files. Thus, "Short" is chosen for 2^{**}, "Middle" is chosen for 1^{**}, 3^{**}, 4^{**}, and "Long" is chosen for 5^{**}.

Step 10: The counts of the regions selected in Step 9 are checked against the predefined minimum support value α. Assuming that α is set at 2 in this example. Since the count value of 2^{**}.Short, 3^{**}.Middle, 4^{**}.Middle are larger than 2, these regions are put in L_1.

Step 11: Since L_1 is not null, the next step is calculated.

Step 12: Set $r = 1$, where r is used to represent the length of sequential patterns currently kept.

Step 13: The candidate 2-sequence C_2 is generated from L_1 as follows: (2^{**}.Short, 2^{**}.Short), (2^{**}.Short, 3^{**}.Middle), (3^{**}.Middle, 2^{**}.Short),...., (3^{**}.

Table 16.3. Codes of file names

Code	File Name	Code	File Name
111	Executive.htm	1^{**}	Government.htm
211	News.htm	2^{**}	Entertainment.htm
212	Discovery.htm	3^{**}	Business.htm
313	Stock.htm	4^{**}	Sport.htm
321	Found.htm	5^{**}	Education.htm
411	Football.htm	11^*	Central.htm
412	Golf.htm	12^*	Local.htm
421	Tiger Woods.htm	21^*	TV.htm
511	University.htm	22^*	Music.htm
512	Highshool.htm	31^*	Investment.htm
		32^*	Management.htm
		41^*	Ball.htm
		42^*	Stellar.htm
		51^*	School.htm
		52^*	Extension.htm

Table 16.4. The web pages browsed with their duration

Client-ID	(Web page, duration)
1	(211, 30)
1	(511, 42)
1	(412, 98)
1	(313, 91)
2	(412, 62)
2	(211, 31)
2	(421,102)
3	(111, 92)
3	(411, 89)
4	(211, 20)
4	(313, 101)
4	(512, 118)
4	(212, 11)
4	(321, 42)
5	(412, 64)
5	(211, 29)
5	(313, 74)
6	(421, 80)
6	(313, 61)
6	(511, 122)
6	(212, 17)

Table 16.5. The browsing sequences formed from Table 16.4

Client ID	Browsing Sequences
1	(211, 30), (511, 42), (412, 98), (313, 91)
2	(412, 62), (211, 31), (421, 102)
3	(111, 92), (411, 89)
4	(412, 64), (211, 29), (512, 118), (212, 11), (321, 42)
5	(412, 64), (211, 29), (313, 74)
6	(421, 80), (313, 61), (511, 122), (212, 17)

Table 16.6. The re-encoded browsing sequences for level 1

Client ID	Browsing Sequences
1	$(2^{**}, 30), (5^{**}, 42), (4^{**}, 98), (3^{**}, 91)$
2	$(4^{**}, 62), (2^{**}, 31), (4^{**}, 102)$
3	$(1^{**}, 92), (4^{**}, 89)$
4	$(4^{**}, 64), (2^{**}, 29), (5^{**}, 118), (2^{**}, 11), (3^{**}, 42)$
5	$(4^{**}, 64), (2^{**}, 29), (3^{**}, 74)$
6	$(4^{**}, 80), (3^{**}, 61), (5^{**}, 122), (2^{**}, 17)$

Table 16.7. The fuzzy sets transformed from the browsing sequences

CID	Fuzzy Set
1	$\left(\frac{0.8}{2^{**}.Short} + \frac{0.2}{2^{**}.Middle}\right) \left(\frac{0.6}{5^{**}.Short} + \frac{0.4}{5^{**}.Middle}\right)$
	$\left(\frac{0.6}{4^{**}.Middle} + \frac{0.4}{4^{**}.Long}\right) \left(\frac{0.8}{3^{**}.Middle} + \frac{0.2}{3^{**}.Long}\right)$
2	$\left(\frac{0.2}{4^{**}.Short} + \frac{0.8}{4^{**}.Middle}\right) \left(\frac{0.8}{2^{**}.Short} + \frac{0.2}{2^{**}.Middle}\right)$
	$\left(\frac{0.6}{4^{**}.Middle} + \frac{0.4}{4^{**}.Long}\right)$
3	$\left(\frac{0.8}{1^{**}.Middle} + \frac{0.2}{1^{**}.Long}\right) \left(\frac{0.8}{4^{**}.Middle} + \frac{0.2}{4^{**}.Long}\right)$
4	$\left(\frac{1}{2^{**}.Short}\right) \left(\frac{0.6}{3^{**}.Middle} + \frac{0.4}{3^{**}.Long}\right)$
	$\left(\frac{0.2}{5^{**}.Middle} + \frac{0.8}{5^{**}.Long}\right) \left(\frac{1}{2^{**}.Short}\right) \left(\frac{0.6}{3^{**}.Short} + \frac{0.4}{3^{**}.Middle}\right)$
5	$\left(\frac{1}{4^{**}.Middle}\right) \left(\frac{0.8}{2^{**}.Short} + \frac{0.2}{2^{**}.Middle}\right) \left(\frac{1}{3^{**}.Middle}\right)$
6	$\left(\frac{1}{4^{**}.Middle}\right) \left(\frac{0.2}{3^{**}.Short} + \frac{0.8}{3^{**}.Middle}\right)$
	$\left(\frac{0.2}{5^{**}.Middle} + \frac{0.8}{5^{**}.Long}\right) \left(\frac{1}{2^{**}.Short}\right)$

Table 16.8. The counts of fuzzy regions for level $k = 1$

Item	Region	Cid-1	Cid-2	Cid-3	Cid-4	Cid-5	Cid-6	Total Count
1**	Short							0.0
	Middle			0.8				0.8
	Long			0.2				0.2
2**	Short	0.8	0.8		1	0.8	1	4.4
	Middle	0.2	0.2			0.2		0.6
	Long							0.0
3**	Short				0.6		0.2	0.8
	Middle	0.8			0.6		0.8	3.2
	Long	0.2			0.4			0.6
4**	Short		0.2					0.2
	Middle	0.6	0.8	0.8		1	1	4.2
	Long	0.4	0.4	0.2				1
5**	Short	0.6						0.6
	Middle	0.4			0.2		0.2	0.8
	Long				0.8		0.8	1.6

Middle, 4**. Middle), (4**. Middle, 3**.Middle), (4**. Middle, 4**. Middle). The results are shown in Table 16.9.

Step 14: The following substeps are done for each newly formed candidate 2-sequence in C_2:

(a) The fuzzy membership value of each candidate 2-sequence in each browsing sequence is calculated. Here, the minimum operator is used for the intersection. Take the sequence (2**.Short, 3**.Middle) as an example. Its membership value in the fourth browsing sequence is cal-

Table 16.9. The candidate 2-sequence C_2 in this example

	2**.Short	3**.Middle	4**.Middle
2**.Short	(2**.Short), (2**.Short)	(2**.Short), (3**.Middle)	(2**.Short) (4**.Middle)
3**.Middle	(3**.Middle), (2**.Short)	(3**.Middle), (3**.Middle),	(3**.Middle) (4**.Middle)
4**.Middle	(4**.Middle), (2**.Short)	(4**.Middle), (3**.Middle)	(4**.Middle) (4**.Middle),

culated as: max[min(1.0, 0.6), min(1.0, 0.4)] = 0.6. There are two sub-sequences of (2**.Short, 3**.Middle) in that browsing sequence. The results for sequence (2**.Short, 3**.Middle) in all the browsing sequences are shown in Table 16.10.

(b) The scalar cardinality (count) of each candidate 2-sequence in C_2 is calculated. Results for this example are shown in Table 16.11.

(c) Since only the counts of 2-sequences (2**.Short, 3**.Middle), (4**. Middle, 2**.Short) and (4**.Middle, 3**.Middle) are larger than the predefined minimum support value 2, they are thus kept in L_2.

Table 16.10. The membership values for sequence (2**.Short, 3**.Middle)

C-ID	Membership Value
1	0.8
2	0.0
3	0.0
4	0.6
5	0.8
6	0.0

Table 16.11. The fuzzy counts of the candidate 2-sequence in C_2

	C-ID						Total Count
	1	2	3	4	5	6	
(2**.Short), (2**.Short)				1			1
(2**.Short), (3**.Middle)	0.8			0.6	0.8		2.2
(2**.Short), (4**.Middle)	0.6	0.6					1.2
(3**.Middle), (2**.Short)				0.6		0.8	1.4
(3**.Middle), (3**.Middle)				0.4			0.4
(3**.Middle), (4**.Middle)							0
(4**.Middle), (2**.Short)		0.8			0.8	1	2.6
(4**.Middle), (3**.Middle)	0.6				1	0.8	2.4
(4**.Middle), (4**.Middle)		0.6					0.6

Step 15: Since L_2 is not null, set $r = r+1 = 2$. Steps 13–15 are then repeated to find L_3. C_3 is first generated from L_2, and the sequence (4**.Middle, 2**.Short, 3**.Middle) is generated. Since its count is 0.8, smaller than 2.0, it is thus not put in L_3. L_3 is an empty set.

Step 16: Set $k = k + 1 = 2$ as new level and repeat Step 5 through Step 15 to find browsing patterns in this level. Repeat this process until no new level is found.

Step 17: The browsing patterns discovered through Step 16 are as follows:

2**.Short→3**.Middle
4**.Middle→2**.Short
4**.Middle→3**.Middle
21*.Short→31*.Middle
211.Short→313.Middle

However, only the first three browsing patterns will be output, as they do not have ancestor patterns.

Step 18: For the fourth and fifth browsing patterns in Step 17, their support interest measures are:

$$I_{support}(21*.Short \rightarrow 31*.Middle)$$
$$= \frac{21*.Short \cap 31*.Middle}{\frac{21*.Short}{2**.Short} \times \frac{31*.Middle}{3**.Middle} \times 21*.Short \cap 31*Middle}$$
$$I_{support}(211.Short \rightarrow 313.Middle)$$
$$= \frac{211.Short \cap 313.Middle}{\frac{211.Short}{21*.Short} \times \frac{313.Middle}{31*.Middle} \times 211.Short \cap 313Middle}$$

These values are smaller than the predefined interest support threshold 1.5. They are not considered as interesting browsing patterns. In this example, the three browsing patterns (2**.Short→3**.Middle), (4**.Middle→2**.Short), (4**.Middle→ 3**.Middle) are output as meta-knowledge concerning the given log data.

16.6 Conclusion

In this work, we have proposed a novel web mining algorithm that can process web server logs to discover fuzzy multiple-level web browsing patterns. The duration time of a web page is considered and processed using fuzzy set concepts to form linguistic terms. The adoption of linguistic terms to express the discovered patterns is more natural and understandable for human beings. In addition, the inclusion of concept hierarchy (taxonomy) of web pages produces browsing patterns of different granularity. This allows the views of users' browsing behavior from various levels and perspectives.

Although the proposed method works well in fuzzy web mining from log data, it is just a beginning. There is still much work to be done in this field. Our method assumes that the membership functions are known in advance. In [16, 17, 19], we also proposed some fuzzy learning methods to automatically derive the membership functions. In the future, we will attempt to dynamically adjust the membership functions in the proposed web mining algorithm to avoid inappropriate choice of membership functions. We will also attempt to design specific web mining models for various problem domains.

References

1. R. Agrawal and R. Srikant (1995). "Mining Sequential Patterns", Proc. of the 11th International Conference on Data Engineering, pp. 3–14.
2. F. Blishun (1987) Fuzzy learning models in expert systems. Fuzzy Sets and Systems, 22, pp. 57–70.
3. L.M. de Campos and S. Moral (1993) Learning rules for a fuzzy inference model. Fuzzy Sets and Systems, 59, pp. 247–257.
4. K.C.C. Chan and W.H. Au (1997) Mining fuzzy association rules. The Sixth ACM International Conference on Information and Knowledge Management, pp. 209–215.
5. R.L.P. Chang and T. Pavliddis (1977) Fuzzy decision tree algorithms. IEEE Transactions on Systems, Man and Cybernetics, 7, pp. 28–35.
6. N. Chen, A. Chen (1999) Discovery of Multiple-Level Sequential Patterns from Large Database. Proc. of the International Symposium on Future Software Technology. Nanjing, China, pp. 169–174.
7. M.S. Chen, J.S. Park and P.S. Yu (1998) Efficient Data Mining for Path Traversal Patterns. IEEE Transactions on Knowledge and Data Engineering, 10, pp. 209–221.
8. L. Chen, K. Sycara (1998) WebMate: A Personal Agent for Browsing and Searching. The Second International Conference on Autonomous Agents, ACM.
9. C. Clair, C. Liu and N. Pissinou (1998) Attribute weighting: a method of applying domain knowledge in the decision tree process. The Seventh International Conference on Information and Knowledge Management, pp. 259–266.
10. P. Clark and T. Niblett (1989) The CN2 induction algorithm. Machine Learning, 3, pp. 261–283.
11. E. Cohen, B. Krishnamurthy and J. Rexford (1999) Efficient Algorithms for Predicting Requests to Web Servers. The Eighteenth IEEE Annual Joint Conference on Computer and Communications Societies, 1, pp. 284–293.
12. R. Cooley, B. Mobasher and J. Srivastava (1997) Grouping Web Page References into Transactions for Mining World Wide Web Browsing Patterns. Knowledge and Data Engineering Exchange Workshop, pp. 2–9.
13. R. Cosala, H. Blockleel (2000) Web Mining Research: A Survey. ACM SIGKDD, Vol. 2(1), pp. 1–15.
14. M. Delgado and A. Gonzalez (1993) An inductive learning procedure to identify fuzzy systems. Fuzzy Sets and Systems, 55, pp. 121–132.
15. Gonzalez (1995) A learning methodology in uncertain and imprecise environments. International Journal of Intelligent Systems, 10, pp. 57–371.
16. T.P. Hong and J.B. Chen (1999) Finding relevant attributes and membership functions. Fuzzy Sets and Systems, 103(3), pp. 389–404.
17. T.P. Hong and J.B. Chen (2000) Processing individual fuzzy attributes for fuzzy rule induction. Fuzzy Sets and Systems, 112(1), pp. 127–1400.
18. T.P. Hong, C.S. Kuo and S.C. Chi (1999) A data mining algorithm for transaction data with quantitative values. Intelligent Data Analysis, 3(5), pp. 363–376.
19. T.P. Hong and C.Y. Lee (1996) Induction of fuzzy rules and membership functions from training examples. Fuzzy Sets and Systems, 84, pp. 33–47.
20. T.P. Hong, K.Y. Lin, and S.L. Wang (2001) Web Mining for Browsing Patterns. Proc. of the 5th International Conference on Knowledge-based Intelligent Information Engineering Systems. Osaka, Japan, September, pp. 495–499.

21. T.P. Hong and S.S. Tseng (1997) A generalized version space learning algorithm for noisy and uncertain data, IEEE Transactions on Knowledge and Data Engineering, 9(2), pp. 336–340.
22. F. Hoppner, F. Klawonn, R. Kruse and T. Runkler (1999) Fuzzy Cluster Analysis. John Wiley & Sons.
23. R.H. Hou, T.P. Hong, S.S. Tseng and S.Y. Kuo (1997) A new probabilistic induction method. Journal of Automatic Reasoning, 18, pp. 5–24.
24. D. Li, J. Han, X. Shi and M.C. Chan (1998) Knowledge representation and discovery based on linguistic atoms. Knowledge-Based Systems, 10, pp. 431–440.
25. M. Maddouri, S. Elloumi and A. Jaoua, (1998) An incremental learning system for imprecise and uncertain knowledge discovery. Journal of Information Sciences, 109, pp. 149–164.
26. H. Mannila and H. Toivonen (1996) Discovering Generalized Episodes Using Minimal Occurrences. Proc. of the 2nd International Conference on Knowledge Discovery and Data Mining. pp. 146–151.
27. T. Oates, et al. (1997) A Family of Algorithms for Finding Temporal Structure in Data. Proc. of the 6th International Workshop on AI and Statistics, Mar 1997, pp. 371–378.
28. S.K. Pal, V. Talwar, and P. Mitra (2002) Web Mining in Soft Computing Framework: Relevance, State of the Art and Future Directions. IEEE Transactions on Neural Network, 13(5), pp. 1163–1177.
29. J. Pei, J.W. Han, B. Mortazavi-Asl, H. Pinto, Q. Chen, U. Dayal and M.C. Hsu (2001) Prefixspan: Mining Sequential Patterns by Prefix-Projected Growth. Proc. of the 17th IEEE International Conference on Data Engineering. Heidelberg, Germany, April.
30. G. Piategsky-Shapiro (1991) Discovery, Analysis and Presentation of Strong Rules. Knowledge Discovery in Databases. AAAI/MIT press, pp. 229–248.
31. H. Pinto (2001) Multiple-Dimensional Sequential Patterns Mining. Master Thesis, University of Lethebridge, Alberta, Canada,.
32. J.R. Quinlan (1993) C4.5: Programs for Machine Learning, Morgan Kaufmann. San Mateo, CA.
33. M. Spliliopoulou, L.C. Faulstich (1998) WUM: A Web Utilization miner. Workshop on the Web and Data Base (WEBKDD), pp. 109–115.
34. R. Srikant and R. Agrawal (1996) Mining Sequential Patterns: Generalizations and Performance Improvements. Proc. of the 5th International Conference on Extending Database Technology. March 1996, pp. 3–17.
35. J. Srivastava, R. Cooley, M. Deshpande, P.N. Tan (2000) Web Usage Mining: Discovery and Applications of Usage Patterns from Web Data. SIGKDD Explorations, 1(2), pp. 12–23.
36. C.H. Wang, T.P. Hong and S.S. Tseng (1996) Inductive learning from fuzzy examples. The fifth IEEE International Conference on Fuzzy Systems. New Orleans, pp. 13–18.
37. S.L. Wang, C.Y. Kuo, T.P. Hong (2001) Mining Similar Sequential Patterns from Transaction Databases. Proc. of the 9th National Conference on Fuzzy Theory and Its Application. Taiwan, November 2001, pp. 624–628.
38. S.L. Wang, W.S. Lo, T.P. Hong (2002) Discovery of Cross-Level Sequential Patterns from Transaction Databases. Proc. of the 6th International Conference on Knowledge-based Intelligent Information Engineering Systems. Italy, September, pp. 683–687.

39. L.A. Zadeh (1965) Fuzzy Sets. Information and Control 8(3), pp. 338–353.
40. L.A. Zadeh (1971) Similarity Relations and Fuzzy Orderings. Inform Sci. 3(1), pp. 177–200.
41. M.J. Zaki (1998) Efficient Enumeration of Frequent Sequences. Proc. of the 7th International Conference on Information and knowledge Management. Washington DC, pp. 68–75.
42. M. Zhang, B. Kao, C.L. Yip, and D. Cheung (2001) A GSP-Based Efficient Algorithm for Mining Frequent Sequences. Proc. of the IC-AI' 2001. Las Vegas, Nevada.

17

Ontology-based Fuzzy Decision Agent and Its Application to Meeting Scheduling Support System

Chang-Shing Lee[1], Hei-Chia Wang[2], and Meng-Ju Chang[1]

[1]Department of Information Management, Chang Jung University, Tainan, 711, Taiwan
leecs@mail.cju.edu.tw/leecs@ismp.csie.ncku.edu.tw
[2]Institute of Information Management, National Cheng Kung University, Tainan, 701, Taiwan
hcwang@mail.ncku.edu.tw

Abstract. A *Fuzzy Decision Agent* (*FDA*) based on personal ontology for *Meeting Scheduling Support System* (*MSSS*) is proposed in this chapter. In this system, when an organization member requests a meeting, the *FDA* can immediately send a meeting request with required information to *MSSS* and *Fuzzy Inference Engine* (*FIE*). When *MSSS* receives the meeting information from *FDA*, it obtains the invitees' information from the *Personal Meeting Scheduling Ontology* (*PMSO*) *Repository* and sends a response to *FDA*. Meanwhile, *FIE* utilizes these invitees' information along with the *Fuzzy Rule Base* and meeting information to infer the suitable meeting time slots. Therefore, *FDA* can analyze each invitee's attendance probability based on soft-computing technology. The experimental result shows that the proposed *FDA* is feasible, efficient and usable for a meeting scheduling support system.

Key words: Personal Ontology, Fuzzy Inference, Meeting Scheduling, Agent

17.1 Introduction

An ontology is a collection of key concepts and their interrelationships collectively providing an abstract view of an application domain. With the support of ontology, both user and system can communicate with each other by the shared and common understanding of a domain [15]. Moreover, an ontology is a computational model of some portions of the world. It is often captured in a semantic network and represented by a graph whose nodes are concepts or individual objects and whose arcs represent relationships or associations among the concepts [7]. On the other hand, an agent is a program that performs unique tasks without direct human supervision. J. Ferber [4] gives another definition of an agent such as "an agent is capable of acting in an environment and can communicate directly with other agents". An intelligent agent is more powerful than an agent because of its reasoning and learning capabilities [3]. Employees in an enterprise may spend much of their time scheduling and attending meetings. The process of searching for a commonly available meeting time can be complicated by communication delayed and by other concurrently scheduled meetings. Automating meeting scheduling can not only save the user time and effort, but also lead to more efficient schedulers and improvements in how the enterprise exchanges information.

S. Sen [14] proposes a software system that uses intelligent meeting-scheduling agents that can negotiate with other agents without compromising their user-specified constraints. In addition, K. Sugihara et al. [16] propose a meeting scheduler for office automation. They consider the priorities on persons and meetings in a real office environment, and use a heuristic algorithm for timetable rearrangement. T. Haynes et al. [6] propose an automated meeting scheduling system that utilizes user preferences. They highlight the usage of user preferences and priorities by the scheduling agent with regards to time of day, day of week, status of other invitees, topic of the meeting, and so on. W. S. Jeong et al. [8] focus their research on how the meeting-scheduling agent can reduce failures when there is no common time-slot. They solve the failure condition by utilizing the cooperation and the rescheduling strategy. E. Mynatt et al. [11] present a calendar system extension that uses a Bayesian model to predict the likelihood of one's attendance at the events listed on his or her schedule. A. Ashir et al. [1] propose a multi-agent based decision mechanism for distributed meeting-scheduling system. They use the concept of quorum to make the scheduling system more flexible and efficient. J. A. Pino et al. [13] accommodate users' availability according to their own preferences and restrictions to schedule meetings. F. Bergenti et al. [2] describe an agent-based computer-supported cooperative work system designed to promote the productivity of distributed meetings by means of agent. C. Glezer [5] proposes and evaluates a comprehensive agent-based architecture for an inter-organizational intelligent meeting-scheduler.

A *Fuzzy Decision Agent* (*FDA*) based on personal ontology for a *Meeting Scheduling Support System* (*MSSS*) is proposed in this chapter. When an organization member requests a meeting, the *FDA* immediately sends the meeting information to the *MSSS* and *Fuzzy Inference Engine* (*FIE*). *MSSS* receives the meeting information from *FDA*, and gets the invitees' information through the *Personal Meeting Scheduling Ontology* (*PMSO*) *Repository*. *FIE* utilizes the invitees' information of *FDA*, the *Fuzzy Rule Base* and meeting information to infer the suitable meeting time slots. Therefore, *FDA* can analyze each invitee's attendance probability based on soft-computing technology. The experimental result shows that the proposed *FDA* is feasible, efficient and usable for meeting scheduling support system. This chapter is organized as follows. In Sect. 17.2, we apply the concept of personal ontology to a meeting scheduling support system, and propose a four-layer object-oriented ontology for this application. Section 17.3 introduces the architecture of meeting scheduling support system. The system implementation of meeting scheduling support system is presented in Sect. 17.4. Finally, the conclusion and discussion are given in Sect. 17.5.

17.2 Personal Ontology
for a Meeting Scheduling Support System

With an ontology, we can organize keywords and database concepts by capturing the semantic relationship among the keywords or among the tables and fields in databases [7]. The semantic relationship can provide an abstract view of the information space for our schedules. T.R. Rayne et al. [12] propose a distributed meeting scheduling agent, called RETSINA Calendar Agent (RCal), that processes schedules marked up on the Semantic Web, and imports them into the user's personal information manager. In this chapter, we will apply the concept of personal ontology [7] to the meeting scheduling support system, and extend the four-layered object-oriented ontology [10] to *Personal Meeting Scheduling Ontology* (*PMSO*) for the *MSSS*.

Figure 17.1 shows the structure of four-layered object-oriented ontology. We define four layers including *Domain Layer*, *Category Layer*, *Class Layer*, and *Instance Layer* in the ontology. There are four kinds of relationships including association, generalization, aggregation, and instance-of in the structure of four-layered object-oriented ontology. The association relation belongs to non-taxonomic relation. On the other hand, the generation, aggregation, and instance-of belong to taxonomic relation. Besides, the aggregation is a whole-part relationship.

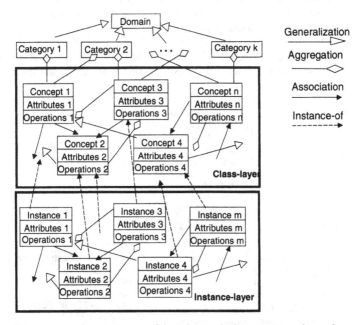

Fig. 17.1. The structure of four-layered object-oriented ontology

Figure 17.2 shows an example of association and aggregation relationships for MSSS. For instance, the association relationships for the concept pairs (*Teacher, Postgraduate*), (*Teacher, Undergraduate*), (*Teacher, Course Information*), (*Teacher, Laboratory Meeting*), and (*Teacher, School Meeting*) are "*Guide*", "*Guide*", "*Teach*",

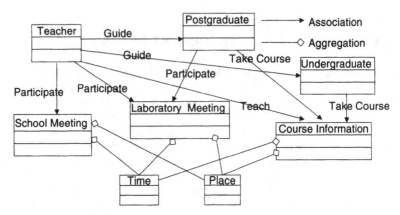

Fig. 17.2. An example of association and aggregation relationship

"*Participate*", and "*Participate*", respectively. In addition, aggregation relations exist for the concept pairs (*Laboratory Meeting, Time*) and (*Laboratory Meeting, Place*) in Fig. 17.2. An example of instance-of relationship is presented in Fig. 17.3. In this example, the concept "*Postgraduate*" in *Class Layer* can be seen, meanwhile, there are two corresponding instances "Meng-Ju Chang: Postgraduate" and "Chih-Wei Chien: Postgraduate" in *Instance Layer*.

Figure 17.4 shows the structure of *Personal Meeting Scheduling Ontology (PMSO)*. We classify the *Category Layer* into "*Teacher Category*", "*Research Assistance Category*", and "*Student Category*" for *MSSS*. Furthermore, we divide the concepts of

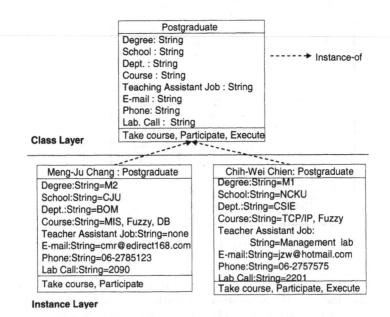

Fig. 17.3. An example of instance-of relationship

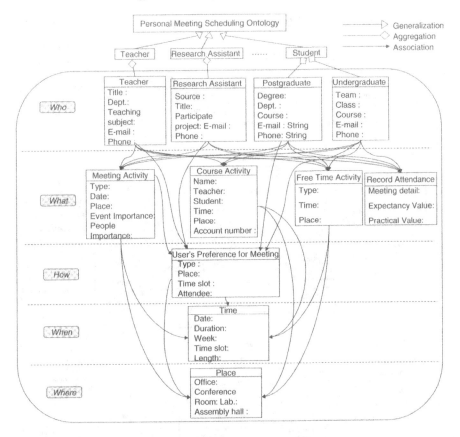

Fig. 17.4. The structure of Personal Meeting Scheduling Ontology (PMSO)

Class Layer into five groups including *Who Concept* group, *What Concept* group, *How Concept* group, *When Concept* group and *Where Concept* group. For example, the concepts, *"Teacher"*, *"Research Assistance"*, *"Postgraduate"* and *"Undergraduate"*, belong to *Who Concept* group. The concepts, *"Meeting Activity"*, *"Course Activity"*, *"Free Time Activity"* and *"Record Attendance"*, belong to *What Concept* group. The concepts *"User's Preference for Meeting"*, *"Time"* and *"Place"*, belong to *How Concept* group, *When Concept* group and *Where Concept* group, respectively. Figure 17.5 shows a part of *Personal Meeting Scheduling Ontology (PMSO)s* for "Decision Support & Artificial Intelligence (DSAI)" Lab. at Chang Jung University in Taiwan. For example, there are two *Who Concepts* "Chang-Shing Lee: Teacher" and "Meng-Ju Chang: Postgraduate" in *Instance Layer*. Association relations *"Participate"* and *"Guide"* exist for the concept pairs (*Teacher*, *Department Meeting*) and (*Teacher*, *Postgraduate*), respectively.

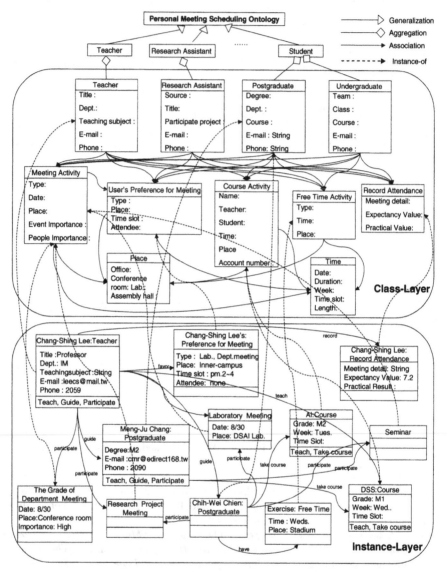

Fig. 17.5. A part of Personal Meeting Scheduling Ontology (PMSO) for DSAI Lab

17.3 The Architecture
of the Meeting Scheduling Support System

17.3.1 The Architecture of the Meeting Scheduling Support System

Figure 17.6 shows the architecture of the meeting scheduling support system. There are two main function blocks including *Fuzzy Decision Agent (FDA)* and *Fuzzy*

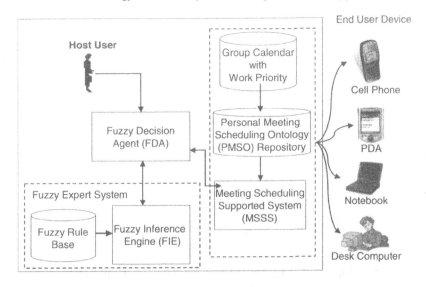

Fig. 17.6. The architecture of meeting scheduling support system

Expert System embedded in this system. At the beginning, *FDA* can receive the meeting information from the host user and send this information to *MSSS* for getting the invitees' information through the *Personal Meeting Scheduling Ontology (PMSO) Repository* and *Group Calendar*. Then, *FDA* further transfers the invitees' information to *Fuzzy Expert System*. The *Fuzzy Expert System* contains a *Fuzzy Inference Engine (FIE)* and a *Fuzzy Rule Base* to perform the parallel fuzzy inference mechanism. In this architecture, the devices of end users include cell phone, PDA, notebook and desk computer, etc. Hence, the invitees can get the meeting information by various platforms. The *FIE* can analyze whether the invitees would attend this meeting or not, then decide the suitable meeting time. Further operations will be described in the next section.

17.3.2 Fuzzy Decision Agent for Fuzzy Expert System

Figure 17.7 shows the flow chart of the *Fuzzy Decision Agent*. When a host user requests a meeting, *FDA* will try to find a suitable meeting time slot by comparing with all the invitees' free time based on the *Group Calendar* of *MSSS*. If a free time slot is found for all invitees, the *Fuzzy Expert System* will perform the parallel fuzzy inference mechanism to search the most suitable meeting time slot and provide the result to *MSSS*. Otherwise, the *FDA* will add the harmonization time slots of all the invitees to find the possible time slots again. If it can be found, the *Fuzzy Expert System* will execute its task to infer the suitable time slot. If there are still not free time slots after harmonizing, then *FDA* will send a failure message to the host user.

Figure 17.8 shows the architecture of the *Fuzzy Expert System*. The structure consists of premise layer, rule layer and conclusion layer [9]. There are two kinds of nodes in this model: fuzzy linguistic nodes and rule nodes. A fuzzy linguistic node represents a fuzzy variable and manipulates the information related to that linguistic

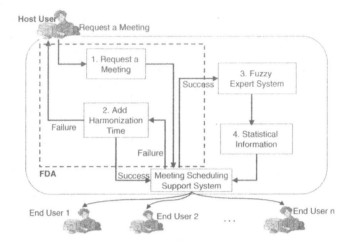

Fig. 17.7. The flow chart of the Fuzzy Decision Agent

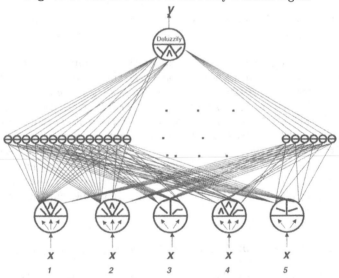

Fig. 17.8. Fuzzy Expert System for the Meeting Scheduling Support System

variable. A rule node represents a rule and decides the final firing strength of that rule during inferring.

Now, we describe these layers in detail.

• Premise layer:

As shown in Fig. 17.8, the first layer is called a premise layer, which represents the premise part of the fuzzy system. Each fuzzy variable appearing in the premise part is represented with a condition neuron. Each of the outputs of the condition node is connected to some node in the second layer to constitute a condition specified in some rules. Note that the output links must be emitted from proper linguistic terms as specified in the fuzzy rules. The premise layer performs the first inference step to compute matching degrees. The input vector is $x = (x_1, x_2, \ldots, x_n)$, where x_i is denoted as the input value of ith linguistic node. Then, the output vector of the premise layer will be

$$\mu^1 = ((u^1_{11}, u^1_{21}, \ldots, u^1_{N_1 1}), (u^1_{12}, u^1_{22}, \ldots, u^1_{N_2 2}), \ldots, (u^1_{1n}, u^1_{2n}, \ldots, u^1_{N_n n})),$$

where u^1_{ij} is the matching degree of the j-th linguistic term in the i-th condition node. In this chapter, triangular function and trapezoidal function are adopted as the membership functions of linguistic terms. The triangular and trapezoidal normal membership functions can be realized by the following formulas:

$$f_{\text{triangle}}(x : a, b, c) = \begin{cases} 0 & x < a \\ (x-a)/(b-a) & a \le x \le b \\ (c-x)/(c-b) & b \le x \le c \\ 0 & x > c \end{cases} \tag{17.1}$$

$$f_{\text{trapezoidal}}(x : a, b, c) = \begin{cases} 0 & x < a \\ (x-a)/(b-a) & a \le x \le b \\ 1 & b \le x \le c \\ (d-x)/(d-c) & c \le x < d \\ 0 & x \ge d \end{cases} \tag{17.2}$$

$$f^1_{ij} = \begin{cases} f_{\text{triangular}} & j \ne 1 \text{ or } n \\ f_{\text{trapezoidal}} & j = 1 \text{ or } n \end{cases} \tag{17.3}$$

where n is the numbers of linguistic term for i-th linguistic node. Therefore, for each element μ^1_{ij} of output vector μ^1 is

$$\mu^1_{ij} = f^1_{ij}(x) \tag{17.4}$$

• Rule layer:

The second layer is called a rule layer where each node is a rule node to represent a fuzzy rule. The links in this layer are used to perform precondition matching of fuzzy logic rules. The output of a rule node in rule layer will be linked with associated linguistic node in the third layer. In our model, the rules are previously defined by expert knowledge, and are shown in Table 17.1.

• Conclusion layer:

The third layer is called conclusion layer. The defuzzification process of conclusion node is shown as follows.

$$CrispOutput = \frac{\sum_{i=1}^{r} \sum_{j=1}^{c} y^k_{ij} w^k_{ij} V_{ij}}{\sum_{i=1}^{r} \sum_{j=1}^{c} y^k_{ij} w^k_{ij}} \tag{17.5}$$

where $w^k = \frac{\sum_{i=1}^{n} \mu^1_i}{n}$, V_{ij} is the center of gravity, r is the numbers of corresponding rule nodes, c is the numbers of linguistic terms of output node, n is the numbers of the fuzzy variable in premise layer and k represents in k-th layer.

Table 17.1. Fuzzy Rules Base for Fuzzy Expert System (L:Low, M:Medium, H:High, Lg:Long, S:Short)

Rule	UP	MEP	MTL	MTP (1)	MTP (2)	AMP	Rule	UP	MEP	MTL	MTP (1)	MTP (2)	AMP
1	L	L	Lg	L	L	L	55	M	M	S	L	L	M
2	L	L	Lg	L	H	L	56	M	M	S	L	H	H
3	L	L	Lg	M	L	L	57	M	M	S	M	L	M
4	L	L	Lg	M	H	L	58	M	M	S	M	H	H
5	L	L	Lg	H	L	L	59	M	M	S	H	L	H
6	L	L	Lg	H	H	M	60	M	M	S	H	H	H
7	L	L	S	L	L	L	61	M	H	Lg	L	L	H
8	L	L	S	L	H	M	62	M	H	Lg	L	H	H
9	L	L	S	M	L	L	63	M	H	Lg	M	L	H
10	L	L	S	M	H	M	64	M	H	Lg	M	H	H
11	L	L	S	H	L	M	65	M	H	Lg	H	L	H
12	L	L	S	H	H	M	66	M	H	Lg	H	H	H
13	L	M	Lg	L	L	L	67	M	H	S	L	L	H
14	L	M	Lg	L	H	L	68	M	H	S	L	H	H
15	L	M	Lg	M	L	L	69	M	H	S	M	L	H
16	L	M	Lg	M	H	M	70	M	H	S	M	H	H
17	L	M	Lg	H	L	L	71	M	H	S	H	L	H
18	L	M	Lg	H	H	M	72	M	H	S	H	H	H
19	L	M	S	L	L	L	73	H	L	Lg	L	L	L
20	L	M	S	L	H	M	74	H	L	Lg	L	H	M
21	L	M	S	M	L	M	75	H	L	Lg	M	L	L
22	L	M	S	M	H	M	76	H	L	Lg	M	H	M
23	L	M	S	H	L	M	77	H	L	Lg	H	L	M
24	L	M	S	H	H	H	78	H	L	Lg	H	H	M
25	L	H	Lg	L	L	M	79	H	L	S	L	L	M
26	L	H	Lg	L	H	M	80	H	L	S	L	H	M
27	L	H	Lg	M	L	M	81	H	L	S	M	L	M
28	L	H	Lg	M	H	M	82	H	L	S	M	H	H
29	L	H	Lg	H	L	M	83	H	L	S	H	L	M
30	L	H	Lg	H	H	M	84	H	L	S	H	H	H
31	L	H	S	L	L	H	85	H	M	Lg	L	L	M
32	L	H	S	L	H	M	86	H	M	Lg	L	H	M
33	L	H	S	M	L	M	87	H	M	Lg	M	L	M
34	L	H	S	M	H	H	88	H	M	Lg	M	H	H
35	L	H	S	H	L	M	89	H	M	Lg	H	L	M
36	L	H	S	H	H	H	90	H	M	Lg	H	H	H
37	M	L	Lg	L	L	L	91	H	M	S	L	L	M
38	M	L	Lg	L	H	L	92	H	M	S	L	H	H
39	M	L	Lg	M	L	L	93	H	M	S	M	L	M
40	M	L	Lg	M	H	M	94	H	M	S	M	H	H
41	M	L	Lg	H	L	L	95	H	M	S	H	L	H
42	M	L	Lg	H	H	M	96	H	M	S	H	H	H
43	M	L	S	L	L	L	97	H	H	Lg	L	L	H
44	M	L	S	L	H	M	98	H	H	Lg	L	H	H
45	M	L	S	M	L	M	99	H	H	Lg	M	L	H
46	M	L	S	M	H	M	100	H	H	Lg	M	H	H
47	M	L	S	H	L	M	101	H	H	Lg	H	L	H
48	M	L	S	H	H	H	102	H	H	Lg	H	H	H
49	M	M	Lg	L	L	L	103	H	H	S	L	L	H
50	M	M	Lg	L	H	M	104	H	H	S	L	H	H
51	M	M	Lg	M	L	L	105	H	H	S	M	L	H
52	M	M	Lg	M	H	M	106	H	H	S	M	H	H
53	M	M	Lg	H	L	M	107	H	H	S	H	L	H
54	M	M	Lg	H	H	M	108	H	H	S	H	H	H

There are five input fuzzy variables used in the *Fuzzy Expert System* for *MSSS*. They are the *User_Priority* (*UP*) fuzzy variable denoting the importance of invitees to the meeting, the *Meeting_Event_Priority* (*MEP*) fuzzy variable denoting the importance of the meeting, the *Meeting_Time_Length* (*MTL*) fuzzy variable denoting the length of the meeting, the *Meeting_Time_Preference_1* (*MTP1*) fuzzy variable, and the *Meeting_Time_Preference_2* (*MTP2*) fuzzy variable. *MTP1* fuzzy variable denotes the preference of meeting time for invitees. For example, if most people like to attend the meeting in the morning at 10:00 AM, and the meeting time is 9:30AM, then *FES* may infer the possibility of attending this meeting for all invitees as high. The *MTP2* fuzzy variable is used to consider the work priority of each invitee's schedule. For example, if a meeting event priority for user A is low, and A has a high priority work to do at the pre-decided meeting time, then A may not attend the meeting at this time. Figure 17-9(a-e) show the fuzzy sets of fuzzy variables *UP*, *MEP*, *MTL*, *MTP1* and *MTP2*, respectively.

The output fuzzy variable for *FES* is *Attend_Meeting_Possibility* (*AMP*), which denotes the possibility of attending the meeting for each invitee. Figure 17.10 shows the fuzzy set of *AMP*.

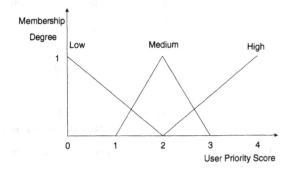

Fig. 17.9a. Fuzzy sets for the User Priority fuzzy variable

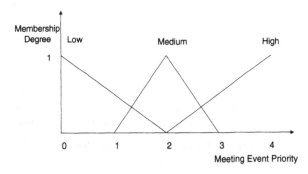

Fig. 17.9b. Fuzzy sets for the Meeting Event Priority fuzzy variable

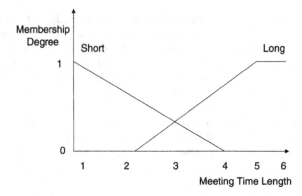

Fig. 17.9c. Fuzzy sets for the Meeting Time Length fuzzy variable

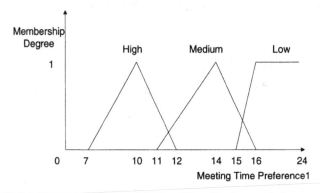

Fig. 17.9d. Fuzzy sets for the Meeting Time Preference 1 fuzzy variable

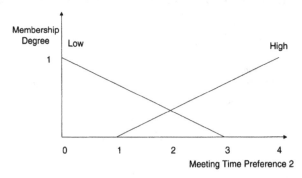

Fig. 17.9e. Fuzzy sets for the Meeting Time Preference 2 fuzzy variable

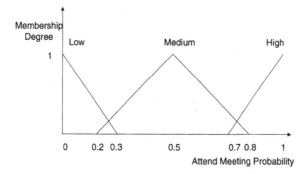

Fig. 17.10. Fuzzy sets for the Attend Meeting Probability fuzzy variable

17.4 System Implementation of the Meeting Scheduling Support System

To evaluate the effect of *FDA*, we constructed an experimental website for a meeting scheduling support system at DSAI Lab. (Decision Support & Artificial Intelligence Laboratory) of Chang Jung University. Figure 17.11 shows a meeting case of this system. The meeting host "Gary" will invite 10 Lab members to attend the meeting. The length of the meeting is 4 hours and the priority of the meeting event is Low.

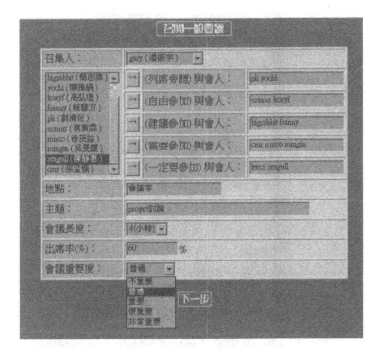

Fig. 17.11. A meeting case of DSAI Lab. at Chang Jung University

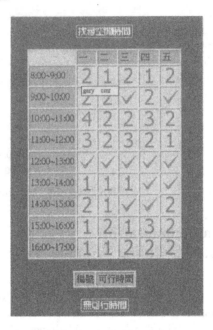

Fig. 17.12a. Common free time slots based on invitees' calendar

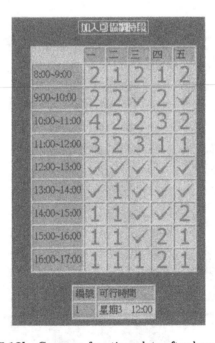

Fig. 17.12b. Common free time slots after harmonizing

Figure 17.12a shows the common free time slots for ten invitees. Because there are no common 4 hours free time slots in this case, *FDA* will consider finding a harmonization time and researching a suitable time slot again. Figure 17.12b shows the result after harmonizing process. Figure 17.13 shows the attending meeting possibility for each invitee.

編號	可行時間	與會人	與會可能性
1	星期3 12:00	pli	18.42%
1	星期3 12:00	yochi	18.42%
1	星期3 12:00	uranus	18.69%
1	星期3 12:00	koayf	18.69%
1	星期3 12:00	bigrabbit	50.0%
1	星期3 12:00	franny	50.0%
1	星期3 12:00	cmr	91.49%
1	星期3 12:00	mirco	91.49%
1	星期3 12:00	mingin	91.49%
1	星期3 12:00	leecs	91.729996%
1	星期3 12:00	seagull	91.729996%
預估出席率		45.454548%	
結果		無法達到要求之出席率	

Fig. 17.13. Attending meeting possibility for each meeting invitee

The experimental result is reasonable, because the meeting event priority is low and the meeting time is too long. Hence, the attending possibility is low for the invitee whose priority is not high.

17.5 Conclusions and Discussion

A *Fuzzy Decision Agent (FDA)* based on personal ontology for a meeting scheduling support system is proposed in this chapter. An organization's member can propose a meeting request and send the meeting information to *FDA*, then *FDA* immediately sends the information to *MSSS* and *FIE*. *MSSS* gets the invitees' information through the *Personal Meeting Scheduling Ontology (PMSO)*. *FDA* will retrieve the invitees' information via *MSSS*, then send the meeting information and invitee's information to *FIE*. Finally, *FIE* infers the suitable meeting time slot based on *Fuzzy Rule Base* and the above information. We have constructed an experimental website at Chang Jung University to evaluate the proposed method. The experimental results exhibit that the *Fuzzy Expert System* can infer the suitable meeting time successfully.

References

1. Ashir A, Joo KH, Kinoshita T, Shiratori N (1997) Multi-agent based decision mechanism for distributed meeting scheduling system. Proc. of the international conference on parallel and distributed systems, pp. 275–280
2. Bergenti F, Poggi A (2000) An agent-based approach to manage negotiation protocols in flexible CSCW systems. Agents 2000 barcelona spain, pp. 267–268
3. Chen H, Yen J (1996) Toward intelligent meeting agents. IEEE computer 29:62–70
4. Ferber J (1999) Multi-agent systems. Addison-wesley, New York
5. Glezer C (2003) A conceptual model of an interorganizational intelligent meeting-scheduler (IIMS). Journal of strategic information systems 12:47–70
6. Haynes T, Sen S, Arora N, Nadella R (1997) An automated meeting scheduling system that utilizes user preferences. Proc. of the first international conference on autonomous agents, pp. 308–315
7. Huhns MN, Stephens LM (1999) Personal ontologies. IEEE Internet Computer 3:85-87
8. Jeong WS, Yun JS, Jo GS (1999) Cooperation in multi-agent system for meeting scheduling. Proc. of the region 10 conference on IEEE Tencon, pp. 832–835
9. Kuo YH, Hsu JP, Wang CH (1998) A parallel fuzzy inference model with distributed prediction scheme for reinforcement learning. IEEE Transactions 28:160–172
10. Lee CS, Kao YF, Kuo YH, Meng IH (2003) An episode-based fuzzy inference mechanism for chinese news ontology construction. The 7th world multiconference on systemics, Cybernetics and informatics, USA, pp. 453–458
11. Mynatt E, Tullio J (2001) Inferring calendar event attendance. Proc. of the 6th international conference on intelligent user interfaces, pp. 21–128
12. Payne TR, Singh R, Sycara K (2002) Processing schedules using distributed ontologies on the semantic web. Springer-Verlag Berlin Heidelberg, Web services, E-business and the semantic web, pp. 203–212
13. Pino JA, Mora HA (1997) Scheduling meeting with guests' approval. The XVII IEEE international conference of the chilean on computer science society, pp. 182–189
14. Sen S (1997) Developing an automated distributed meeting scheduler. IEEE Expert 12:41–45
15. Soo VW, Lin CY (2001) Ontology-based information retrieval in a multi-agent system for digital library. Proc. of the sixth conference on artificial intelligence and applications, pp. 241–246
16. Sugihara K, Kikuno T, Yoshida N (1989) A meeting scheduler for office automation. IEEE Transaction on software engineering 15:1141–1146

18

A Longitudinal Comparison of Supervised and Unsupervised Learning Approaches to Iso-Resource Grouping for Acute Healthcare in Australia

Eu-Gene Siew[1a], Kate A. Smith[1b], Leonid Churilov[1c], and Jeff Wassertheil[2]

[1]School of Business Systems, Building 63, Monash University, Vic 3800
[a]Eu-Gene.Siew@infotech.monash.edu.au [b]Kate.Smith@infotech.monash.edu.au
[c]Leonid.Churilov@infotech.monash.edu.au
[2] Peninsula Health,Hastings Road, P.O. Box 52, Frankston Vic 3199
JWassertheil@phcn.vic.gov.au

Abstract. Estimating resource consumption of hospital patients is important for various tasks such as hospital funding, and management and allocation of resources. The common approach is to group patients based on their diagnostic characteristic and infer their resource consumption based on their group membership. This research looks at two alternative forms of grouping of patients based on supervised (classification trees) and unsupervised (self organising map) learning methods. This research is a longitudinal comparison of the effect of supervised and unsupervised learning methods on the groupings of patients. The results for the four-year study indicate that the learning paradigms appear to group patients similarly according to their resource consumption.

Key words: Classification trees, self organising map, iso-resource groupings, acute healthcare

18.1 Introduction

Knowledge about a patient's resource consumption is important in funding, management, and administration of hospitals. Currently, "Casemix" funding for patients are used in most developed countries including Australia [1]. Patients are grouped based on their clinical diagnostic categories (DRGs) and their resource consumption is inferred based on their group membership.

However, there are many problems with "Casemix" implementation in Victoria, Australia [2], such as when "Casemix" is used to predict resource consumption in groups that have presence of co-morbidities or other complicated clinical factors in their medical diagnosis.

Eu-Gene Siew et al.: *A Longitudinal Comparison of Supervised and Unsupervised Learning Approaches to iso-resource Grouping for Acute Healthcare in Australia*, Studies in Computational Intelligence (SCI) **4**, 283–303 (2005)
www.springerlink.com

An alternative to grouping patients based on their diagnostic category is to group patients based on their demographic and hospital admission and discharge information [3, 4]. Once the groups have formed, the average length of stay (LOS) of each group can be inspected, and used as a proxy for the predicted resource consumption.

LOS is used as a resource consumption proxy because directly measuring consumption is difficult, costly, and time consuming. According to Ridley et al. [5], LOS seems the best available proxy for resource consumption because of its ease of measurement. Furthermore, since LOS correlates well with patient injury (Iezzoni et al. [6]), it could be used for resource consumption planning (Walczak et al. [7]).

In addition, LOS is adopted in research and funding for resource consumption. LOS has also been used as a proxy for resource consumption in Australia to fund public hospitals. Hospitals are funded for their resource consumption based on their patients' DRG, which is attributed to a corresponding average LOS. As mentioned above, Ridley et al. [5] have also used classification trees as an alternative method for grouping iso-resource consumption based on LOS. This involves using LOS as the dependent variable, and training the tree to recognize different subsets of patients based on average LOS. Since our neural clustering approach is an unsupervised learning method, and classification trees are a supervised learning method, it is interesting to determine the effect of the learning paradigm on the resulting grouping of patients and predicted resource consumption.

The objective of this research is to look at:

- cluster profiles that are generated by supervised and unsupervised learning methods.
- four-year comparison of the different learning paradigms to establish the effect of the different methods on the groupings of patients.

This paper is organised as follows: Section 18.2 presents the case study settings and the data description; Sect. 18.3 describes the research methodology on the comparisons of methods and the methods used to cluster the data; Sect. 18.4 presents analysis and discussion; and the summary and conclusions are discussed in Sect. 18.5.

18.2 Case Study

The case study data is obtained from Frankston Hospital, a medium sized health unit in Melbourne, Australia. It is part of a network of hospitals called Peninsula Health, which serves nearly 290,000 people, with a seasonal influx of visitors of up to 100,000. The hospital is one of the main providers of hospital services in Mornington Peninsula. It has over 300 beds, providing acute and non-urgent medical and surgical services, mental health, maternity, and pediatric services [8].

The population for this study is selected in the following way: it includes all cases from the acute inpatients database for a given year (July-June) that arrived to the hospital through the emergency department, were subsequently admitted as inpatients into one of the acute care wards or into the hospital in the home ward, and were discharged from Frankston hospital. Thus, cases of patient transfer to a different hospital, as well as elective cases were explicitly excluded from this study, as their resource consumption patterns can differ significantly from those of the inpatients under consideration.

The resulting data consists of four-year period of about 64,000 records of admitted patients from 1 July 1997 to 30 June 2001. It contains information about patients' demographics, admission and discharge ward information, and their LOS. Inpatients to the hospital can be admitted either to the emergency department (EMG) (if they spend more than four hours there), or, after a short stay in the emergency department, be admitted as inpatients to children's ward (CHW), emergency department ward (EDW), cardiac care (C/C), short stay unit (SSU), post natal and gynecology unit (PNA), health home unit (HHU), intensive care unit (ICU), and to general wards located in various geographical areas of the hospital (4&5 SN, 3N, and 1&2 WA). EDW differs from EMG in that EDW it is used as an observation and recovery ward. It consumes some emergency department resources as patients are looked after by emergency staff. HHU refers to patients being looked after by hospital staff at home. SSU is a multi diagnostic observation unit whose focus is on the care of patients that have low diagnostic and therapeutic complexity.

The last discharge ward refers to the final ward from where the patient is discharged. Once patients are treated, they may be transferred to other non-emergency wards for observation and are discharged when their condition has improved. The system at the Frankston hospital records only the initial ward the patient is admitted to and the last ward the patient is discharge from. The intermediate wards to which the patient is transferred are not recorded in the database. Patient LOS is determined by subtracting the time of patient admission from the time of patient discharge.

Before the data could be used, it needed to be preprocessed. As the data mining based model could only accept numeric data, the variable gender was converted to binary form, "1" being male and "0" female. Likewise, other variables such as admission ward and discharge ward required columns containing binary numbers, "1" indicating the patient had entered the ward and "0" indicating they had not.

Table 18.1 gives some insight into the characteristics of the data. There is a large proportion of elderly patients admitted into Frankston Hospital. Elderly patients tend to consume more than the "average" amount of resources compared to other patients in the same DRG groupings and there is an equal distribution of males and females.

The variables "LOS" and "time in emergency" appear to be widely distributed. For example, in 1999, the average LOS of patients is 3.5 days, but its standard deviation is 9 days and 17 hours. LOS of patients has been steadily falling since

Table 18.1. Data Characteristics

Data	Gender	AGE				LOS		Time in Emergency	
		Mean	Quartile 1	Quartile 2	Quartile 3	Mean	Std. Dev.	Mean	Std. Dev.
00–01	0.49	50	28	52	74	3.03	5.84	0.34	0.24
99–00	0.5	50	27	53	75	3.55	9.7	0.32	0.24
98–99	0.5	50	28	53	74	3.66	6.52	0.21	0.74
97–98	0.49	49	26	52	73	3.61	6.65	0.22	0.216

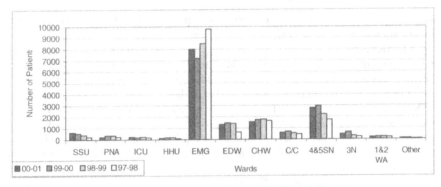

Fig. 18.1. Wards of patients initially admitted

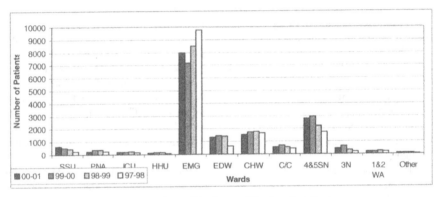

Fig. 18.2. Last Discharge Wards of patients

1997, from about 3 days and 14 hours in 1997 to 3 days in 2001. Time in emergency has slightly increased, but it too is widely distributed.

Figure 18.1 and Fig. 18.2 show the total number of patients that constitute the various admissions and discharges. There are 4 major wards that most patients are admitted to: EMG, EDW, CHW, and 4&5 SN. About half of all the patients are initially admitted to and last discharged from the emergency department. There is a significant drop in patients initially admitted to the emergency department from above 60% to below 50% and an increase in patients admitted to EDW after 1997.

The last discharge wards from Fig. 18.2 shows that there is a small yearly increase in people being discharged from the EMG wards. This might explain the decrease in LOS, as more people are discharged straight from emergency rather than going to other wards for recovery. There is also a slight increase in SSU admissions since SSU is a newly introduced unit in Frankston hospital and its use is becoming more accepted. The increase in discharges from those wards comes from the drop in discharges from CHW, C/C, 4&5 SN, and 3N wards.

Below, we demonstrate how various data mining approaches can be used for eliciting valuable information from this data.

18.3 Methodology

18.3.1 Research Design

The research objective is a comparison of supervised and unsupervised learning methods to establish the effect of the learning paradigms on the groupings of patients over a four-year period. The purpose of clustering is to group patients who have similar attributes. By clustering patients according to their demographics and admission and discharge wards, it is hoped that their resource consumption could be predicted.

Two different data mining techniques, which are Classification and Regression Trees (CART) and Self Organising Maps (SOM) were used to group the patients. Both of these techniques use different learning paradigms. CART requires an independent variable to split the data up, while SOM clusters the data independently without the need for any prior knowledge or guidance.

Quantitative and qualitative criteria were used together to indicate the optimal patient clusters. The quantitative criteria used are the "lowest error" measure produced in the data mining tools. The quantitative measure is to help find an initial clustering by providing a starting point for the optimal cluster size. A qualitative measure was used to determine how reasonable the optimal cluster is. The qualitative criteria were:

- The clusters themselves need to be distinct in terms of LOS. If two groups could independently come up with two different average LOS, then chances are these two groups are different from one another.
- The variables that belong to each cluster must make sense; they should be distinct and carry some information of their own. When each cluster is analyzed, its profile should be unique and meaningful.
- The total size of each cluster needs to be monitored. If a cluster is too large then it is possible that more groups that are distinct could lie within the cluster. Likewise, if it is too small, then there is high probability that the cluster is artificial.

Once the clusters were formed and examined, comparisons were made between CART and SOM to see the effect of the learning paradigms on the clusters formed.

To make the comparison, a table was used to contrast the CART and SOM clusters. In the table, SOM clusters are arranged in an increasing LOS horizontally while CART clusters are arranged in increasing LOS vertically. Clusters obtained were also renumbered in the order of ascending LOS, 1 being the shortest.

If both unsupervised and supervised learning methods produced groupings of the data that are indeed similar, one would expect to see a diagonal pattern emerge from such a matrix, that is those patients grouped in a low LOS cluster should also appear in a low LOS terminal node. As the predicted LOS increases for one method, we should see it also increase for the other method. If the two methods produced different groupings of the data, random patterns would emerge.

The total number and percentage of cluster in each terminal node for the four years are shown in Table 18.6 to Table 18.9. The percentage of SOM cluster in each terminal node is calculated as shown in the equation below:

$$P_{ij} = \frac{|S_i \cap C_j|}{|S_i|} \times 100\% \tag{18.1}$$

Where,

> |A| is a power (number of elements) in the set A
> P_{ij} is the percentage of SOM cluster i in CART terminal node j
> S_i is number of patients in SOM cluster i
> C_j is number of patients in CART terminal node j

The percentage of cluster in each terminal node shown in Table 18.6 to Table 18.9 is rounded up and if the percentage is not significant, that is, less than 1%, it is not shown in the comparison table.

18.3.2 An Unsupervised Learning Approach (Neural Clustering)

in the unsupervised approach, LOS was not used as input. The objective was to establish the extent to which this unsupervised learning approach is able to distinguish the data into clusters using only demographic, ward admission and discharge information.

The term "unsupervised learning" refers to a learning method which comes up with its own clustering of patients without any direction during learning. It is useful for cases where the user does not know which clusters and relationships to expect.

The unsupervised learning approach used in this paper is based on Kohonen's SOM [9, 10]. SOM is a type of neural network that specializes in grouping of data without any prior knowledge of the data relationships. The basic principle of identifying those hidden relationships is that if input patterns are similar then they should be grouped together. Two inputs are similar if the distance between the two inputs is close to each other.

A software package called Viscovery SOMine [10] was used to model the data. It reduces the data complexity by mapping high dimensional inputs into low dimensional outputs. The results of the SOM clustering are presented in 0.

18.3.3 A Supervised Learning Approach (Classification and Regression Trees)

In contrast to the unsupervised approach described in the previous section, classification and regression trees (CART) is a supervised learning data analysis tool that explores the effects of the independent variables on the dependent variable [11]. CART was introduced by Leo Breiman, Jerome Friedman, Richard Olshen, and Charles Stone in 1984.

CART is a useful tool in trying to find complex and hidden relationships in the data, while at the same time provide rules to determine each of the classification or prediction methods. It is a non-parametric and non-linear estimation and classification method. It does not require the theoretical and distributional assumptions required by traditional methods [12].

At each stage of the tree generation process, the CART algorithm chooses the best independent variable that will allow it to split the data into two groups on the basis of reducing the misclassification costs of the classification predictor or the least square error on regression predictor. It keeps splitting data into two groups with new independent variables until no further splits are possible. After finding the maximal tree, CART prunes it until finding the best tree, based both on predictive

Table 18.2. The cluster mean for each period based on SOM

Period		1	2	3	4	5	6	7	8	9	10
	LOS	0.34	0.37	1.53	4.04	5.26	6.6	6.9	7.98	9.98	10.91
	Emerg T	0.3	0.33	0.17	0.16	0.32	0.42	0.48	0.5	0.33	0.41
2000–2001	Gender	0.5	0.5	0.6	0.4	0.5	0.5	0.5	0.5	0.5	0.4
	Age	28	73	6	43	52	62	64	66	59	59
	Size	4000	3974	1544	7	2289	2631	535	793	264	144
	LOS	0.35	1.61	2.98	5.62	7.18	8.28	9.80			
	Emerg T	0.31	0.15	0.17	0.35	0.41	0.31	1.18			
1999–2000	Gender	0.5	0.6	0.5	0.5	0.5	0.5	0.5			
	Age	52	7	47	58	61	45	62			
	Size	7146	1753	10	2583	2847	635	1235			
	LOS	0.34	1.73	1.77	6.23	7.32	7.41	8.57	10.13		
	Emerg T	0.29	0.14	0.06	0.15	0.16	0.06	0.20	0.14		
1998–1999	Gender	0.5	0.6	0.6	0.5	0.5	0.4	0.5	0.5		
	Age	52	7	17	54	61	61	64	54		
	Size	6940	1765	61	2585	2188	1507	944	313		
	LOS	0.38	1.69	6.86	7.16	7.57	8.59				
	Emerg T	0.32	0.13	0.18	0.2	0.06	0.06				
1997–1998	Gender	0.5	0.6	0.5	0.5	0.5	0.4				
	Age	51	7	54	59	61	57				
	Size	7002	1651	1694	2302	1763	1017				

accuracy and on a penalty applied to large trees. The optimal tree is the one that generates the lowest misclassification costs or least square error.

To form a complete instruction one needs to trace a path down the tree. A branch of the tree is a terminal node if there are no more branches that spring out of the node. The terminal node is considered to be the end point of all the instructions to obtain a certain classification or prediction. Figure 18.3 to Fig. 18.6 shows the terminal nodes obtained from CART for the four-year periods.

18.4 Results

18.4.1 SOM Profiles

This section presents cluster by cluster profiles generated by the unsupervised neural clustering approach across the four-year period. Table 18.2 presents the cluster mean attribute of each period while Table 18.3 presents the initial admission wards and the last discharge wards. The numbers in brackets in the table indicate the number of patients last discharged from the wards, while those numbers without brackets indicate the number of patients admitted to the wards.

Clusters 1 and 2 (2000–2001) and Cluster 1 (1997–2000) represent patients who are initially admitted to and last discharged from the emergency department. These clusters have the lowest LOS and the largest of all patient clusters. In all periods except for 2000–2001, these patients are represented as one cluster. In the 2000–2001 period, these patients are split equally into two clusters based on age, with Cluster 1 having the average age of 28, and Cluster 2 having the average age of 73. The

Table 18.3. Number of patients admitted and discharged from the wards based on SOM. Numbers inside brackets indicate discharge, while numbers outside shows admission

Period	Clusters	SSU	PNA	ICU	HHU	EMG	EDW	CHW	C/C	4&5 SN	3N	1&2 WA	Other
2000–2001	1					4000(4000)							
	2					3974(3974)							
	3			1				1537(1537)					3(3)
	4	(1)			(1)	6(1)		1(1)			(2)		(1)
	5	595(566)	229(245)	124(96)	124(160)			2(2)	535(518)		432(449)	189(188)	59(65)
	6									2631(2631)			
	7	(45)	(11)	(2)	(38)	(2)	535(175)	(1)	(16)		(228)		(17)
	8						793			(793)			
	9	55	18	91	5	3		1	39	(264)	21		31
	10	(24)	(12)	(13)	(48)	(2)	(1)		(7)	144	(16)		(21)
1999–2000	1					7146(7146)	4	1739(1741)	1				
	2	6(9)	(2)	2				(2)					1(1)
	3	(2)			(3)	9(1)			(1)				1(1)
	4	457(499)	343(383)	20(3)	1	1	470(142)	2(1)	626(641)	40	605(912)		18(2)
	5	(13)	(4)	(15)			(1)	(1)		2847(2812)			(1)
	6	5	6	85(90)	161(291)		44	1	10	52	10	221(221)	40(33)
	7	61	28	84(16)	2	7	939	6(2)	54	(1213)	17		37(4)
1998–1999	1	1	1		(1)	6936(6940)	4	1765(1759)					2
	2	(3)	(1)										(1)
	3	3				49		(61)		3			2
	4	345(362)	327(359)	142(113)	172(215)	(2)	462(136)	3	509(543)	33	315(580)	242(242)	35(33)
	5	(9)	(10)	(12)	(30)	(7)				2188(2119)			(1)
	6	(73)	(77)	(18)	(29)	1507	(62)		(82)	(956)	(198)		(12)
	7						944			944			
	8	75	31	95	1		7	17	59	(313)	11		24
1997–1998	1		1			6989(7002)				5			
	2	7	(1)		(5)	(7)	6	1634(1638)		3			1
	3	178(180)	216(280)	85(63)	88(116)	(3)	229(73)	2	406(421)	85	191(341)	194(196)	20(21)
	4	63	21	90(9)	3		398	16(1)	27	1649(2291)	21	1	13(1)
	5					1763				(1763)			
	6	(65)	(146)	(23)	(70)	1017	(42)	(115)	(134)		(405)	(2)	(15)

Table 18.4. Terminal Node attributes represented by their node averages

Period	Variable	Terminal Node																
		1	2	3	4	5	6	7	8	9	10	11	12	13	14	15	16	17
2000–2001	LOS	0.37	1.64	1.90	2.86	3.93	4.93	6.61	6.73	7.71	8.25	8.49	8.52	10.62	14.63	15.24	17.05	34.87
	Emerg T	0.32	0.18	0.31	0.36	0.32	0.43	0.43	0.26	0.36	0.35	0.25	0.44	0.47	0.44	0.29	0.12	0.61
	Gender	0.5	0.6	0.5	0.4	0.6	0.5	0.5	0.4	0.6	0.5	0.5	0.4	0.5	0.5	0.6	0.5	0.4
	Age	51	7	33	35	69	37	60	70	36	71	32	80	77	67	70	61	64
	Size	7983	1711	413	547	816	1064	665	59	95	50	147	1840	611	89	51	20	17
1999–2000	LOS	0.35	1.60	2.73	2.83	3.16	3.84	4.68	5.31	7.24	7.55	7.95	9.75	19.47	20.56	22.89	93.81	
	Emerg t	0.31	0.15	0.44	0.25	0.32	0.35	0.27	0.25	0.27	0.41	0.37	0.42	0.03	0.25	0.35	88.83	
	Gender	0.5	0.6	0.4	0.0	0.5	0.5	0.7	0.5	0.6	0.5	0.5	0.4	0.6	0.6	0.5	0.4	
	Age	52	6	77	72	68	34	68	76	31	58	35	78	32	73	68	47	
	Size	7144	1744	83	38	143	2394	491	39	176	684	158	3021	23	39	21	11	
1998–1999	LOS	0.34	1.73	1.96	4.52	4.73	9.00	9.09	9.67	12.71	18.39	18.54						
	Emerg t	0.29	0.14	0.14	0.11	0.14	0.15	0.10	0.15	0.17	0.09	0.09						
	Gender	0.5	0.6	0.7	0.5	0.5	0.5	0.6	0.6	0.4	0.6	0.7						
	Age	52	7	31	69	36	75	32	68	79	70	69						
	Size	6949	1813	340	822	2482	2818	220	86	702	49	22						
1997–1998	LOS	0.38	2.08	4.41	5.03	6.15	8.90	10.35	11.51	15.24	15.32	16.83	23.03					
	Emerg t	0.32	0.13	0.14	0.13	0.15	0.14	0.24	0.11	0.11	0.14	0.13	0.24					
	Gender	0.5	0.5	0.6	0.5	0.5	0.4	0.4	0.5	0.4	0.6	0.4	0.5					
	Age	51	13	28	63	52	77	79	29	78	66	52	80					
	Size	7012	2395	963	1050	890	2134	183	132	505	67	61	37					

Table 18.5. Number of patients admitted and discharged from the wards based on CART

Period	Clusters	SSU	PNA	ICU	HHU	EMG	EDW	CHW	C/C	4&5 SN	3N	1&2 WA	Other
2000-2001	1	(1)	(1)	(1)	(1)	7983(7974)	(7)	(1)	(1)	(3)	(2)	(1)	(1)
	2	66(63)	20(24)	4(1)	4(4)	(2)	20(7)	1527(1530)	2(1)	56(65)	(1)	5(5)	7(8)
	3	362(413)	4	1	2		24	2		9			9
	4	26	169(202)	56(46)		(1)	91(57)	9(8)	99(111)	19	58(80)		20(42)
	5	136(147)	31(38)	62(51)		(1)	158(111)	2(2)	395(424)		10		22(42)
	6	29	14	38			183		1	789(1064)	1		9
	7	10	3	2	2		158		11	473(665)	2		6
	8				59(59)								
	9	1	3		59(95)		10	1	3	16	2		
	10	(12)	(4)	(13)		(1)	(1)		(5)	50		147(147)	(14)
	11												
	12	13	1		3		446		27	1316(1840)	18		16
	13	2	1	4			210		21	15	357(611)		1
	14	5	1		(89)		28		15	32	5		3
	15			51						(51)	(1)		
	16			2					2			20(19)	
	17			1					1			17(17)	
1999-2000	1	6				7143(7144)	5		2	1			1
	2					2	83(83)	1725(1744)					1
	3						4	1					
	4	115(143)	31(38)			1	20			4			
	5	361(380)	298(312)	74(46)		3	395(60)	16	155(151)	961(1296)	87(123)		43(26)
	6	1		6		1	13		455(491)	14	1		
	7												
	8			39(39)	39(39)								
	9	17	38(39)	30(18)	30(54)	2	138	2(2)	11	348(482)	58(84)	176(176)	10(5)
	10	5	4		95(158)	2	22	1	2	22	5		
	11	22	6	(21)	38(80)	6	777	3(1)	65	1585(2210)	479(698)		40(11)
	12			39	(1)							23(23)	
	13									(33)			
	14										(5)		
	15											21(21)	
	16	1			(1)	3(3)				4(4)	2(2)	1(1)	

Table 18.5. cont.

Period	Clusters	SSU	PNA	ICU	HHU	EMG	EDW	CHW	C/C	4&5 SN	3N	1&2 WA	Other
	1	1	1			6936(6949)	2	1755(1813)		7			2
	2	3		3		46	4	5	1	3			2
	3	242(340)	7			56	19			6	1		
	4	76(107)	44(68)	68(76)		136	104(99)	4(7)	355(447)	22	1		12(18)
	5	69	297(379)	113(67)	125(189)	480	394(99)	21	150(178)	770(1465)	33(76)		30(29)
1998–1999	6	28	8			652	681		50	1374(2818)	10		15
	7											220(220)	
	8	1	1		48(86)	7	11		3	15			
	9	4	1	4		179	195		9	27	281(702)		2
	10			49						(49)			
	11											22(22)	
	1		1			6989(7012)	7	7	3	5			
	2	120(154)	156(294)	24(22)	35(76)	310	66(33)	1618(1737)	26(36)	22	6(21)	(5)	12(17)
	3	26	8	23		371	66	9		456(963)	2		2
	4	62(91)	57(133)	56(73)	52(115)	312	75(82)	11(17)	373(519)	40	2	1	9(20)
	5	18	4		3	404	88	6	4	360(890)	1	1	1
1997–1998	6	19	9			988	244	1	23	822(2134)	18		10
	7										183(183)	132(132)	
	8												
	9	3	3	5	1	395	94		4		(505)		
	10			67						(67)			
	11											61(61)	
	12									37	(37)		

Table 18.6. Total number and percentage of cluster in each terminal node for 2000–2001

Terminal Node	CLUSTER									
	1	2	3	4	5	6	7	8	9	10
1	4000(100%)									
2		3974(100%)	1529(99%)	6(86%)	99(4%)	53(2%)	14(3%)	6(1%)	3(1%)	3(2%)
3			9(1%)	1(14%)	380(17%)		24(4%)		6(2%)	9(6%)
4			2		428(19%)		91(17%)			19(13%)
5					656(29%)		158(30%)			
6						789(30%)		183(23%)	92(35%)	
7						473(18%)		158(20%)	34(13%)	
8					59(3%)					
9			1		68(3%)		10(2%)			16(11%)
10										50(35%)
11					147(6%)					
12						1316(50%)		446(56%)	78(30%)	
13					386(17%)		210(39%)			15(10%)
14					29(1%)		28(5%)			32(22%)
15									51(1%)	
16					20(1%)					
17					17(1%)					
TOTAL	4000(100%)	3974100%	1541(100%)	7(100%)	2289(100%)	2631(100%)	535(100%)	793(100%)	264(100%)	144(100%)

sizes of these clusters are quite consistent throughout the four yearly periods, and contain approximately 7000–7900 patients yearly.

Children tend to recover the quickest and, on average, with the least complications. They are represented in cluster 3 (2000–2001) and cluster 2 (1997–2000). They have one of the lowest times spent in emergency, as well as one of the shortest LOS. The children have an average age of 7 years and are more likely to be male.

Cluster 4 (2000–2001), Cluster 3 (1998–2000), Cluster 6 (1997–1999), and Cluster 5 (1997–1998) are patients initially admitted to emergency and then transferred to other wards for recovery. There was a change in hospital policy, and after 1999, a substantial drop in numbers and LOS of these types of patients is evident. In 1997–1998, there were 2800 of these patients, in 1998–1999, there were 1550 patients, and in 1999–2001 there were less than 10.

For the entire four-year period, except for 1999–2000, SOM does not distinguish whether patients were initially admitted and last discharged from SSU, PNA, ICU, HHU, 3N and C/C wards. SOM seems to cluster these types of patients together into Cluster 5 (2000–2001), Cluster 4 (1998–1999), and Cluster 3 (1997–1998). These types of patients seem to have an average LOS of 5 to 7 days and contain approximately 1700 to 2600 patients.

However, in 1999–2000, if these patients were admitted to or last discharged from HHU or the geographical area of 1&2 WA, they are classified as Cluster 6, while otherwise they are grouped as Cluster 4. Cluster 6 (1999–2000) has a longer LOS and is less numerous than Cluster 4.

The rest of the other clusters are grouped according to the way in which the patients are transferred to EDW and 4&5 SN:

- If patients are initially admitted to and last discharged from the geographical ward 4&5 SN they are likely to be grouped as Cluster 6 (2000–2001) and Cluster 5 (1998–2000). They have an average LOS of about $6\frac{1}{2}$ to 7 days, an average age of 50 years, and a cluster size of over 2200 patients.
- If patients are initially admitted to wards other than 4&5 SN and last discharged from the same ward, they are likely to be grouped as Cluster 9 (2000–2001), Cluster 7 (1999–2000) and Cluster 8 (1998–1999). They have an average LOS of about 10 days.
- If patients are initially admitted to wards 4&5 SN and last discharged from other wards, they are likely to be grouped as Cluster 10 (2000–2001). They have the longest LOS with an average of 11 days.
- If patients are initially admitted to EDW and last discharged from ward 4&5 SN, they are likely to be grouped as Cluster 8 (2000–2001) and Cluster 7 (1998–1999). They have an average LOS of about $7\frac{1}{2}$ to 8 days and an average age of 65 years.
- If patients are admitted to EDW and last discharged from other wards, they are likely to be grouped as Cluster 7 (1999–2000). They have an average LOS of 7 days and an average age of 64 years.
- The 1997–1998 period combines the above mentioned clusters into one cluster, which is Cluster 4. It has all the above mentioned attributes with an average LOS of 7 days and average age of 59 years.

Overall, the SOM clusters produce an interesting picture. Ward information of patients seems to have a major impact on the way the clusters are formed. Age attribute seems to have some impact while gender and time in emergency appear to have minimal effect on the clusters formed.

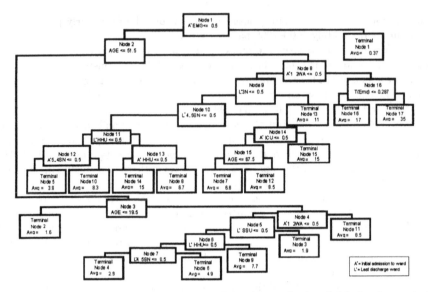

Fig. 18.3. Classification and Regression Tree for predicting LOS for the period 2000–2001

18.4.2 CART Profiles

This section will provide a brief description of the terminal nodes' profiles obtained from CART. The trees generated by CART are shown in Fig. 18.3 to Fig. 18.6, and a summary of the mean of terminal nodes is shown in Table 18.4. Table 18.5 shows the wards the patients are initially admitted to and discharged from.

Terminal Node 1 of all periods represents patients who are initially admitted to and last discharged from the emergency department. These nodes have the lowest LOS and contain the most patients of all the nodes. They have an average LOS of just 9 hours. The size of these nodes is quite consistent throughout the periods, which is about 7000–7900 patients yearly.

Terminal Node 2 represents child patients. It is one of the largest groups. These patients tend to have the quickest recovery and, therefore, one of the shortest lengths of stay.

For the rest of the other terminal nodes, there were a number of similarities across all the periods:

- If patients are initially admitted to and last discharged from SSU, PNA, ICU, HHU, and C/C wards, they tend to be grouped into Terminal Nodes 4 and 5 (2000–2001), 7 and 11 (1999–2000), 3 and 4 (1998–1999), and 4 (1997–1998), they tend to have moderate resource consumption. They have a LOS of between 4 to 8 days.
- If patients are initially admitted to and last discharged from wards EDW, 4&5 SN and 3N only, such as Terminal Nodes 6, 7, 12, 13 and 14 (2000–2001), 14 (1998–1999), 6, 9, 10 (1998–1999), and 7, 9, 10 and 12 (1997–1998), they seem to have a higher LOS.

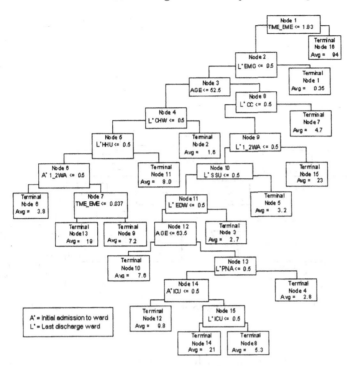

Fig. 18.4. Classification and Regression Tree for predicting LOS for the period 1999–2000

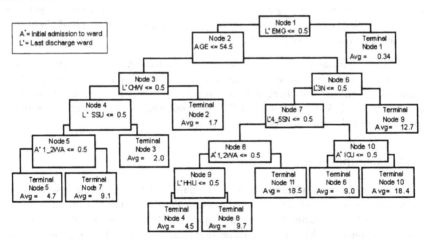

Fig. 18.5. Classification and Regression Tree for predicting LOS for the period 1998–1999

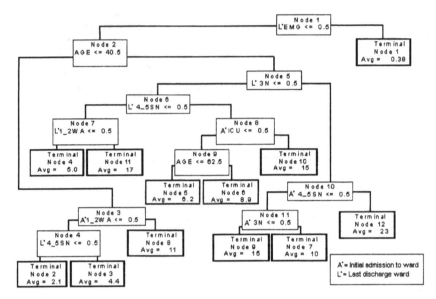

Fig. 18.6. Classification and Regression Tree for predicting LOS for the period 1997-1998

- However, if patients are initially admitted to other wards and last discharged from wards 4&5 SN, such as Terminal Nodes 6, 10 and 12 (1999–2000), 5 (1998–1999), and 3, 5 and 6 (1997–1998), they seem to have moderate resource consumption.
- The geographical ward 1&2 WA seems to have patients with the longest LOS. Terminal Nodes 16 and 17 (2000–2001), 15 (1999–2000), and 11 (1997–1999) have an average LOS of over 20 days.

In general, if one were to compare the profiles for CART with SOM one would notice many similarities between them. They both classify patients who are initially admitted to and last discharged from emergency, a children's group, a combined group of SSU, PNA, ICU, HHU, and C/C patients, and one or more groups admitted to or discharged from 4&5 SN.

18.4.3 Comparison

This section presents a comparison of the supervised and unsupervised learning paradigms based on their grouping of patients. The unsupervised learning paradigm did not use LOS as input whereas the supervised learning paradigm used LOS as a predictor output. Table 18.6 to Table 18.9 shows the period comparison between supervised and unsupervised learning paradigms. Looking at these tables, there is a pattern of similarity between supervised and unsupervised learning. For the period 2000–2001, the majority of patients lie in clusters and terminal nodes across the diagonal of the matrix (these cells have been highlighted). In fact, 79% of the patients fall in clusters or CART terminal nodes that produce consistent indications of likely resource consumption via LOS in period 2000–2001. Clusters 3, 4 and 6 appear to contain some patients with inconsistent predictions across the two learning

Table 18.7. Total number and percentage of cluster in each terminal node for 1999–2000

Terminal Node	CLUSTER						
	1	2	3	4	5	6	7
1	7143 (100%)		1 (10%)	1	1		1
2		1739 (99%)	2 (20%)		1		1
3				83 (3%)			
4				38 (1%)	4		
5			1 (10%)	138 (5%)			
6		11 (1%)	1 (10%)	1003 (39%)	957 (34%)	56 (9%)	366 (30%)
7			1 (10%)	490 (19%)			
8						39 (6%)	
9						176 (28%)	
10		2		124 (5%)	334 (12%)	69 (11%)	155 (13%)
11			2 (20%)			156 (25%)	
12		1	2 (20%)	699 (27%)	1548 (54%)	92 (14%)	679 (55%)
13						23 (4%)	
14				5		1	33 (2%)
15						21 (3%)	
16	3			2	3	2	1
TOTAL	7146 (100%)	1753 (100%)	10	2583 (100%)	2847 (100%)	635 (100%)	1235 (100%)

Table 18.8. Total number and percentage of cluster in each terminal node for 1998–1999

Terminal Node	CLUSTER							
	1	2	3	4	5	6	7	8
1	6940 (100%)			2				
2		1755 (99%)	58 (95%)		7			
3		3		275 (11%)	6	56 (4%)		
4		4	3 (5%)	666 (26%)	16 (1%)	133 (9%)		
5		3		813 (31%)	770 (35%)	480 (32%)	263 (28%)	153 (49%)
6					1374 (63%)	652 (43%)	681 (72%)	111 (35%)
7				220 (9%)		7		
8				64 (7%)	15 (1%)			
9				523 (20%)		179 (12%)		
10								49 (16%)
11				22 (1%)				
TOTAL	6940 (100%)	1765 (100%)	61 (100%)	2585 (100%)	2188 (100%)	1507 (100%)	944 (100%)	313 (100%)

Table 18.9. Total number and percentage of cluster in each terminal node for 1997–1998

Terminal Node	CLUSTER					
	1	2	3	4	5	6
1	7002 (100%)	7	3			
2		1635 (99%)	450 (27%)			310 (30%)
3			3	592 (26%)	371 (21%)	
4		9 (1%)	718 (42%)	11		312 (31%)
5				486 (21%)	404 (23%)	
6				1146 (50%)	988 (56%)	
7			183 (11%)			
8			132 (8%)			
9			110 (6%)			395 (39%)
10				67 (3%)		
11			61 (4%)			
12			37 (2%)			
TOTAL	7002 (100%)	1644 (100%)	1694 (100%)	2302 (100%)	1763 (100%)	1017 (100%)

paradigms. In answer to the research question on the effect of the two learning paradigms, they appear to produce similar LOS prediction and groupings of the data.

The same analysis is done over the other periods to establish whether there has been any change in the comparison conclusion. Table 18.7, Table 18.8 and Table 18.9 show these results. For the 1999–2000 period, 75% of patients are similarly grouped. For the 1998–1999 and 1997–1998 periods, the percentage of patients similarly grouped is even higher (83%). However, Clusters 1,2 and 3 of period 1999–2000, Clusters 3 and 4 of period 1997–1998 and Clusters 2 and 5 of period 1997–1998 do not show consistent supervised and unsupervised patient groupings.

Overall, these inconsistent cluster numbers are small and there is no change in the comparison conclusion. Thus, this four-year period seems to confirm that for the case study under question both supervised and unsupervised learning method grouped patients similarly in terms of LOS.

18.5 Conclusion

This paper shows that there are other alternatives to estimating resource consumption besides using DRGs. CART is a rule-based system that generates rules to differentiate one group from the other. CART is a supervised learning method, which uses LOS as a basis to produce the tree nodes. SOM, on the other hand, is an unsupervised learning method, which separates the data into clusters without any guidance during learning. Even though those techniques use different approaches, they achieved comparable results in this case study for the past four yearly periods. In SOM, LOS was not used as input yet it achieved similar groupings of the data to those obtained when LOS was used as a dependent variable in CART. Evidently, there is enough quality in the demographic and hospital admission and discharge information to reflect patients' resource consumption.

References

1. Sanderson H, Anthony P, and Mountney L (1998) Casemix for All. Radcliff Medical Press, Abingdon, Oxon, UK
2. Auditor-General of Victoria (1998) Acute Health Services Under Casemix. A Case of Mixed Priorities. Special Report No. 56. Victorian Government Printer, Melbourne
3. Siew E-G, Smith KA, Churilov L, and Ibrahim M (2002) A Neural Clustering Approach to Iso-Resource for Acute Healthcare in Australia. HICSS-35, 35th Hawaii International Conference on System Sciences. IEEE Computing Society, Hawaii, pp. 154–164
4. Siew E-G, Smith KA, Churilov L, and Wassertheil J (2003) A Comparison of Patient Classification Using Data Mining in Acute Health Care. In: Abraham A et al. (Eds.) Intelligent Systems and Design Applications. Springer-verlag, Tulsa, Oklahoma, USA, pp. 599–608
5. Ridley S, Jones S, Shahani A, Brampton W, Nielsen M, and Rowan K (1998) Classification trees. A possible method for iso-resource grouping in intensive care. Anaesthesia 53(9):833–840

6. Iezzoni LI, Schwartz M, Ash AS, and Mackiernan YD (1996) Does severity explain differences in hospital length of stay for pneumonia patients? Journal of Health Services Reserch and Policy 1(2):65–76
7. Walczak S, Pofahl WE, and Scorpioc RJ (2003) A decision support tool for allocating hospital bed resources and determining required acuity of care. Decision Support Systems 34(4):445–456
8. Peninsula Health (2000) Annual Report. Frankston, Australia
9. Deboeck G and Kohonen T (1998) Visual Explorations in Finance with Self-Organizing Maps. Springer-Verlag, London
10. GMBH ES (1999) Viscovery SOMine 3.0 User Manual. www.eudaptics.com
11. Steinberg D and Phillip C (1997) CART–Classification and Regression Trees., Salford Systems, San Diego, CA
12. Martin K (2003) An Overview of the CART Methodology. www.laurus.pl/DM_Robo/Trees/CART/CART_Algorithm/CART_Algorithm.htm

Data Mining of Missing Persons Data

K. Blackmore[1], T. Bossomaier[1], S. Foy[2] and D. Thomson[2]

[1] School of Information Technology, Charles Sturt University, Bathurst, NSW
2795, Australia
kblackmore@csu.edu.au
[2] School of Social Sciences and Liberal Studies, Charles Sturt University, Bathurst,
NSW 2795, Australia
sfoy01@postoffice.csu.edu.au

Abstract. This paper presents the results of analysis to evaluate the effectiveness
of data mining techniques to predict the outcome for missing persons cases. A rule-
based system is used to derive augmentations to supplement police officer intuition.
Results indicate that rule-based systems can effectively identify variables for prediction.

Key words: Classification; Rules; Missing Values; Uncertainty

19.1 Introduction

Many people go missing in NSW, at a rate of approximately 7000 per year. Fortunately, around 99% of these missing persons are quickly located, with 86% returning
home or being found by family or friends within a week. In some cases, however,
the person who has been reported as missing has become the victim of foul play
or has committed suicide. Whilst these high priority cases occur infrequently, the
fact remains that to date there exist no adequate guidelines for the police to utilize
when making a decision about the likelihood that a person is at risk of harm. In particular, inexperienced police officers may find it difficult to disregard stereotypical
views of why the missing person is missing [1]. In order to standardize the process of
assigning possible risk to missing person reports, a system is needed that is reliable,
not too complicated to use and efficient.

There are several powerful advantages in being able to deepen the inferences
that can be drawn about the circumstances of a missing person's disappearance.

- The police response can be tuned to make the most efficient use of resources.
  While foul play and suicide as the reason for going missing will need an immediate
  response, a repeat runaway who is of low risk for foul play or suicide might be
  assigned lower priority.
- The police investigation may be focused: it is more important to ask the right
  questions and collect the most useful data than to try, and fail, to collect every-
  thing which might possibly be useful. Again this is a resource issue.

K. Blackmore et al.: *Data Mining of Missing Persons Data* , Studies in Computational Intelligence (SCI) **4**, 305–314 (2005)
www.springerlink.com

- Where foul play is suspected, some inferences about the criminal *modus operandi* may be possible. Other disappearances, which may be separated over time and space, may have important linking features that may help in apprehending the offender or in warning the public of specific dangers.

Thus the NSW Police are interested in a system that can both utilize artificial intelligence and be available across NSW for all police officers to use. Towards this end, case histories of disappearances with known outcomes falling into the categories of *runaway*, *suicide* and *foul play* were made available to the investigators.

We approach the problem in the first instance through a search for a rule-based classification system. Given the largely nominal data quality of our first extract from the police data, rule systems are a natural choice. To eliminate the necessity for software training for police officers, some heuristics to support computational results are highly desirable. The rule-based approach generates simple augmentations to intuition that may be used by police anywhere at any time.

The aim of this work is to show that given the uncertain nature of the data, rules can be derived which can be used to predict the outcome for future missing persons cases. Examples exist within the dataset that escape rules, containing random patterns which result in examples that do not match classes. What were sought were not a holistic analysis of all patterns, but rather a subset of patterns, presented as rules, that could be used reliably to supplement police intuition. The potential also exists for these patterns to be used to isolate links and possible repeat or serial "offenders".

19.2 Collection and Parameterisation of Data

19.2.1 Data Source

This research was funded by the Maurice Bryers Foundation and is in collaboration with the Missing Person's Unit (MPU) of the New South Wales Police Service. The Missing Person's Unit in Parramatta, Sydney, is the central location for the collation of all missing persons reports that are made within New South Wales, Australia. This study was conducted as a systematic investigation of these archival police files which were originally collected upon the missing person report being made to the police, and during the investigation of the person's whereabouts. The information used in this study was extracted from records held in a centralised database (COPS), as well as from photocopies of records and other relevant information stored within files held at the Missing Persons Unit. Also included were relevant files archived within the Homicide Library of the NSW Police Service.

The criteria for cases selected for inclusion in this study were the circumstances surrounding the disappearance of the missing person, the theoretical and practical relevancy of the case to the research, and the quality of the information contained in the police file. Only persons who were known to have run away, attempted or completed suicide, or fallen victim to foul play were included in the study. Persons who had gone missing because of an accident or through being lost, and persons reported missing because of misunderstanding, were not included in the study. All the files were treated with the strictest confidentiality, in accordance with the Charles

Sturt University Animal and Human Ethics Committee, The Australian Psychological Society Code of Ethics and the New South Wales Police Service confidentiality practices.

19.2.2 Criterion Selection and Data Reduction

All aspects of the information contained within the files were considered for their capacity to predict type of missing person. Determining what aspects of the data were suitable and obtainable required combing through the files a number of times. The priority in the initial stages of data collection was to maximise both the type of information included as well as the coverage of that information. Ultimately, deciding what variables were important to preserve, and which could be discarded was a matter of judgment combined with the frequency with which the information was available.

The procedure adopted for the collection and the coding of data comprised (a) listing those variables that were deemed desirable for the research, (b) listing those variables that were observed to be available; (c) inspecting case files for the quality of the records and to determine what variables were and were not usable in the study; and (d) reviewing the suitability of the variables in light of theoretical and practical relevance.

19.2.3 Data Quality

Technically, the dataset contained no missing values. This was due to the fact that a response option for "not relevant" or "not known" was included for most variables because of the difficulty in finding files with all of the information relevant to the study. Missing data in the police file was taken to imply that the information of interest was either (a) not relevant to that particular case; (b) not known by the reporting person; (c) not considered for its possible relevance by the police officer; or (d) explored by the officer but not noted in the COPS narrative. Rather than scoring the response for that variable as missing data, it was more practical to categorise the absence of knowledge as "not relevant or not known" because of the non-specific nature of lack of information.

Systematic randomisation of cases was not possible due to the nature of the data and issues of availability. However, all of the cases included in this research were from New South Wales. All the runaway cases occurred during 2000, and the suicide and foul play occurred between 1980 and 2000. Cases prior to 1980 were not included in the study due to the influence of different temporal factors such as differences in the availability of forensic evidence [2], as well as changes to police recording methods.

Because of the practical constraints on the data collection it was not possible to validate the criterion measures by inter-rater reliability. Therefore, care was taken to define variables in a clear, precise and consistent manner. Also, in this study an effort was made to minimise the impact of a retrospective design. Only cases where the missing person had been missing for three months and under before being reported as a missing person were included in the research. This condition aimed to reduce the influence of errors in the reporting person's memory recall when the report was made to the police [3].

Table 19.1. Variables based on information in police files describing missing persons

• Does missing person have any dependents	• Is there a past history of suicide attempt or ideation
• Residential status	• Any known mental health problems
• Time of day when last seen	• Any known drug and alcohol issues
• Day of week when last seen	• Any known short term stressors
• Season of year when last seen	• Any known long term stressors
• Last seen in public	• Method of suicide
• Is this episode out of character for the missing person	• Was the perpetrator known or a stranger to the victim
• What does the reporting person suspect has happened	• Was missing person alive, deceased or hospitalized when located
• Any known risk factors for foul play	
• Is missing person known to be socially deviant or rebellious	
• Is there a past history of running away	

By the completion of the data collection a total of 26 input variables and 1 output variable described the sample of 357 finalised missing persons cases. Information present in every file, which was deemed relevant, included age, gender, nationality by appearance, residential address, occupation, marital status, date last seen, and who the reporting person was. Below (Table 19.1) are the remaining variables that were usually dependent on information contained in the files free narrative, which was reported by the officer in charge of the case.

19.3 Developing Rule Systems

19.3.1 Participants

A total of 357 case files were used in this research. The sample ranged in age from 9 years to 77 years with a mean age of 28 years (SD = 15 years). There were 184 females (51.5%) in the entire sample, and 173 (48.5%) males. Those who appeared to be Caucasian were the most frequently reported missing in this sample, and comprised 85.4% of the entire sample. Those who fell into "all other categories" comprised 7% of the sample, with Asian accounting for 4.8% of the missing persons sample.

Those appearing to be Aboriginal were the least reported as missing persons, comprising 2.8% of the entire sample. There were 250 (70%) runaways, 54 (15.1%) victims of suicide and 53 (14.8%) persons missing due to foul play included in the present study.

19.3.2 Rule Systems

Structural patterns in data can be discovered and expressed as rules, which are easily interpreted logical expressions. Covering algorithms use a "separate and conquer"

Table 19.2. Data Summary

Data Summary						
Gender		Age		Outcome		
Female	Male	<18	>18	R	S	F
157	168	107	218	188	59	75
48%	52%	33%	67%	58%	18%	23%

(R = runaway; S = suicide; F = foulplay)

Total cases: 325

Age Range = 9–77 years   Mean Age = 27.41

strategy to determine the rule that explains or covers the most examples in the data set. The examples covered by this rule are then separated from the data set and the procedure repeats on the remaining examples [4]. Decision trees are a popular "divide and conquer" method to discover knowledge from data [5]. Decision trees represent variables and variable values as trees, branches and leaves from which rules must be transformed.

Earlier work [6] using datasets containing missing values focused on the problems associated with classifying and defining rules for data with missing values, which is well recognised and has been widely discussed in the literature [7, 8, 9, 10]. Given that the occurrence of missing values in the dataset carries no significance, a suitable approach to develop rules in this study was to apply a covering algorithm to a decision tree derived from the training dataset. Algorithms that derive rule sets from decision trees first generate the tree, and then transform it into a simplified set of rules [4]. Based on results from previous research [3], the *WEKA J48.PART* [21] algorithm was selected to derive and evaluate rules sets from the training data. This is a standard algorithm that is widely used for practical machine learning. The algorithm is an implementation of C4.5 release 8 [9] that produces partial decision trees and immediately converts them into the corresponding rule. The methods used were found to be robust against missing values.

Classification algorithms provide statistical measures of fit to assess the effectiveness of the derived rules to predict outcomes. In this study, consistency was used as a measure of the derived rules predictive ability. Under this model, a rule appearing in all folds for all samples is considered a consistent and accurate predictor. Statistical measures of effectiveness can then be ascertained by averaging the individual measures.

19.3.3 Methodology

Evaluation of an algorithm's predictive ability is best carried out by testing on data not used to derive rules [21], thus all training was carried out using tenfold cross-validation. Cross-validation is a standard method for obtaining an error rate of the learning scheme on the data [21]. Tenfold cross-validation splits the data into a number of blocks equal to the chosen number of folds (in this case 10). Each block contains approximately the same number of cases and the same distribution of classes. Each case in the data is used just once as a test case and the error rate of the classifier produced from all the cases is estimated as the ratio of the total number of errors on the holdout cases to the total number of cases. Overall error of the classification is then reported as an average of the error obtained during the test phase of each fold.

Variations in results for each iteration in a cross-validation occur depending on the cases used in the training and holdout folds, which can lead to differences in overall results. Given the uneven distribution of cases in the dataset and the possible influence of case selection in folds, the tenfold cross-validation was repeated 10 times on randomised data. The following results section will refer to each individual cross-validation as a "run".

19.4 Results

Preliminary analysis of variables was used to evaluate the worth of a subset of variables by considering the individual predictive ability of each feature along with the degree of redundancy between them [9]. Although significant correlations between the suspicion variable and the outcome were not identified using statistical analysis (Table 19.3), the variable was removed from the analysis due to possible conflicts. Reasoning for the removal of the suspicion variable is provided in the discussion.

Table 19.3. Correlation analysis of suspicion and outcome status

	Runaway	Suicide	Foulplay
Suspicion – runaway	0.57	−0.31	−0.43
Suspicion – suicide	−0.27	0.57	−0.23
Suspicion – foulplay	−0.30	−0.20	0.59

The status variable, which identifies if the missing person was alive, deceased or hospitalized when located, was removed from the dataset due to high correlation with the outcome. The remaining 24 variables were used for analysis.

A "best first" classifier subset evaluation, using the *J48.PART* algorithm, was used to estimate the most promising set of attributes. From the entire dataset, the following variables were identified as having the highest predictive ability; age category; gender; marital status; residential status, appearance, person(s) reporting, previous history of suicide, mental health status, and whether their behaviour was typical behaviour. No variables were identified as redundant.

19.4.1 Rule Determination

The *WEKA J48.PART* classifier derived 22 rules for each of 10 repetitions of the tenfold cross-validation. Although the number of rules generated was consistent, the accuracy measures were inconsistent Fig. 19.1. On average, 71% or 253 of the 357 cases were classified correctly.

The confusion matrices for each classification, which were combined, are shown in Table 19.4 and detail the number of correctly and incorrectly classified cases in

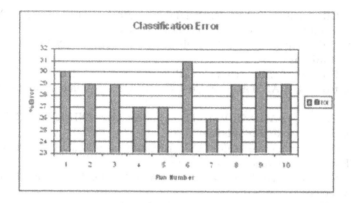

Fig. 19.1. Error for ten runs of a tenfold cross-validated classification using J48.PART

Table 19.4. Classification confusion matrix

Actual	R	R	R	S	S	S	F	F	F
Classed	R	S	F	R	S	F	R	S	F
Run 1	207	25	18	32	20	2	23	7	23
Run 2	210	25	15	31	19	4	27	1	25
Run 3	216	22	12	34	16	4	25	5	23
Run 4	215	16	19	30	21	3	23	4	26
Run 5	215	16	19	30	21	3	23	4	26
Run 6	202	30	18	34	18	2	21	6	26
Run 7	210	23	17	22	28	4	24	3	26
Run 8	206	28	16	28	23	3	25	5	23
Run 9	201	25	24	27	22	5	23	3	27
Run 10	207	28	15	32	21	1	22	5	26
Avg	209	23	17	30	21	3	23	4.3	25
% Error	16				61				53

each sample for both algorithms. While the overall predictive accuracy was 71%, the confusion matrix shows that the *J48.PART* algorithm correctly classifies 84% of runaway (R) cases, and only 39% of suicide (S) and 47% of foulplay (F) cases.

An analysis of the rules derived during each training fold of a single run identified a total of 195 distinctly different rules and only 1 rule appeared consistently in all folds.

In general, although the number of derived rules from the repeated cross-validations was consistent, the constitution of rules was inconsistent. Variations in the number variables constituting a rule and the structure of the underlying decision trees produced inconsistencies. Similarly to the issues of rule consistency occurring within folds of a single classification run, a total of 120 distinctly different rules were identified across all 10 runs. Interestingly though, 10 of the 22 rules from each run appeared consistently across all runs, providing a valid subset which are listed below.

- PHSUICID = unknown AND REPORTIN = immediate_family AND MENTAL_R = no_mental_health_problems_experienced:>> **runaway**
- PHSUICID = unknown AND REPORTIN = not_reported AND PUBLIC = public: >> **foulplay**
- PHSUICID = yes AND DEVIANTR = no_deviancy AND REGION = Sydney_metro AND MARITAL = single AND REPORTIN = immediate_family: **runaway**
- PHSUICID = yes AND DEVIANTR = no_deviancy AND RESIDENT = sharing_married_defacto: >> **suicide**
- SHORTERM = mental_health AND REGION = regional_NSW ANDCHARACTE = out_of_character: >> **suicide**
- PUBLIC = home AND DEPENDEN = no AND REPORTIN = immediate_family: >> **runaway**
- DEVIANTR = social_externalised AND REPORTIN = immediate_family AND ALCODRUG = no_drug_or_alcohol_problems_mentioned: >> **runaway**
- PHSUICID = unknown AND MARITAL = married_defacto AND PASTRUN = first_time: >> **suicide**
- PHSUICID = no AND ALCODRUG = no_drug_or_alcohol_problems_mentioned AND PUBLIC = public AND AGE_BEE = 18_25: >> **foulplay**
- ALCODRUG = no_drug_or_alcohol_problems_mentioned AND MENTAL_R = no_mental_health_problems_experienced AND PUBLIC = public AND AGE_BEE = 17_under: >> **foulplay**

19.5 Discussion

Although consistency was not achieved in the derived rules, pruning resulted in several valid rules that appeared consistently in all of the 10 cross-validations performed. The consistency of these rules makes them valuable supplements to police intuition. The variables identified in the subset evaluation as having predictive ability and those appearing consistently in derived rules were flagged as priorities for missing value reduction and re-evaluation for coding inconsistencies.

The presence of consistent predictor variables supports the need for quality control in data collection. Although data quality is a typical problem, police officers investigating missing persons often deal with conflicting, scarce and emotionally charged information. Knowledge of the key factors affecting the outcome provides a guide for police interviews and data capture. Although the rules derived are generally inconsistent, the consistent variables identified in this study are more useful perhaps than soft-computing in this regard.

The data obtained by police officers investigating reported missing persons consists of facts, judgments and model-based attitudes. Factual information such as gender, marital status and location last seen present few issues for data capture. Judgments are required when collecting information on variables like mental health problems, deviancy and drug and alcohol problems, and a more structured approach to the capture of this information may improve data quality. For instance, a field-based tool to prompt officers to explore specific areas of concern, such as the duration or type of drug abuse, may provide richer data for analysis and ultimately more comprehensive and effective prediction of outcomes. The suspicion variable is a model-based attitude, in itself reflecting an individual's prediction of outcome based on rules derived from the underlying variables. For this reason, suspicion is considered correlated with the outcome category and exists as a stand-alone prediction rather than a factor that may affect the outcome. Mental health problems also appear to be valid indicators although may not be as simple as a binary answer. Their predictive ability may be improved by a scaled system to indicate the severity of illness.

The algorithm implemented by the *WEKA J48.PART* produced inconsistent results. Given that the most consistent rules were sought, correct information may have been discarded to achieve consistency. Given this, recent work [12, 13] has considered the use the soft-computing methodologies for the missing persons problem. The predictive accuracy of the rule-based classifier was compared with an artificial neural network (ANN). The ANN achieved superior accuracy over the WEKA J48.PART rule based classifier, correctly predicting outcomes for 99% of cases. This issue is particularly pertinent given the vast array of "off-the-shelf" data mining software applications currently being applied to a diverse range of problems. Clearly, algorithms produce different results given the same training data and care must be taken to ensure the correct method is "on the job".

Although in this case ANNs offer improved predictive accuracy over rule-based classifiers, the nature of the problem domain requires rules or insights to support police officer intuition. To this end, a method of rule extraction from ANNs using a genetic algorithm has been used [12] to extract heuristics from the trained network. The resulting rule set was found to "cover" or explain 86% of cases in the dataset. Generalisation issues exist with the rule extraction method used, however, combining the use of an ANN and genetic algorithm appears to be a more appropriate approach to the missing persons problem.

Rules must be reliable in order to supplement police intuition and general "rules of thumb". Interesting results arose which tend to go against general intuition. Reliable associations between missing persons reported by a member of their immediate family (implying runaway) and missing persons married or in a de facto relationship (implying suicide) are unexpected patterns, as too is the limited predictability of time of day of the disappearance.

This study has provided insight into variables that have potential to accurately predict outcomes for missing persons cases and highlighted issues pertaining to data capture, preprocessing and rule determination. Data capture and quality may be improved through the use of structured systems, such as forms or computer collection of data using palm-tops, to prompt and guide police officers to ensure all pertinent data is consistently collected. There are indications that the missing persons problem is an instance of a "random problem" [14], a problem of a high Kolmogorov complexity [15].

Acknowledgement

This research was carried out under a grant from the Sir Maurice Byers Research Fellowship of the NSW Police Service. Foy thanks the NSW Police Service Missing Persons Unit, notably Geoff Emery and Jane Suttcliffe, for their assistance with data capture.

References

1. Newiss G (1999) The Police response to missing persons, Police Research Series Paper 114, Home Office
2. Salfati CG (2000) The Nature of Expressiveness and Instrumentality in Homicide: Implications for Offender Profiling, Homicide Studies, 4(3):265–293
3. Conwell Y, Duberstein PR, Cox C, Herrmann JH (1996) Relationships of age and axis I diagnoses in victims of completed suicide: A psychological autopsy study. *American Journal of Psychiatry.* 153(8):1001–1008
4. Quinlan JR (1993) *C4.5: Programs for machine learning.* Morgan Kaufmann, San Mateo, CA
5. Crémilleux B (2000) Decision trees as a data mining tool, *Computing and Information Systems,* 7(3):91–97, University of Paisley
6. Blackmore K, Bossomaier T, Foy S, Thomson D (2002) *Data Mining of missing persons data.* In Proceedings of 1st International Conference on Fuzzy Systems and Knowledge Discovery (FSKD02), Orchid Country Club, Singapore, 18–22 November, 2002
7. Ragel A, Crémilleux B (1998) *Treatment of missing values for association rules,* In proceedings of the Second Pacific-Asia Conference on Knowledge Discovery and Data Mining (PAKDD-98), Melbourne, Australia, 258–270
8. Crémilleux B, Ragel A, Bosson JL (1999) *An interactive and understandable method to treat missing values: application to a medical data set.* In proceedings of the 5th International Conference on Information Systems Analysis and Synthesis (ISAS/SCI 99), Torres M Sanchez B, Wills E (Eds.), Orlando, FL, 137–144
9. Witten I, Frank E (2000) Data mining: Practical machine learning tools and techniques with Java implementations. Morgan Kaufmann: San Francisco
10. Ripley BD (1996) Pattern recognition and neural networks. Cambridge University Press
11. Frank E, Witten IH (1998) *Generating accurate rule sets without global optimisation.* In Proceedings ICML'98–International Conference on Machine Learning, Shavlik J (Ed.), Morgan Kaufmann, Madison, Wisconsin, pp. 144–151
12. Blackmore KL and Bossomaier TRJ (2003) *Using a Neural Network and Genetic Algorithm to Extract Decision Rules.* In proceedings of the 8th Australian and New Zealand Conference on Intelligent Information Systems, 10-12 December 2003, Macquarie University, Sydney, Australia
13. Blackmore KL and Bossomaier TRJ (2003) Soft computing methodologies for mining missing person data. *International Journal of Knowledge-based Intelligent Engineering Systems.* Howlett RJ and Jain LC (Eds). 7(3), UK
14. Abu-Mostafa Y (1986). Complexity of random problems, in *Complexity in Information Theory,* Abu-Mostafa Y (Ed.), Springer-Verlag
15. Li M and Vitanyi P (1997) An introduction to Kolmogorov complexity and its applications, 2nd ed., Springer

Centralised Strategies for Cluster Formation in Sensor Networks

Malka N. Halgamuge[1], Siddeswara M. Guru[2] and Andrew Jennings[3]

[1] ARC Special Research Centre for Ultra-Broadband Information Network, Department of Electrical & Electronic Engineering, University of Melbourne, VIC 3010, Australia
malka.nisha@ee.unimelb.edu.au
[2] Department of Mechanical and Manufacturing Engineering, Mechatronics Research Group, University of Melbourne, Australia
s.guru@pgrad.unimelb.edu.au
[3] School of Electrical and Computer Systems Engineering, RMIT University, Melbourne, Australia
ajennings@rmit.edu.au

Abstract. Cluster based communication protocols for sensor networks are useful if the cluster formation is energy efficient. Centralised cluster formation can be applied when the sensor network is hybrid, fully wireless but less mobile, or fully wireless with a known sensor location. In a cluster based communication protocol, sensors within a cluster are expected to be communicating with a cluster head only. The cluster heads summarise and process sensor data from the clusters and maintain the link with the base station. The clustering is driven by the minimisation of energy for all the sensors. The paper extends the energy equations to hybrid and wireless sensor networks. Recent developments in generic clustering methods as well as customized optimisation methods based on Genetic Algorithms are used for centralised cluster formation. The paper compares the simulation results of 4 clustering algorithms.

Key words: Energy optimisation, wireless sensor networks, hybrid sensor networks, genetic algorithms, Fuzzy C-mean, clustering.

20.1 Introduction

A collaboration of large number of sensors in a network to perceive events in the environment is considered as the major strength of sensor networks. The emergence of wireless networks and ad hoc networks has influenced the research in this area [1, 6]. Various aspects of sensor networks such as data aggregation or fusion [3, 5] packet size optimisation [19], target localisation [25], network protocols [11, 13, 16, 20, 24] are discussed in the literature with respect to crucial energy limitations.

Generally, there are three methods that can be considered as possible networking protocols: direct communication, multi-hop routing, and clustering. As direct

Malka N. Halgamuge et al.: *Centralised Strategies for Cluster Formation in Sensor Networks*, Studies in Computational Intelligence (SCI) **4**, 315–331 (2005)
www.springerlink.com

communication between the base and a large number of sensors is extremely energy consuming, and the multi-hop routine is considered globally inefficient, clustering seems to be the appropriate method to use.

In order to send information from a very high number of sensor nodes to the base station, it is necessary and economical to group sensors into clusters. Each cluster will contain a cluster head (CH). Each CH gathers and sends data, from its group of sensors, to the base station [13].

Given the parameters for variation of energy consumption in the nodes, there are three main problems: how many sensors should be connected to each CH, how many clusters are needed, and where each CH should be positioned. The clusters of sensors can be overlapping in a realistic scenario.

Low Energy Adaptive Clustering Hierarchy (LEACH) [14] is a cluster based communication protocol where cluster heads collect information from the sensors, process and transmit further to the base stations. LEACH estimates cluster heads in a distributed manner without being aware of the location of the sensors, i.e., no information is exchanged between nodes regarding cluster formation or base stations. LEACH-C (C for centralised) [12] presents a clustering procedure using the location awareness of each sensor at the base station. Each node transmits the location information to the base station where an optimal cluster formation is performed and the decision is sent back to the sensors. The advantage of centralized clustering is that data processing at the base station can be conducted using substantial energy resources, as energy is no longer dependent on batteries. The disadvantage of centralised cluster formation is that it can only be used when the sensor network is hybrid, fully wireless but less mobile, or fully wireless with a known sensor location.

The name "sensor network" is synonymously used to refer a fully wireless sensor network. Real applications sometimes require them to be hybrid, i.e., wired and wireless combined. A possible application of a hybrid sensor network is in an automated factory where all the robots contain networked sensors. In such a situation, it may be possible to have wired connections for all the sensors within the robot (a sensor cluster), and wireless connections between the robots via CH. As the number of sensors used in this application can be very large and costly (macro sensors), it should be attempted to make sensors more economical. This can be achieved by not building each sensor with a battery, radio, a powerful processor and memory, but by having a few powerful wireless enabled sensors and more unsophisticated sensors that can only be networked using wired connections. The powerful sensors can share the processing capability, memory and power with the less sophisticated sensors. Sensors should be clustered to make it more economical, and powerful CHs should be associated with each cluster. CHs will have the capability to organise themselves in an ad hoc way, communicate with the remote system and to process the data, thus reducing the cost of having large numbers of powerful sensors. The role of clustering in hybrid sensor networks is to establish the wired clusters and the placement of the wireless sensors or cluster heads.

In most real applications sensor field in 3-D can be converted into a 2-D field. Therefore, in this study we assume that the nodes are randomly distributed in a 2-D plane. The clustering algorithms presented in Sect. 20.2 describe the various options of centralised cluster formations. Section 20.3 presents the energy equations for consideration in wireless and hybrid sensor systems. Section 20.4 describes the implementation options for various clustering algorithms. Section 20.5 compares the results of 4 clustering algorithms and Sect. 20.6 concludes the paper.

20.2 Centralised Clustering of Sensors

Remote management and control of sensors have received attention recently due to the possible use in E-commerce to allow potential customers to experience or inspect the products using the Internet. Sensors can provide feedback on the behaviour of the product itself and the environment when the remote user instructions are carried out. When many such sensors are used (for example) in a large mechanical structure, the supply of power to the sensors, which are ideally distributed throughout, become a limitation. It is therefore necessary to form clusters of sensors with a CH associated to each cluster. The CH should be ideally placed in the optimised cluster centre, which is not necessarily the geometrically optimised cluster centre. The latter may be located inside the mechanical structure making it impossible to place the CH and it is also likely that other environment related factors must be taken into account in selecting the location of CH for a particular cluster of sensors.

Cluster formation for a collection of randomly scattered sensors is discussed in this section. In the first set of techniques discussed in this paper (Sect. 20.2.1), the clustering occurs according to generic cluster formation techniques and are mainly a function of the location of the sensors. These clustering algorithms are well established and a few of them are even available as functions in MATLAB [17]. For our purpose, clustering should be driven by the minimisation of energy for all the sensors. Due to the relationship between energy minimisation and the location of sensors, it is important to investigate those traditional clustering algorithms often neglected by the research community in the networking area.

Generic optimisation techniques can be customised for efficient formation of clusters. In Sect. 20.2.2 a Genetic Algorithms (GA) based method is discussed.

20.2.1 Generic Clustering Methods Based on Sensor Location

Fuzzy C-Mean Algorithm

Bezdek [2] introduced the fuzzy C-mean (FCM) algorithm. The FCM based algorithm is a data clustering technique wherein each data point belongs to a cluster to some degree that is specified by a membership grade. The FCM allows each input vector to belong to every cluster with a fuzzy truth-value (between 0 and 1) computed [15]. The algorithm assigns an input to a cluster according to the maximum weight of the input over all clusters. For example an input vector with equal distance to two clusters will have the same truth-value. Fuzzy c-means algorithm uses the reciprocal of distances to decide about the cluster centers. This representation reflects the distance between a given input and the cluster center but does not differentiate the distribution of the clusters [7].

K-Mean Algorithm

In the K-mean algorithm the goal is to divide the N input samples into K clusters, minimizing some metrics relative to the cluster center. Various metrics can be minimized such as

- Distance from the cluster center to object

• Total distance between all objects and cluster centre

The clusters are defined by their centres c_i (centre unit). First a random data assignment is made, and then the goal is to partition the data in sets S_i to minimize the Euclidean distance between the data partition N_i and the cluster centres c_i: [1]

$$J = \sum_{i=1}^{k} \sum_{n \in S_i} |X_n - c_i|, \qquad (20.1)$$

where X_n is the current pattern. The data centers c_i are defined by

$$C_i = \frac{1}{N_i} \sum_{n \in S_i} x_n$$

K-mean clustering requires a batch operation when the samples are moved from cluster to cluster to minimize (20.1) [18].

Subtractive Clustering Algorithm

Subtractive clustering assumes each node is a potential cluster center and calculates the measure of the likelihood that each node defines the cluster center, based on the density of surrounding nodes.

The algorithm:

• Selects the node with the highest potential to be the first cluster center
• Removes all nodes near to the first cluster center, in order to determine the next node cluster and its center location
• Iterates this process until the entire nodes are within radii of a cluster.

The subtractive clustering method is an extension of the mountain clustering method proposed by R. Yager [23].

If one is not certain about the number of clusters needed for a given set of nodes, then the subtractive clustering method is an effective way of finding the number of clusters and their centers.

20.2.2 Customised Optimisation Algorithms for Cluster Formation

The GA is based on the collective learning process within a population of individuals, each of which represent a search point in the space of potential solutions to a given optimization problem. The population initialised arbitrarily is moving towards better regions of the search space by means of randomised processes of selection, mutation and recombination.

The major advantage of GA lies in its ability to solve many large complex problems where other methods have not been satisfactory. Its ability to search populations with many local optima makes it increasingly popular for use in many applications but primarily in optimisation and machine learning. The areas of application include function optimisation, image processing, system identification and control, robotics, communication networks, and clustering.

A typical Genetic Algorithm is based on three operations known as selection, crossover and mutation and a cost function known as the fitness function. The selection operator's purpose is to attempt to promote the information contained in most fit individuals while discarding the weaker individuals. Crossover allows the exchange of information similar to that of a natural organism undergoing reproduction. The pair of individuals promoted by the selection operator swap information. This produces a new individual made up of parts of the parent individuals.

The main idea of cluster fitting is to assume the shape of a possible cluster boundary in advance and attempt to fit a number of such predetermined clusters to cover the range of sensors. The generalised model suggested was a simple ellipse with five parameters that allowed it to change position, size and rotate. These five parameters are encoded to allow manipulation by the optimisation algorithm. Any number of these ellipses could be used to obtain the best approximation. The optimisation algorithm also requires an evaluation function that returns a value representing the fitness of each model. Evolutionary algorithms GA and Particle Swarm Optimisation (PSO) algorithm were used as the optimisation algorithm for fitting cluster models for binary images [4].

The algorithm then initialises the optimisation procedures with the appropriate utilities and an initial random population of encoded models. The next step is to begin the iteration of the optimisation algorithm. The encoded models are then manipulated to find the model with the highest evaluation.

Ellipses were used as they are reasonably simple to represent yet provide significant flexibility. Each set of model ellipses is represented by an array of parameters, [Cluster 1, Cluster 2 ... Cluster n]. The ellipse is described by 5 parameters $[w, h, x, y, \theta]$, where w and h are width and height of the ellipse, x and y are coordinates of the centre of ellipse, and θ is the angle of rotation. Obviously, an ellipse will become a circle with appropriate selection of parameters, if an omni directional transmission range is assumed for the cluster head.

In order for the optimisation algorithms to be able to determine the better fitting individuals in the population from the poorly fitting individuals, each individual needs a measure of its fitness evaluated. The fitness function should reward placing clusters in areas with higher density and offer penalties for leaving sensors without a cluster head.

Both GA and PSO based algorithms were recently developed based on the Incremental Optimisation method suggested in [11]. The idea is inspired with the cluster formation in sensor networks in mind. However, the first applications of Incremental Optimisation are in Image Clustering and Segmentation. Incremental Optimisation uses an optimisation algorithm, such as a GA or PSO, to fit a number of models of a possible cluster to a distribution of sensors or an original image.

Rather than trying to manipulate a large number of parameters that encode a number of ellipses either defined by the user or automatically determined through encoding into the bit string (Chromosome), this algorithm incrementally introduces elliptic models, after a series of algorithm steps in improving the solution. For instance, during the first set of iterations of the algorithm, the model will have a single ellipse. The optimisation program evolves a population of these models until a satisfactory solution that minimises the fitness function is evolved. It will stop when an initial threshold is reached or the maximum number of iterations has been performed. When either of these conditions has been met another ellipse is added to the model.

A new population of the new model is created using the fittest model from the previous iteration preserving the parameters of this model. The new population consists of models that all have the fittest from the previous iteration as well as a random additional ellipse. The optimisation routine then evolves this population until a predetermined increase is achieved upon the fittest model from the previous iteration. This process continues until the desired number of ellipses has been added to the model.

The total number of evolutions (i.e. sets of iterations without incrementing the number of models) performed by the optimisation algorithm is limited to the number of ellipses multiplied by the maximum number of evolutions.

It can also be seen that generally, the incremental version, irrespective of whether GA or PSO is used, works better than the regular version [9]. In this study the incremental version of GA (IGA) has been used.

20.3 Energy Equation for Wireless and Hybrid Sensor Networks

Recent advances in sensor technology based on Micro-Electromechanical Systems (MEMS), low power electronics and Radio Frequency Designs have paved the way to the development of relatively inexpensive low power wireless microsensors [21]. It is not expected that those microsensors are reliable and accurate as their expensive macrosensor counterparts. But their smaller size and low cost allow them to be used in large numbers, and collective distributed decision making in an ad hoc way enables the network to be fault tolerant. The energy model developed will be for clustering supported communication for fully wireless operation (such as in a microsensor network) and for a hybrid network.

Hardware architecture of a sensor node consists of a sensor module that includes the sensor itself and the A/D converter, the data and control module responsible for data preprocessing if necessary, the radio communication module, and a power module that provides energy for the other three modules. Sensors will be able to communicate as long as the battery in the power module is not dead. Therefore, it is important to consider the energy minimisation in the formation of clusters.

The energy model used in this work is adapted from [22], and summarised in this section. Energy consumption of a wireless sensor node transmitting and receiving data from another node at a distance d can be divided into two main components: Energy used to transmit, receive, and amplify data (E_T) and energy used for processing the data (E_p), mainly by the microcontroller. Assuming the transfer of b-bit packets by the sensor unit, and using the values: transmit amplifier constant as $0.1\,\mathrm{nJ/bit/square}$ meter, and transmission/receiving energy loss as $50\,\mathrm{nJ/bit}$, adapted from [12], E_T can be derived as

$$E_T = \left(100 + 0.1 * d^2\right) b \, .$$

E_p has two components: the energy loss due to switching and the energy loss due to leakage. It can be shown that E_p is a function of the supply voltage V and the frequency f associated with the sensor. Microcontroller data sheets can provide the average capacitance C_{avg} switched per cycle, and the average number of clock cycles

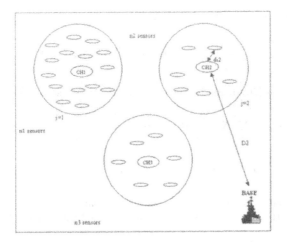

Fig. 20.1. Networked Sensor Clusters ($k = 3$)

needed for the task N_{cyc} is assumed as known. Therefore total capacitance switched is $N_{\text{cyc}} * C_{\text{avg}}$. Assuming I_0 as the leakage current:

$$E_p = N_{\text{cyc}} * C_{\text{avg}} V^2 + V(I_0 e^{v/nv_t}) \frac{N_{\text{cyc}}}{f} ,$$

the total energy loss (E) of the sensor is:

$$E = \left(100 + 0.1 * d^2\right) b + N_{\text{cyc}} * C_{\text{avg}} V^2 + V(I_0 e^{v/nv_t}) \frac{N_{\text{cyc}}}{f} . \qquad (20.2)$$

Assume a k set of sensor clusters (see Fig. 20.1) with $n_j + 1$ sensors in the cluster C_j, where $1 < j < k$, as the energy optimised network architecture of a randomly distributed n_{total} number of sensors, where

$$n_{\text{total}} = \sum_{J=1}^{k} (n_j + 1) . \qquad (20.3)$$

Considering the cluster C_j, in which a sensor S_{ij} has the distance d_{ij} to the cluster head CH_j, and CH_j has the distance D_j to the base station, the energy loss of all the sensors can be derived using (20.2). Leakage currents I_0 can be as large as a few mA for microcontrollers, and the effect of leakage currents can be ignored for higher frequencies and lower supply voltages. Assuming the leakage current I_0 as negligible, (20.2) can be simplified further. It is also assumed that transmission and receiving energy used by a CH is r times stronger than that for a sensor, and additional data processing/aggregation tasks associated with a CH cause a q times stronger average clock cycles needed for a task in comparison to a sensor, where $r, q > 1$.

Energy used by all k cluster heads E_k is:

$$E_k = \sum_{j=1}^{k} ((100r + 0.1 D_j^2) b + q N_{\text{cyc}} C_{\text{avg}} V^2) . \qquad (20.4)$$

Energy used by all the other sensors E_s is:

$$E_s = \sum_{j=1}^{k} \sum_{i=1}^{n_j} ((100 + 0.1d_{ij}^2)b + N_{\text{cyc}}C_{\text{avg}}V^2) \,. \tag{20.5}$$

From (20.4) and (20.5), the total energy loss for the sensor system (E_{total}) is:

$$E_{\text{total}} = 0.1b \sum_{j=1}^{k} \sum_{i=1}^{n} \left(\left(d_{ij}^2 + \frac{D_i^2}{n_j} \right) + (n_{\text{total}} - k)(100b + N_{\text{cyc}}C_{\text{avg}}V^2) \right.$$
$$\left. + K(100rb + qN_{\text{cyc}}C_{\text{avg}}V^2) \right) \,. \tag{20.6}$$

In order to study the optimisation with the first part of (20.6), it was assumed that $r = q = 1$ for the simulation purpose. The simplified total energy loss (E_{total}) is:

$$E_{\text{total}} = 0.1b \sum_{j=1}^{k} \sum_{i=1}^{n} \left(\left(d_{ij}^2 + \frac{D_i^2}{n_j} \right) + (100b + N_{\text{cyc}}C_{\text{avg}}V^2)n_{\text{total}} \right) \,. \tag{20.7}$$

It is clear that the first part of (20.6) and (20.7) or energy due to distances (E_{dd}.) is the only component that can be optimised independently from parameters related to the micro-controller and the supply voltage used. Consequently, E_{dd} was used as the energy loss for formation of clusters:

$$E_{dd} = \sum_{j=1}^{k} \sum_{i=1}^{n} \left(\left(d_{ij}^2 + \frac{D_i^2}{n_j} \right) \right) \,. \tag{20.8}$$

Energy consumption for a wired sensor could be derived from the energy equation $E = \int_0^1 I^2 R \, d\tau$ where R is the resistance of wire connecting the sensor to CH and I is the current of operation, $\tau =$ time variable and t is the total time spent. R can be defined as $R = \frac{\rho l}{A}$, and applied in energy equation. Assuming that energy loss due to inductive and capacitive energy is negligible, the energy consumption of each sensor node will be:

$$E = b^2 \rho^1 lt \,, \tag{20.9}$$

where $\rho^1 = \frac{\rho*\text{const}^2}{A}$ and $I = \text{const} * b$.

Assuming that all distances d_{ij} in Fig. 20.1 are wired connections, and considering the cluster C_j, the energy of all the sensors can be derived from above equations.

Energy used by all the other sensors E_{sw} derived from (20.9) and (20.6):

$$E_{sw} = \sum_{j=1}^{k} \sum_{i=1}^{n_{ij}} [b^2 \rho^1 d_{ij} t + N_{\text{cyc}}C_{\text{avg}}V^2] \,. \tag{20.10}$$

Energy used by all k cluster heads E_{kw} is the same as in (20.4). From (20.4) and (20.10) the total energy loss for the hybrid sensor system (E_{htotal}) is:

$$E_{\text{htotal}} = \sum_{j=1}^{k} \sum_{i=1}^{n_j} \left(\left(0.1b \frac{D_j^2}{n_j} + b^2 \rho^1 d_{ij} t \right) + N_{\text{cyc}}C_{\text{avg}}V^2 n_{\text{total}} + 100bk \right) \,. \tag{20.11}$$

It is clear from the above equation that the first and the third components depend on the cluster and distance related parameters, and therefore assuming $r = 1, E_{hdd}$ the energy component that can be used for optimisation of clusters can be written as:

$$E_{hdd} = \sum_{j=1}^{k} \sum_{i=1}^{n_j} \left(0.1 \frac{D_j^2}{n_j} + b\rho^1 d_{ij} t \right) + 100\, k \,. \qquad (20.12)$$

It can be observed that for hybrid derivation of "interesting" components of energy equation has the influence of the number of clusters formed very strongly, and the number of bits used for transmission will also have an effect on the cluster formation.

20.4 Implementation Options

20.4.1 GA Based Implementations for Wireless Sensor Networks

The proposed algorithm for cluster formation for a fully wireless network based on IGA works over a few steps or evolutions. In the first evolution, it starts with the minimum possible number of clusters acceptable for the network, which is assumed here as 1 cluster. After few iterative steps, and evaluating the fitness function, it may reach the point that the best individual cannot be significantly improved. Then, a new cluster is added while keeping the parameters of the old cluster settings. The second evolution begins with an initialised random setting for the new cluster and the best possible settings found for the old clusters. The incrementing cluster process continues until the fitness function is satisfactory, and the best possible solution is reached.

For wireless transmission, it is assumed that transmission is omni-directional, and therefore the clusters are circular (as opposed to the elliptic model used in the previous application of IGA). Further, it is possible to set a radial parameter R, which is the maximum distance the radio transmission can reach within the cluster as close as possible to a user set parameter r.

The encoding for the first evolution has 3 strings; each string can be of 8 bits. The three fields represent R, $X1$, and $Y1$. R is the radius of the cluster, and $(X1, Y1)$ is the centre point of the cluster on XY plane. All the sensors within the radius of R around $(X1, Y1)$ belong to cluster 1. Consequently, there are five fields for the second evolution. The five fields represent R, $X1$, $Y1$, $X2$, and $Y2$. R is the radius of each cluster, which is assumed to be the same for all the clusters, but the algorithm can handle a different radius for each cluster, if it is needed. $(X1, Y1)$ and $(X2, Y2)$ are the centre points of each cluster on XY plane, where R, and $(X1, Y1)$ are the best settings obtained from evolution 1. The encoding will be extended in a similar way to 7, 9, 11... fields as the incrementing progresses.

There will be an unavoidable overlapping of clusters in this algorithm. However, it is possible to allow a P percentage of overlapping between two clusters,

R	X1	Y1	X2	Y2
8 bits	8 bits	8 bits	8 bits	8 bits

Fig. 20.2. Encoding for the second evolution for a chromosome with 5 parameters

and strongly discourage the overlapping of 3 or more clusters. Assigning each sensor a counter to calculate the number of CHs it is connected, and processing $\{n_0, n_1, n_2, n_3\}$, where

1. n_0 = number of sensors without any connection to a CH
2. n_1 = number of sensors with connection only to a single CH
3. n_2 = number of sensors with connection only to two CHs
4. n_3 = number of sensors with connection to three or more CHs

The fitness function should guide the optimisation process and include penalty for having high numbers of 3 or more overlapping of clusters (n_3) and no cluster membership (n_0).

The fitness function suggested is:

$$E_{dd} + 2100(n_0 + n_3) + \left(\text{If } (n_1 = 0) \Rightarrow 500(n_2) , \right.$$

$$\left. \text{else} \left(\frac{n_2 * 100}{n_1 + n_2} - P \right) \right) + 10(R - r) . \tag{20.13}$$

The first term is the distance-based component of the energy equation derived for the cluster based wireless sensor network discussed in the previous section. The second term is the penalty term to discourage high n_0 and n_3. The third term is to make sure that overlapping of two clusters is not too far away from the desired percentage of P. It was anticipated that at the beginning of IGA, there may be no clusters containing sensors $n_1 = n_2 = 0$, in which case this term will be zero. It is also possible that $n_1 = 0$ while n_2 is non zero, in which case a considerable penalty based on n_2 is added to the fitness function. The last term shows the penalty, if the automatically set cluster radius "R" is too far away from the desired range "r". All constants, e.g. 2100 and 500, are experimentally set for this problem.

If the radius of the cluster needs to be fixed, the last term can be deleted, and the encoding of R deleted. For the rest of calculations, it can be assumed that all clusters are of R radius. It is also possible to use the GA if the number of cluster heads is predetermined.

20.4.2 GA Based Implementation for Hybrid Sensor Networks

For hybrid networks, there is no need to limit the clusters to the circular model, and therefore encoding will be for elliptic clusters. Therefore, the chromosome size is the number of ellipses used in each evolution × 5 (parameters) *times* 8 bits (or number of ellipses *times* 40 bits). There will be no R parameter needed.

The fitness function can be calculated according to (20.14).:

$$E_{hdd} + 2100(n_0 + n_3) + \left(\text{If } (n_1) = 0 \Rightarrow 500(n_2), \text{ELSE} \left(\frac{n_2 * 100}{n_1 + n_2} - P \right) \right) . \tag{20.14}$$

Note that the distance-dependent energy component is now dependent on the number of bits b as well. Further, there is no need to set a desired value for R. However, it is quite easy to add a term to reflect the cost relationship between wires and CHs, if cost should be minimised at the same time, as described in the next section.

20.4.3 Implementation of Generic Cluster Algorithms

Matlab is used for simulation of the fully wireless sensor networks. The graphical user interface with a 2-D visualisation allows the simulation of up to 100 nodes. The user selects the total number of sensor nodes to be randomly distributed and the cluster algorithm out of Fuzzy C-means, K-mean, Subtractive, and GA based solutions.

For the same number of nodes the position of the node may be different at different instance of selection. E_{dd} is used for the evaluation of the quality of the result [10].

Few generic clustering techniques have been used in evaluating the cost effectiveness of generated clusters for hybrid sensor networks. The GU Interface and the cost effective implementation was developed by [86] using the cost minimisation equation:

$$\text{TOTAL\_COST} = \text{Cost\_of\_wire\_unit\_length}*$$
$$(\cos t\_ratio * k + \text{total\_length\_of\_wires}) .$$

Where cost_ratio (Cr) is defined as the cost of a CH unit over the cost of unit length of wire (Cw), and k is the number of clusters generated:

$$\text{TOTAL\_COST} = c_w \left(c_r * k + \sum_{j=1}^{k} \sum_{i=1}^{n_j} (d_{ij}) \right) . \qquad (20.15)$$

The user can set the number of nodes and the ratio of cost of CH to the cost of wiring per unit length (cost ratio) as shown in Fig. 1. The cost of wiring is based on the actual cost of wiring from the center unit to all the sensor nodes (in the cluster) per unit length. Cost of the centre unit is the cost of installing a centre unit for each cluster.

The K-mean based algorithm was used with a single CH ($K = 1$) and the cost of wiring and cost of CU are calculated for a given cost ratio, then one more cluster is added (now $K = 2$) and the nodes are grouped. If the total cost is greater then the total cost when $K = 1$, then condition of convergence is achieved, and the optimum cost is at $K = 1$, otherwise continue with $K = 3$.

Fuzzy C-mean based solution is implemented in a similar manner, and the subtractive clustering is also implemented (without the need to indicate the number of clusters). In most cases [8] either k-mean or Fuzzy-C-mean algorithms provide the best solution.

The fitness function for energy optimisation in hybrid networks can be extended to include the cost consideration. As C_w is a parameter that can be taken out, the new fitness function for energy cost optimisation is:

$$E_{hdd} + 10(n_0 + n_3) + (\text{if } (n_1) = 0 \Rightarrow 500(n_2) ,$$
$$\text{else} \left(\frac{n_2 * 100}{n_1 + n_2} - P \right) + \frac{\text{TOTAL\_COST}}{c_w} . \qquad (20.16)$$

20.5 Simulation Results

Simulations were conducted for centralised cluster formation of wireless sensor networks. We used 100-node networks that are randomly placed in a 2-Dimensional problem space [0:10, 0:10]. The base station is placed at location (5, 15), i.e. outside, but not far away from the sensor field. Both generic algorithms (Fuzzy C-mean, K-mean, and Subtractive) as well as customised GA are applied to several sets of nodes. As the results are consistent, we present here only for two sets of 100-node data.

We have used the simple radio model as discussed in Sects. 20.3 and 20.4 to calculate the energy dissipation to transmit and receive data from the node. Once the cluster-head (CH) and the position of the cluster-head is identified, if there is no node in that position then a node nearest to the cluster-head position will become a cluster-head. For our experiment, we have restricted the transmission range of each node to 3 units. This will help in enhancing the life span of each node. However the node, which will become a cluster-head, will not have any restriction on the transmission range.

In our experiment, the area of the problem space is 100 sq. units. Considering the transmission range of each sensor as 3 units and omni-directional, the minimum threshold number of clusters needed to connect the entire sensor field is 4. However, the actual number of cluster heads generated could be more than the minimum number, if the base station is not too far away from the sensor field, as is the case here. Otherwise, it is possible to get a solution containing a number of cluster heads much lower than the minimum threshold calculated above.

In K-mean based clustering, we randomly place the cluster-head and each cluster-head will form a cluster based on the Euclidian distance measure. The algorithm will be terminated when there is no significant improvement in the position of the cluster-head. The selection of K during the start of the algorithm depends on the criterion explained in the previous section. For our experiment, $K \geq 4$.

In FCM-based clustering, the cluster-heads are randomly distributed and each node will be assigned a membership grade for each cluster-head. Iteratively the cluster-head will move to the right direction improving the membership grade.

In subtractive algorithm based clustering, each node is set with an influence range. Here each node is considered as a potential cluster-head. The algorithm calculates the measure of the likelihood that each node would define a cluster. The node with the highest surrounding density will be a cluster-head.

For K-mean, and FCM clustering we choose 4, 6, and 8 clusters to be formed for the given example of 100 nodes distributed in the problem space. However, the number of minimum number of clusters needed to group all the nodes for the example network is 4. The Genetic Algorithm based method uses the fitness function described in (20.13) with $P = 0$ and $R - r = 0$. Convergence of the GA as the algorithm progresses is given in Fig. 20.3. According to Figs. 20.4 and 20.5, the GA based method has the least intra-cluster distance.

Experiments were conducted to find the total energy loss based on the energy model discussed in this paper. The results are tabulated in the tables shown in this section. The results from Tables 20.1 to 20.3 are for one set of nodes and results from Tables 20.4 to 20.6 are for another set of nodes. The entries for GA have the energy loss E_{dd} as well as the value from the fitness function. Clearly, for both datasets, the expected best results are provided in Tables 20.1 and 20.4 where the minimum numbers of clusters are generated, and in both datasets the GA based solution is the best one (highlighted). The GA does not give good results for clusters six and eight because of overlapping of clusters. The number of nodes involved in 2 or more clusters is greater when six and eight clusters. So intra-cluster distance of each

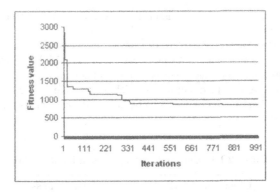

Fig. 20.3. Convergence of the GA as the algorithm progresses

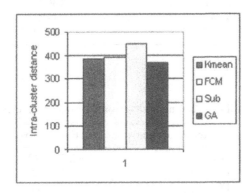

Fig. 20.4. Intracluster distance for 4 cluster solution

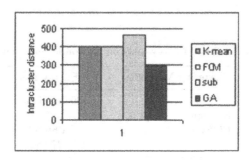

Fig. 20.5. Intra-cluster distance for 4 cluster solution using alternative data

clusters are more than the intra-cluster distance in a generic cluster where each node belongs to a single cluster. Thus E_{dd} is higher in the GA based algorithm as the number of clusters increases. For other tables, the Fuzzy C-mean as well as K-mean provides better results. This is to be expected as the GA based method has more directed fitness function ideally suited for the optimum number of clusters. If the user needs to have specified number of CHs other than the minimum threshold, the fitness function of the GA needs to be adjusted, or a generic clustering method should be used.

Table 20.1. Energy loss comparison for four clusters

Clustering Methods	Energy Loss
GA	**608.36 (766.13)**
K-mean	636.13
FCM	619.44
Sub	668.51

Table 20.2. Energy loss comparison for six clusters

Clustering Methods	Energy Loss
GA	800.71 (812.90)
K-mean	660.82
FCM	642.46
Sub	**755.19**

Table 20.3. Energy loss comparison for eight clusters

Clustering Methods	Energy Loss
GA	1094.07/1145.03
K-mean	691.25
FCM	**688.52**
Sub	779.82

20.6 Conclusion and Future Work

Clustering algorithms that can be used for sensor cluster formation in hybrid and wireless networks can be categorised into three types:

Table 20.4. Energy loss comparison for four clusters using alternative data

Clustering Method	Energy Loss
GA	**640.9 (814.88)**
K-mean	641.53
FCM	660.56
Sub	678.77

Table 20.5. Energy loss comparison for Six clusters using alternative data

Clustering Method	Energy Loss
GA	797.09 (875.52)
K-mean	**664.58**
Sub	733.31
FCM	668.11

Table 20.6. Energy loss comparison for eight clusters using alternative data

Clustering Method	Energy loss
GA	1005.74(1135.22)
K-mean	706.39
Sub	752.05
FCM	696.92

1. Clustering algorithms known in data mining (e.g. K-mean, Fuzzy C-mean, Subtractive)
2. Use of optimisation algorithms such as GA and PSO for cluster formation
3. Clustering algorithms known in networking research, typically in association with the routing protocol.

Most of the developments in category 3 are limited to fully wireless operation. Authors believe that there are many useful and practical applications of hybrid sensor networks in industry. In particular they are common for tension monitoring in large engineering equipment such as bridges, rockets, cars etc., and therefore it is worth considering the development of clustering algorithms for hybrid networks as well. Obviously, in hybrid networks, wired components within the cluster are fixed and no mobility is expected other than the possible movement of the cluster as a whole. In this work 3 algorithms in category 1 and the GA based algorithm in category 2 are compared for the best energy efficient solution.

This paper extends formulae for energy optimisation for fully wireless and hybrid networks. It also demonstrated the capabilities of generic clustering techniques using those formulae. Simulation results show that they perform reasonably well on cluster formation for sensor network examples considered in this paper. However, this may not provide satisfactory results. For example, if there are obstacles between the sensor nodes, and the relationship between the physical distance between the nodes and the energy dissipation due to transmission is no longer straightforward. In this work, two new methods are proposed to generate clusters directly based on energy considerations, and one such method is implemented using a GA.

If the base station is very close to the sensor field, the energy minimisation component E_{dd} encourages the formation of more clusters than if the base station is far away from the sensor field. Obviously, if the number of bits per packet increases the energy loss is higher, but it is unlikely to influence the cluster formation process. It can be generally concluded that Fuzzy C-mean and K-mean algorithm provide a good solution in all cases, and GA can be tailored to outperform those algorithms, either with the assumption that the user wants the optimum number of CHs or by including the exact number of CHs the user intends to have in a network into (20.13).

Acknowledgement

This work was supported by the Australian Research Council (ARC). The first author is with the ARC Special Research Center for Ultra-Broadband Information Networks, an affiliated program of National ICT Australia, Department of Electrical and Electronic Engineering, The University of Melbourne, Melbourne, VIC 3010, Australia.

References

1. Akyildiz IF, Su W, Sankarasubramaniam Y and Cayirci E (2002) A survey on sensor networks. Communications Magazine, IEEE 8:102–114
2. Bezdek JC (1981) Pattern recognition with fuzzy objective function algorithm. Plenum Press, USA
3. Boulis A, Ganeriwal S and Srivastava MB (2003) Aggregation in sensor networks: an energy-accuracy trade-off. In: IEEE International Workshop on Sensor Network Protocols and Applications, 2003. pp. 128–138
4. Brewster C, Farmer P, Manners J and Halgamuge MN (2002) An Incremental Genetic Algorithm for Model based Clustering and Segmentation. In: Fuzzy System and Knowledge Discovery, 2002. Singapore,
5. Cayirci E (2003) Data Aggregation and Dilution by Modulus Addressing in Wireless Sensor Networks. IEEE Communications Letters 8:
6. Estrin D, Girod L, Pottie GJ and Srivastava M (2001) Instrumenting the World with Wireless Sensor Networks: A Survey. In: International Conference on Acoustics, Speech, and Signal processing (ICASSp 2001). Salt Lake City, Utah,
7. Fasulo D (2002) An Analysis of recent work on clustering algorithms. Technical report # 01-03-02, University of Washington
8. Guru SM, McRae G, Halgamuge S and Fernando S (2002) Cost effective selection of sensor node Clusters. In: FSKD'02. Singapore, pp.
9. Halgamuge MN (2002) Energy Minimization of Ad-hoc Wireless Sensor Networks. Master Thesis (Minor), RMIT University
10. Halgamuge MN, Guru SM and Jennings A (2003) Energy efficient cluster formation in wireless sensor networks. In: 10th International Conference on Telecommunications, 2003. Tahite, French Polynesia, pp. 1571–1576

11. Heinzelman W, Kulik J and Balakrishnan H (1999) Adaptive Protocols for Information Dissemination in Wireless Sensor Networks. In: ACM MobiCom' 99. Seattle, Washington., pp. 174–185
12. Heinzelman W (2000) Application-Specific Protocol Architectures for Wireless Networks. Ph.D. thesis, Massachusetts Institute of Technology
13. Heinzelman WB, Chandrakasan AP and Balakrishnan H (2002) An application-specific protocol architecture for wireless microsensor networks. IEEE Transactions on Wireless Communications 4:660–670
14. Heinzelman WR, Chandrakasan A and Balakrishnan H (2000) Energy-efficient communication protocol for wireless microsensor networks. In: Proceedings of the 33rd Annual Hawaii International Conference on System Sciences, 2000. Maui, HI, pp. 3005–3014
15. Hutchinson A (1994) Algorithmic Learning. Clarendon Press, Oxford
16. Intanagonwiwat C, Govindan R and Estrin D (2000) Directed Diffusion: A Scalable and Robust Communication Paradigm for Sensor Networks. In: ACM MobiCom'00. Boston, MA, pp. 56–67
17. Palm WJ (2001) Introduction to Matlab 6 for Engineers. McGraw-Hill Publication,
18. Principe JC, Euliano NR and Lefebvre CW (2000) Neural and adaptive systems: Fundamentals through Simulations. John Wiley & sons,
19. Sankarasubramaniam Y, Akyildiz IF and McLaughlin SW (2003) Energy efficiency based packet size optimization in wireless sensor networks. In: IEEE International Workshop on Sensor Network Protocols and Applications, 2003. pp. 1–8
20. Shen C, Srisathapornphat C and Jaikaeo C (2001) Sensor Information Networking Architecture and Applications. IEEE Personal Communications 52–59
21. Wang A and Chandrakasan A (2001) Energy efficient system partitioning for distributed wireless sensor networks. In: IEEE International Conference on Acoustics, Speech, and Signal Processing, 2001. Proceedings. (ICASSP '01). 2001. pp. 905–908 vol. 2
22. Wang A and Chandrakasan A (2002) Energy-efficient DSPs for wireless sensor networks. Signal Processing Magazine, IEEE 4:68–78
23. Yager R and Filev D (1994) Generation of Fuzzy Rules by Mountain Clustering. Journal of Intelligent & Fuzzy Systems 3:209–219
24. Ye W, Heidermann J and Estrin D (2002) An Energy Efficient MAC Protocol for Wireless Networks. In: INFOCOMM'02. New York, pp.
25. Zou Y and Chakrabarty K (2003) Target Localization Based on Energy Considerations in Distributed Sensor Networks. In: IEEE International Workshop on Sensor Network Protocols and Applications.

Adaptive Fuzzy Zone Routing
for Wireless Ad Hoc Networks

Tawan Thongpook

Faculty of Engineering, London South Bank University, 103 Borough Road
London, UK SE10AA
tawan@s-t.au.ac.th

Abstract. In this chapter we introduce how to implement fuzzy logic for ad hoc
networks. Ad hoc networks are characterized by multi-hop wireless connectivity,
frequently changing network topology and the need for efficient dynamic routing
protocols. Designing ad hoc routing protocols is complicated by the fact that every
host is moving, leading to the dynamic nature of the topology. Routing and network
flow information are typically unavailable. Furthermore, incremental delay and residual link capacities are affected by the traffic being carried over other links. Wireless
links further introduce constraints such as limited and unreliable bandwidth, as well
as finite battery power in each mobile node which consequently restricts the transmission range. The decisions made in the design of the routing protocols are also
dependent on the underlying link layer and physical layer technologies.

Many protocols have been proposed to support routing of data from a source to
an arbitrary destination. Existing routing protocols fall into two categories: proactive
or reactive, with some possessing characteristics associated with both of them, also
referred to as hybrid protocols.

Hybrid protocols attempt to take advantages of proactive and reactive protocols.
For example, Zone Routing Protocol (ZRP) [1, 2, 3] divides the network into non-
overlapping routing zones .and runs independent protocols that study within and
between the zones. Intra-zone protocol (IARP) operates within a zone by using
proactive protocol. Inter-zone protocol (IERP) is reactive and a source node finds a
destination node which is not located within the same zone.

We use fuzzy logic by modifying the ZRP. This technique is known as Fuzzy
Rule-Based Zone Routing Protocol (FZRP). Even though ZRP and FZRP share
a similar hybrid routing behavior, the technique to adjust zone radius is different.
Our technique does not just consider the number of hops but also mobility of node
and traffic direction, which has to be taken into account to decide whether or not
to adjust the zone radius for the ZRP routing protocol. It has been demonstrated
that the differences in the protocol mechanism can lead to significant changes in
terms of performance. The performance of the FZRP scheme is compared to the
ZRP routing protocol. The results indicate clearly that FZRP not only outperforms
ZRP by significantly reducing the number of route query messages but also reduces
average end to end delay between pair nodes. Moreover, FZRP increases the average
throughputs of ad hoc networks and the efficiency of network utilizations.

Tawan Thongpook: *Adaptive Fuzzy Zone Routing for Wireless Ad Hoc Networks*, Studies in
Computational Intelligence (SCI) **4**, 333–347 (2005)
www.springerlink.com   © Springer-Verlag Berlin Heidelberg 2005

21.1 Introduction

Wireless networks have become increasingly popular in many areas, especially within the past decade. The term, wireless network, has been most often used to describe either a cellular type network of radio towers (Fig. 21.1) or stationary devices which serve as gateways for mobile hosts. More recently, the term refers to a pair of devices which connect two LANs which are separated by distance not exceeding radio coverage. It remains more feasible to use a wireless solution.

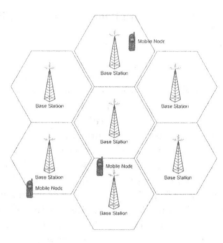

Fig. 21.1. Cellular Networks

The current advances in mobile communication technologies have provided users with various types of services. The primary service has been mobile voice communication by cellular networks and PCS (Personal Communication Services). Mobile users can often meet under circumstances that have not been explicitly planned for. In such situations requiring each user to connect to a wide-area network, such as the Internet, only to communicate with each other, may not be possible due to lack of facilities, or may not be convenient or may be impractical due to associated time or expense required for such connections. These kinds of networks, where mobile hosts communicate directly, have become known as ad hoc networks or infrastructure-less networks.

21.2 What are Ad Hoc Networks?

A mobile ad hoc network (MANET) is a collection of wireless mobile nodes dynamically forming a network without the use of any existing network infrastructure (Fig. 21.2). The mobile hosts are not bound to any centralized control like base stations or mobile switching centers.

Due to the limited transmission range of wireless network interfaces, multiple network hops may be needed for one node to exchange data with another across the

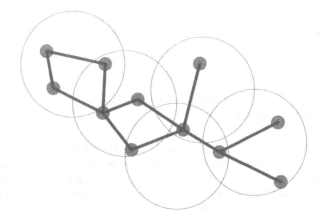

Fig. 21.2. Ad Hoc Networks

network. In such a network, each mobile node operates not only as a host but also as a router, forwarding packets for other mobile nodes in the network that may not be within the direct wireless transmission range of the other. Each node participates in an ad hoc routing protocol that allows it to discover multi hop paths through the network to any other node. The idea of an ad hoc network is sometimes also called an infrastructure-less network, since the mobile nodes in the network dynamically establish routing among themselves to form their own network on the fly.

21.3 Applications of Ad Hoc Networks

The concept of ad hoc networks is attractive due to the following reasons:

1. Ease of deployment
2. Speed of deployment
3. Decreased dependence on infrastructure

The range of applications varies from military to commercial purposes. Some applications could include industrial and commercial applications involving cooperative mobile data exchange. There are many existing and future military networking requirements for robust, IP-compliant data services within mobile wireless communication networks, many of these networks consist of highly dynamic autonomous topology segments. Also, the developing technologies of "wearable" computing and communications may provide applications for these networks. When properly combined with satellite-based information delivery, MANET technology can provide an extremely flexible method for establishing communications for fire/safety/rescue operations or other scenarios requiring rapidly-deployable communications with survivable, efficient dynamic networking. They could also be used by a team of scientists to conduct field studies in remote locations. Ad hoc networks are very useful when a group of people in a conference need to communicate with one another. Sensor networks, which consist of several thousand small low powered nodes with sensing capabilities, is one of the futuristic applications of ad hoc networks. The possibility of using them in commercial applications like Bluetooth is also being explored. There could be other possible applications which may not have been realized or envisioned by the authors.

21.4 Characteristics and Challenges of Ad Hoc Networks

Ad hoc networks consist of mobile nodes which are free to move about arbitrarily. The system may operate in isolation, or may have gateways to and interface with a fixed network. In the latter operational mode, it is typically envisioned to operate as a "stub" network connecting to a fixed inter network. Stub networks carry traffic originating at and/or destined for internal nodes, but do not permit exogenous traffic to "transit" through the stub network. The nodes are equipped with wireless transmitters and receivers using antennas which may be omni-directional (broadcast), highly directional (point-to-point), possibly steerable, or some combination thereof. At a given point in time, depending on the nodes' positions and their transmitter and receiver coverage patterns, transmission power levels and co-channel interference levels, a wireless connectivity in the form of a random, multihop graph or "ad hoc" network exists between the nodes. These networks have several salient characteristics [4]. These features give rise to several challenges:

21.4.1 Bandwidth-Constrained, Variable Capacity Links

Wireless links have significantly lower capacity than their hardwired counterparts. In addition, the actual throughput of wireless communications, after accounting for the effects of multiple access, fading, noise, and interference conditions, etc., is often much lower than a radio's maximum transmission rate. One effect of the relatively low to moderate link capacities is that congestion is typically the norm rather than the exception, i.e. aggregate application demand will likely approach or exceed network capacity quite frequently. As the mobile network is often simply an extension of the fixed network infrastructure, mobile ad hoc users have to demand similar services. These demands are expected to continue to increase as multimedia computing and collaborative networking applications rise.

21.4.2 Dynamic Topologies

Nodes are free to move arbitrarily. Thus, the network topology, which is typically multihop, may change randomly and rapidly at unpredictable times, and may consist of both bidirectional and unidirectional links.

21.4.3 Energy-Constrained Operation

Some or all of the nodes in such networks may rely on batteries or other exhaustible means for their energy. For these nodes, the most important system design criteria for optimization may be energy conservation.

21.4.4 Limited Physical Security

Mobile wireless networks are generally more prone to physical security threats than are fixed-cable nets. An increased possibility of eavesdropping, spoofing, and denial-of-service attacks should be carefully considered. Existing link security techniques are often applied within wireless networks to reduce security threats. As an inherent benefit, the decentralized nature of network control in a MANET provides additional robustness against the single point of failure present in centralized approaches.

In addition, some envisioned networks (e.g. mobile military networks or highway networks) may be relatively large (e.g. tens or hundreds of nodes per routing area) and there is a need for scalability considerations. These characteristics create a set of underlying assumptions and performance concerns for protocol design which extend beyond those guiding the design of routing within the higher speed, semi-static topology of the fixed Internet. Mobility of hosts results in frequent breaks in links. This is an important issue to be considered while routing. Dynamically changing topologies, lack of centralized control, sharing of a single channel between all users and energy constrained operations are the principal challenges. It is therefore difficult to achieve good quality of service. Efforts are ongoing to enhance Quality of Service (QoS) parameters in these networks.

Designing ad hoc routing protocols is complicated by the fact that every host is moving, leading to the dynamic nature of the topology. Routing and network flow information are typically unavailable. Furthermore, incremental delay and residual link capacities are affected by the traffic being carried over other links. Wireless links further introduce constraints such as limited and unreliable bandwidth, as well as finite battery power in each mobile node which consequently restricts the transmission range. The decisions made in the design of the routing protocols are also dependent on the underlying link layer and physical layer technologies.

21.4.5 Routing Protocols

Many protocols have been proposed to support routing of data from a source to an arbitrary destination. Existing routing protocols fall into two categories: proactive or reactive, with some possessing characteristics associated with of both of them, also referred to as hybrid protocols.

Proactive protocols require network nodes (or at least some of them) to maintain up-to-date information of the network topology. Topology information must be propagated through the network periodically or at least immediately after a topological change, and due to the inherently dynamic topology of ad hoc networks, periodic updates are almost guaranteed. Example protocols are: DSDV [5], WRP [6, 7] and CGSR [8]. Reactive routing protocols search for routes only when they are required. The discovered routes are then cached for future use until link failures or topological changes make them invalid and they are then deleted. Example protocols include: DSR [9], AODV [10] and TORA [11]. The advantages of proactive schemes rest on the fact that route information is available whenever a route is required. In reactive schemes, route information may not be available when needed and the delay incurred to determine a route can be quite significant. Furthermore, a significant amount of control traffic is required by the global search process, which coupled with the delay makes purely reactive protocols inappropriate for real time applications. Likewise, purely proactive protocols in an ad hoc environment continuously use a large portion of the network capacity to keep the routing information up-to-date. In highly dynamic environments, most of the routing information is never used, thus creating unnecessary overheads. Hybrid protocols combine the techniques of proactive and reactive protocols in an attempt to an optimal solution. For example, reactive algorithms are used for routing within clusters while proactive algorithms are used for inter-cluster routing. ZRP [1, 2, 3] is an example protocol.

21.5 ZRP

In order to understand how the FZRP protocol works, we must first understand how the Zone Routing Protocol works. ZRP is a hybrid protocol. This protocol divides the network into non-overlapping routing zones and runs independent protocols that study within and between the zones. Intra-zone protocol (IARP) operates within a zone, and learns all the possible routes. So, all nodes within a zone know about its zone topology very well. Protocol which will run in intra-zone is not defined, but can be any proactive protocol such as DSDV. Different zones may operate with different protocols. Inter-zone protocol (IERP) is reactive and a source node finds a destination node which is not located within the same zone, by sending RREQ messages to all border nodes. This continues until A destination is found.

The notions of routing zone and zone radius are the most important terms in the ZRP protocol. Each node defines its zone as the set of all of the nodes whose minimum distance (number of hops) from itself is mostly equal to the zone radius. More precisely, a node's routing zone is defined as a collection of nodes whose minimum distance in hops from the node in question is no greater than a parameter referred to as the zone radius.

Figure 21.3 illustrates the routing zone concept with a routing zone of radius equal to 2 hops. The particular routing zone belongs to node S which we refer to as the central node of routing zone. Node A though H are members of S's routing zone. Node J, however, is three hops away from node S. and is therefore outside of S's routing zone. An important subset of the routing zone nodes is the collection of nodes whose minimum distance to the central node is exactly equal to the zone radius. These modes are aptly named peripheral nodes. In our example, nodes G-K

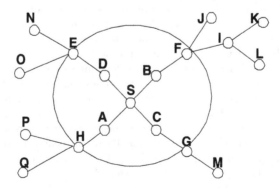

Fig. 21.3. A Routing Zone of Radius 2 hops

are peripheral nodes of node A. We typically illustrate a routing zone as a circle centered around the central node. However, one should keep in mind the zone is not a description of physical distance, but rather nodal connectivity (hops).

Figure 21.4 illustrates the Route Discovery procedure. The source node S prepares to send data to destination D. S first checks whether D is within its routing zone. If so, S already knows the route to node D. Otherwise, S sends a query to all its peripheral nodes (E, F, G and H). Now, in turn, each own zone, forwards the query to its peripheral, F sends the query to I, which is to recognize D as being in its routing zone of I and responds to the query indicating the forwarding path : S-F-L-D.

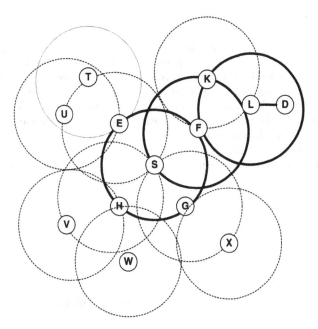

Fig. 21.4. IERP Operation (zone radius = 2 hops)

The architecture of ZRP consists of four major parts on a per-node basis: MAC-level functions, IARP, IERP and BRP. The MAC-level performs a protocol called the Neighbor Discovery/Maintenance (NDM). NDM informs the IARP layer of neighbor nodes found or lost. The NDM component will notify the IARP layers of the new neighbors. A node uses IARP to maintain route information about nodes within its zone. Using the IARP routing table, all nodes maintain information about the nodes in their zones. Therefore, whenever a node wishes to route messages to a node within its zone, it simply gets the next-hop information to that destination from the IARP routing table. The IARP routing table is discussed in further detail in [1]. If a node wishes to route a packet to a node outside its zone, it must first find the best route using IERP. The IERP layer maintains a routing table of routes to destinations outside of the node's zone. When a node wishes to send data to a node outside its zone, the IERP first checks to see if it has a route to that destination within the

IERP routing table. If a route to that destination is not in the table, the IERP initiates a route query and passes control to the BRP layer. BRP uses a message-passing mechanism called bordercasting to transmit route queries and replies across the ad hoc network. Bordercasting gets its name from the fact that route control packets are passed between border nodes. Therefore, BRP traffic does not technically flood the network, since only nodes on the periphery of a zone transmit and receive packets. BRP prevents the looping of route queries and replies by maintaining two tables: the Detected Queries Table (DQT) and the Detected Replies Table (DRT) [1, 2, 3].

21.6 FZRP

In zone routing, each node has its own zone. Intrazone routing (e.g., DSDV and DBF) maintains a communication path within the zone, while Interzone routing determines routes to the destination using flooding technique. In [1, 2, 3] techniques called "min-searching" and "traffic adaptive" have been proposed. This techniques allow individual nodes to identify changes in the network, and appropriately react to the changes in the network's configuration. The modifications are conducted by either increasing or decreasing the number of hops used by each node. Note that the number of hops for a particular node will be increased (or decreased) in all directions depending on the traffic between interzone and intrazone routing (see Fig. 21.5).

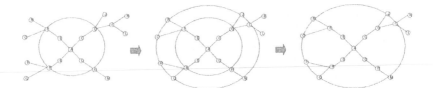

Fig. 21.5. Changes of zone radius in all directions

This introduces changes in the routing table for intrazone routing. What happen is that if the traffic of a network in a zone only occurs in a particular direction, changes in the zone radius for all directions can seriously increase the size of the routing table even though the traffic in other direction is not significant.

We proposed a new technique for adaptively resizing the zone radius by using a fuzzy rule-base and concentrating the changes of network traffic in particular directions depending on the interzone traffic. By periodically monitoring the interzone network traffic, the direction that generates high interzone traffic can be determined. Furthermore, changing the number of hops can also be implemented only for that particular direction. An example of this technique is demonstrated in Fig. 21.6.

Changing in radius only in a particular direction, which has heavily loaded traffic, can significantly reduce the size of the routing table. Therefore, maintaining a small routing table for individual zones is much simpler. Moreover, the amount of control traffic particularly used to update the routing table will also be decreased since the number of nodes after changing zone radius is relatively small compared to the ZRP technique [1, 2, 3].

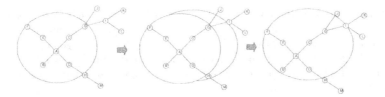

Fig. 21.6. Changes of zone radius in a particular direction

21.7 Methodology

21.7.1 Parameter

In order to adjust the zone radius in a particular direction, there are three parameters that must be taken into account (see Fig. 21.7). The first parameter is network size (number of nodes, the size of the area within the zone that the nodes are moving). The second is traffic network load. This can be characterized by three parameters: packet size, number of connections and the rate that we are sending the packet with. The last parameter is mobility, which is the most important for ad hoc network. This will affect the dynamic topology as links will go up and down.

21.7.2 Mobility

Because mobility is an important metric when evaluating ad hoc networks we need some definition of mobility. There are many definitions of mobility. The CMU Monarch project [12, 13] has for instance used the pause time in the way point as a definition of mobility. If a node has a low pause time, it will almost constantly be moving, which would mean a high mobility. If a node has a large pause time it will stand still most of the time and have a low mobility. We did not think this mobility definition was good enough, because even if the pause time is low and all nodes are

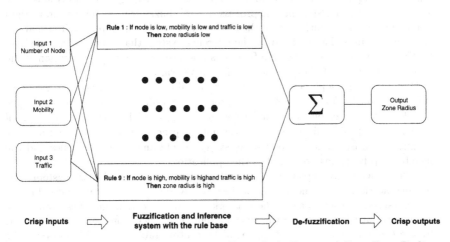

Fig. 21.7. Decision Making by using a Fuzzy Rule Base to Adjust Zone Radius

constantly moving, the nodes could all be moving with a very slow speed in the same area. We have defined mobility slightly differently. Our definition is based on the relative movement of the nodes. This definition gives a very good picture of how the nodes are moving in relation to each other. The definition is as follows:

First of all, the average distance from each node to all other nodes within the zone has to be calculated.

$$A_x(t) = \sum_{i=1}^{n} \mathrm{dist}(N_x, N_y) \qquad (21.1)$$

$\mathrm{dist}(n_x, n_y)t$ = the distance between node x and node y at time t; n = number of node; i = index; $A_x(t)$ = Average distance of node x to all other nodes at time t. Then, the average mobility for that particular node has to be calculated. This is the average change in distance during a whole simulation. The mobility for node x is:

$$M_x = \frac{\sum_{t=0}^{T-\Delta t} |(A_x(t) - A_x(t + \Delta t))|}{T - \Delta t} \qquad (21.2)$$

M_x = average mobility for node x to all other nodes at time t; T = Simulation time; Δt = Granularity, simulation step

Finally, the mobility for the whole scenario is the sum of the mobility for all nodes (7.2) divided by the number of nodes within the zone:

$$R - \mathrm{Mobility} = \frac{\sum_{i=1}^{n} M_i}{n} \qquad (21.3)$$

The unit of the relative mobility factor (7.3) is m/s.

21.8 Fuzzy Rule

Knowledge-base systems such as fuzzy logic have been successfully implemented in multifarious applications where human expertise and dealing with uncertainty play a vital role in the decision making process [14, 15]. Fuzzy logic avoids arbitrary rigid boundaries by taking into account the continuous character of imprecise information. A fuzzy system is characterized by the inference system that contains the rule base for the system, input membership functions that are used for the fuzzification of the input variables and de-fuzzification of the output variables.

Fuzzification is a process where crisp input values are transformed into membership values of the fuzzy sets. After the process of fuzzification, the inference engine calculates the fuzzy output using fuzzy rules which are linguistic in the form of if then rules. De-fuzzification is a mathematical process used to convert the fuzzy output to a crisp value. A good fuzzy system is obtained when the rules and the membership functions are tuned to the application.

The above figure depicts the block diagram of the proposed fuzzy system. The routing metrics, total of nodes within the zone, average traffic and mobility of nodes are measured at each node of the network for each direction and these values are given as input parameters for the fuzzy system. The size of zone radius is computed by the fuzzy system. The ZRP updates the routing tables for each node by using global parameter and size of zone radius.

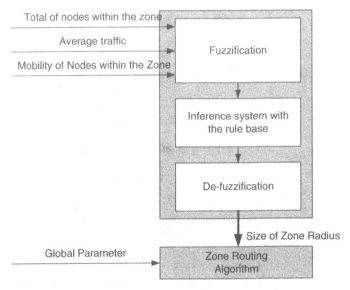

Fig. 21.8. Fuzzy System

In the proposed fuzzy system, The Mandani minimum inference method [16] was used as the fuzzy inference method, where the "and" operation was set to minimum and de-fuzzification was carried out using centroid defuzzifer. Mandani's inference system can be mathematically written as,

$$\max(\min(\mu, \mu\overline{w}(Z))) \quad \text{for all } z \tag{21.4}$$

Where, $\mu\overline{w}(Z)$ is the output membership function and is the combined membership in the rule antecedent.

21.8.1 Membership Functions

Triangular membership functions were used for the linguistic variables that represent number of nodes, mobility of nodes and average traffic. The triangular membership function is specified by a, b and c as shown in Fig. 21.9:

$$\text{Triangle}(x, a, b, c) = \left\{ \frac{(x-a)}{(b-a)}, \frac{(c-x)}{(c-b)} \right\} \tag{21.5}$$

Table 21.1 illustrates the rule base used in the fuzzy inference system. Expert knowledge was used in developing the rule base.

21.9 Simulation Model

In our simulation environment, using MATLAB and NS-2 [13] simulation tools, we assume that the link bandwidth of every node is 11 Mbps. Three types of traffic

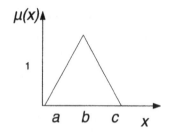

Fig. 21.9. Triangular membership

Table 21.1. Fuzzy Rule

	IF		THEN	
Fuzzy Rules	Node	Mobility	Traffic	Zone Radius
R1	High	High	High	Large
R2	Medium	High	Medium	Large
R3	Low	High	Low	Medium
R4	High	Medium	High	Large
R5	Medium	Medium	Medium	Medium
R6	Low	Medium	Low	Small
R7	High	Low	High	Medium
R8	Medium	Low	Medium	Small
R9	Low	Low	Low	Small

(high, medium and low) are assumed. The traffic is considered to be high, medium and low when it consumes link bandwidth of 50%, 25% and 15%. We varied the number of nodes in the network using this criterion in 9.1:

Number of nodes: The range of the membership function is 0 to 50. The various fuzzy sets used are Small: triangle in (x, [0 12 25]); Medium: triangle (x, [12 25 35]); Large: triangle (x, [25 35 50])

We varied the mobility by randomizing scenario files, because we can not predict the mobility of nodes in scenario which is randomly generated. Thus we must use the maximum speed of mobility to control a scenario. The simulation parameters that have been used for the mobility simulations are shown in Table 21.2.

The scenario is a very crucial part of the simulation. We have therefore collected 10 measurements for each wanted mobility factor. The mobility factors that we simulated on are: 0, 0.5, 1.0, 1.5, 2.0, 2.5, 3.0 and 3.5. Because of the difficulty of getting scenarios that are precise we have used an interval of ± 0.1 for the above mentioned mobility factor.

Mobility of node: The range of the membership function is 0 to 3.5. The various fuzzy sets used are Small: triangle (x, [1 1.05 1.25]); Medium: triangle (x, [0.875 1.75 2.675]);Large: triangle (x, [1.75 2.675 3.5])

Table 21.2. Global Parameter used during mobility simulations

Parameter	Value
Transmitter range	250 m
Bandwidth	11 Mbit/s
Simulation time	250 s
Number of node	50
Pause time	1 s
Environment size	1000 × 1000 m
Traffic type	Constant (CBR)
Bit Rate Packet rate	5 packet/s
Packet size	64 byte
Number of flows	15

21.10 Performance Results

21.10.1 Performance Metrics

Two key performance metrics are evaluated:

1. Number of Query messages
2. Average end-to-end delay of data packets. This includes all possible delays caused by buffering during route discovery latency, queuing at the interface queue, retransmission delays at the MAC, propagation and transfer times.

In Fig. 21.10, the number of query messages from FZRP is smaller than ZRP. This means that FZRP can reduce the number of query messages by using a Fuzzy Rule-base to adjust zone radius.

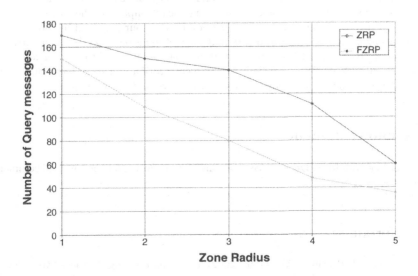

Fig. 21.10. The number of query messages

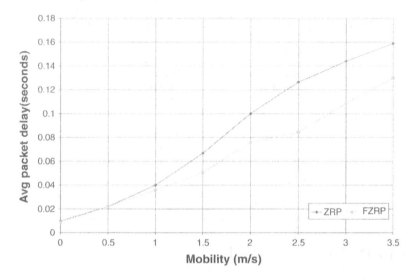

Fig. 21.11. The average number of packet delay

In Fig. 21.11, the average end to end delay of packet from FZRP has lower delay than ZRP when the mobility factor changes from 0 to 3.5.

21.11 Conclusion

The result has shown that the FZRP outperforms ZRP by reducing significantly the number of route query messages and thereby increases the efficiency of the network load. This technique can have a significant impact on the cost of updating the routing table compared to ZRP. However, while our proposed method yields a promising performance, we also note that the method is only suitable for a particular situation. Nevertheless, this promising result may prove useful to another researcher in the fuzzy logic field, especially the implementation of hybrid technique, which combines the advantages of proactive and reactive routing techniques.

Acknowledgments

I am very grateful to Dr. D.A. Batovski and Dr. A. Pervez for their helpful comments on this work.

References

1. Haas ZJ, Pearlman MR (1999–2002) The Zone Routing Protocol (ZRP) for Ad Hoc Networks, In: Internet Draft, draft-ietf-manet-zone-zrp-02.txt

2. Haas ZJ, and Pearlman MR (2001), The Performance of Query Control Schemes for the Zone Routing Protocol, In: ACM/IEEE Transactions on Networking, vol. 9, no. 4, August, pp. 427-438

3. Pearlman MR, Haas ZJ (1999) Determining the optimal configuration for the zone routing protocol, In: IEEE Journal on Selected Areas in Communications 1999; 17: No. 8, pp 1395–1414

4. David Johnson, David Maltz, (1996) Protocols for Adaptive Wireless and Mobile Networking. In: IEEE Personal Communications, Volume 3 Issue 1, February pp. 112–115

5. Perkins CE, Pravin Bhagwat (1994) Highly dynamic Destination-Sequenced Distance-Vector routing (DSDV) for mobile computers. In: Proceedings of the SIGCOMM '94 ACM Conference on Communications Architectures, Protocols and Applications, Aug 1994, pp. 234–244

6. Chiang CC (1997), Routing in Clustered Multihop, Mobile Wireless Networks with Fading Channel. In: Proc. IEEE SICON'97, April, pp. 197–211

7. Iwata A et al. (1999) Scalable Routing Strategies for Ad Hoc Wireless Networks, In: IEEE Journal on Selected Areas in Communications, Special Issue on Ad-Hoc Networks, August pp. 1369–79

8. Murthy S, Garcia-Luna-Aceves J.J. (1996) An Efficient Routing Protocol for Wireless Networks In: ACM Mobile Networks and App. J., Special Issue on Routing in Mobile Communication Networks, October, pp. 183–97.

9. David Johnson (1996) Dynamic Source Routing in Ad Hoc Wireless Networks. In: Mobile Computing Kluwer Academic Publishers, pp. 153–181

10. David A. Maltz (2001). On-Demand Routing in Multi-hop Wireless Ad Hoc Networks. Ph.D. Thesis, School of Computer Science, Carnegie Mellon University, Pittsburgh, PA

11. Vincent DP, M Scott Corson (2001). Temporally-Ordered Routing Algorithm (TORA) version 1: In: Internet Draft, draft-ietf-manet-tora-spec- 03.txt, work in progress

12. The CMU Monarch Project (1998). The CMU Monarch Projects Wireless and Mobility Extensions to ns URL: http://www.monarch.cs.cmu.edu/

13. Kevin Fall, Kannan Varadhan (1998) ns notes and documentation in: The VINT project, UC Berkeley, LBL, USC/ISI, and Xerox PARC Available from URL: http://www-mash.cs.berkeley.edu/ns/

14. Ronal R. Yager, Dimitar P. Filev (1994) Essentials of fuzzy modeling and control. In: A Wiley – Interscience publication

15. Constantin Von Attrock (1997) Fuzzy Logic & NeuroFuzzy Applications Explained. New Jersey: Prentice Hall International Inc.

16. Constantin Von Attrock (1997) Fuzzy Logic & NeuroFuzzy Applications Explained. New Jersey: Prentice Hall International Inc.

A New Approach to Building Fuzzy Classifications in Sociological Research with Survey Data

Yu. N. Blagoveschensky and G.A. Satarov

Information for Democracy Foundation (INDEM), Russia
fond@indem.ru

Abstract. This paper offers a new approach to building classifications in sociological and marketing research. This approach is based on survey data and expert evaluations of conditional probabilities of choice of answers with typical representatives of different classes. In the framework of this approach, we are constructing consistent estimators of class membership functions for all respondents. Besides, we are introducing characteristics of quality for different experts, and solving the problem of dimensionality reduction (the number of questions used for estimating). Theoretical results are illustrated by data from sociological research on corruption in Russia.

Key words: Fuzzy classifications, mathematical modeling, public opinion surveys, experts' evaluation, attitude towards corruption

22.1 Introduction

Public opinion surveys are one of the primary sources of information in sociological and marketing research. The task of dividing respondents into classes that reflect some number of their social and behavioral particularities is one of the most important. Examples of these particularities might be types of motivation, forms of relationships to the choice of product or to evaluation of events, electoral preferences, etc.

The traditional method of dividing respondents into classes is based on their responses to the questions. Respondents who equally questions form a class. However, in this case some methodological problems arise.

Firstly, there is a problem of language ambiguity by which communication between social scientists and respondents is made. Secondly, there is a difference in understanding of the same words by respondents and social scientists. Thirdly, responses on the questions are motivated by several reasons or features, among which a researcher's interest is only one of them. Fourthly, this latent feature can be concerned with confidential spheres of respondents' life and in such situation direct questions in a questionnaire would be useless and inadequate.

Yu. N. Blagoveschensky and G.A. Satarov: *A New Approach to Building Fuzzy Classifications in Sociological Research with Survey Data*, Studies in Computational Intelligence (SCI) **4**, 349–356 (2005)

In every-day communication these obvious factors are compensated by standard methods. The prevalent one is to ask a partner several questions dealing with a problem, which we are interested in, and generalize answers to these questions by a unified statement about this problem.

Finally, the following reasoning is also defensible. The object of sociological (marketing) research is not individual respondents, but different social groups and their features. We investigate these groups through their representatives – respondents. If we make judgments about a group's features based on respondents' answers, then it is natural to think that questions represent features as we think that respondents are representatives of social groups, which we are interested in. In other words, in sociological research we presume respondents are a main part of the sampling, but we deal with questions in another way. Meantime they can be considered as a sampling of the parent population of questions that investigate a particular feature.

Thus, in this method, respondents answer the range of questions. We intend that all these questions deal with the feature being investigated. It means that responses on these questions are determined to high extent by the degree of testament of this feature in respondents. Based on the range of respondent's answers we aimed to decide to what extent each of the respondents has this or that feature. To solve the problem we have to fix a correlation between respondents' answers on the questions from the range and investigating feature. This correlation will be determined by experts' procedures.

We will presume that the degree of testament of the feature being investigated will be described by fuzzy classification. To attribute a feature to a respondent is to determine his function of attitude to the classes of classification. In doing so it is essential to determine the chances of typical representatives of different classes of fuzzy classification to choose this or that answers.

Naturally, such a method solves some subsidiary problems. Among them is experts' evaluation and their selection, synthesis of different experts' estimations and evaluation of questions as indicators of the investigated feature.

22.2 Formal Problem Statement

Let's assume that there is a general assembly of W objects, each of which is characterized by its membership functions

$$F(w) = \{f_1(w), f_2(w), \ldots, f_k(w)\}, \quad w \in W . \tag{22.1}$$

Components of $F(w)$ conform to classes $A_1, A_2, \ldots, A_k$. If we assume that these elements are normalized, that is $f_1(w) + f_2(w) + \ldots + f_k(w) = 1$, the set $F = \{F(w), w \in W\}$ will represent a family of distributions into classes.

In fact, in sociological and marketing research, the primary task consists of determining the structure of this family and establishing connections between the functions $F(w)$ and factors that determine decision making. For this purpose N objects are derived from the set W, and each object is "measured", fixing its characteristics.

$$X_n = \{x_{n1}, x_{n2}, \ldots, x_{nm}\}, \quad 1 \leq n \leq N. \tag{22.2}$$

(Answers of respondents to m questions from the survey). Assuming that X_n is the product of two mechanisms: of the random *variable* ν (the choice of the class number

i by distribution $F(n)$) and of the vector $\xi = (\xi_1, \xi_2, \ldots, \xi_m)$, its realization under the condition that $\nu = i$. It is assumed that the components ξ are independent, and they take on the value $1, 2, \ldots, s(j)$ with the probabilities

$$p_{ij}^{(0)}(s) = \mathbf{P}\left\{\xi_j = s/\nu = i\right\}, \quad 1 \le s \le s(j), \quad 1 \le j \le m, \quad 1 \le i \le k. \quad (22.3)$$

If $\pi_1, \pi_2, \ldots, \pi_k$ is a priori distribution of classes, then it is easy to calculate a posteriori distribution

$$P(n) = \{\pi_1(X_n), \pi_2(X_n), \ldots, \pi_k(X_n)\}$$

under the condition $\xi = X_n$. The distribution $P(n)$ is a sufficient evaluation of membership function $F(n)$ if the number of independent components of the vectors $\xi = (\xi_1, \xi_2, \ldots, \xi_m)$ will grow and probabilities (22.3) will satisfy several simple conditions (ξ_j must contain positive information about the classes, separate from the zero).

In the case of the above-formulated assumptions, we could certainly use the distribution $P(n)$ as an estimator for membership functions $F(n)$. But, they would all be this way if we knew the probabilities $p_{ij}^{(0)}(s) = \mathbf{P}\{\xi_j = s/\nu = i\}$, which are, in fact, unknown. In reality, we only have $p_{ij}^{(r)}(s)$ that is the evaluation of the expert indexed by his personal number $r = 1, 2, \ldots, R$.

In the end, we need to find and substantiate "good" estimators for $F(n)$, having R expert evaluations for probabilities $p_{ij}^{(0)}(s)$. In other words, we need to find a formula to calculate the quality of expert evaluations and a procedure for estimating the informational "power" of the components ξ_j under various hypotheses (classes).

In addition we must construct total estimators $\widehat{P}(n)$ for the membership functions $F(n)$. This project is being realized through our research.

22.3 Method and Results

The key problem is the correct choice of a numerical characteristic to measure quality of separate components of vector ξ in the presence of one or another expert evaluation of distributions.

Let $P_{ij}^{(r)} = \{p_{ij}^{(r)}(s); 1 \le s \le s(j)\}$ is the distribution of components ξ_j when we consider the hypothesis of belonging to the class A_i, $1 \le i \le k$, and evaluations given by the "r" expert, $r = 1, 2, \ldots, R$, at that $r = 0$ corresponds to the case of a priori distribution.

Let us suppose that all of our data is only observations ξ_j and presents only one class A_i, and moreover $P_{ij}^{(r)}$ is the true distribution ξ_j. And then we ask: "How long is the sequence of independent observations $\xi_j(1), \xi_j(2), \ldots, \xi_j(H)$, $H \to +\infty$, which are distributed as ξ_j, in order that a posteriori probability of the class A_i reaches the level β (an analog of reliability, for example, $\beta = 0.8$)?"

The length $\nu_{ij}^{(r)}(\beta)$ of this sequence (the number of members in it) is the random variable, dependent on all conditional distributions $P(j; r) = \{P_{1j}^{(r)}, P_{2j}^{(r)}, \ldots, P_{kj}^{(r)}\}$ wholly, i.e. from the class membership ξ_j. This length can be equal $+\infty$ (for example, when $P_{tj}^{(r)} = P_{ij}^{(r)}$ for $t \ne i$).

Now let's assume that we can calculate the median $\mu_{ij}^{(r)}(\beta)$ for the distribution of $\nu_{ij}^{(r)}(\beta)$, at that $\mu_{ij}^{(r)}(\beta) = +\infty$, if $\mathbf{P}\{\nu_{ij}^{(r)}(\beta) = +\infty\} > 0$. The choice of the median instead of a mathematical expectation is connected with a large asymmetry of these distributions. It should be noted that the values $\mu_{1j}^{(r)}(\beta), \ldots, \mu_{kj}^{(r)}(\beta)$ give an idea of expenditures for classification, and their variability contains an useful information about the potential ratio between classification mistakes. Specifically, the more average the values of $\bar{\mu}_j^{(r)}(\beta)$, the worse the quality of the classification on average. And the greater the range of $\Delta_j^{(r)}(\beta)$, the greater probability that the mistakes of classification will be strongly distorted (as in the following extreme example: all respondents have been put into only one class).

It's clear that the final function of expenditures should be monotone creasing function from the average value and range. For these reasons, we use one of the more simple functions of expenditures:

$$Y(j,r) = \sqrt{\bar{\mu}_j^{(r)}(\beta) \cdot \Delta_j^{(r)}(\beta)} \,. \tag{22.4}$$

To calculate the value $Y(j,r)$, giving a measure of quality of the pair "component – expert", we used the method of mathematical modeling. In fact we simulate $\xi_j(1), \xi_j(2), \ldots$ and step by step calculate a posteriori probability $P_{ij}^{(r)}(1), P_{ij}^{(r)}(2), \ldots$ of class A_i in case of a priori belonging ξ_j to the class A_i, $1 \leq i \leq k$. After that we run up to first moment when $P_{ij}^{(r)}(g)$ will exceed given level β at a step $g = g(j,r)$ or when we reach a given number H out of exceeding β, and then $g(j,r) = H + 1$ by definition. Now we reiterate the procedure of modeling many times and gather all values in one sample $g_1(j,r), g_2(j,r), \ldots, g_L(j,r)$. Finally we calculate median $\widehat{\mu}_{ij}^{(r)}(\beta)$ of the sample as an estimate for $\mu_{ij}^{(r)}(\beta)$.

Basic parameters of the procedure are β, H and L, length and size simulating. It turned out that we have reliable results with $\beta = 0.8$, $H = 600$ and $L = 800$ in every case. These numbers are acceptable because individual realizations of values $P_{ij}^{(r)}(1), P_{ij}^{(r)}(2), \ldots$ exhibit "strange" behavior. We can illustrate this fact in Fig. 22.1 where there are 5 pairs of trajectories. Namely each of 5 classes have been represented by 2 trajectories simulated independently one from another.

The following problem has been concerned that "bad" expert data cause substantial deviations from results given "ideal" expert data $P_{ij}^{(0)}$. Therefore it needs to define what is a "good" expert, and, using (22.4), to separate "assuredly good" experts from the controversial rest.

Let $P = (p_1, p_2, \ldots, p_n)$ – any discrete distribution and, at the same time, a point in n-dimensional space. And let $P^* = (p_1^*, p_2^*, \ldots, p_n^*)$ is an expert evaluation for P. The set of all distributions $Q = (q_1, q_2, \ldots, q_n)$ with $\max_{1 \leq i \leq n} q_i < 1$ is defined as G. Further, let $U_d(P)$ – d-neighbourhood of $P \in G$, at that $P^* \in U_d(P) \subseteq G$.

In the case we will say that P^* is d-close evaluation for P if P^* is the realization of a *random variable* ς^* *distributed into* $U_d(P)$, and $\mathbf{E}\varsigma* = P$.

Now we can give the definition of "good" expert. But at first we choose a little number $\varepsilon > 0$ in such the way that

$$U_\varepsilon(P_{ij}^{(0)}) \subseteq G, 1 \leq i \leq k, 1 \leq j \leq m \,. \tag{22.5}$$

Definition: The "r" expert is a "ε-good" expert if for any pair $(i,j), 1 \leq i \leq k$, $1 \leq j \leq m$, the evaluation $P_{ij}^{(r)}$ is ε-close for $P_{ij}^{(0)}$, $r = 1, 2, \ldots, R$.

In the main our *theoretical* result consists of the following:

If we know that among all experts in number R there are "ε-good" experts in numbers *not less than* $R' < R$ (value R' is known, but it is unknown about experts personally), then we can produce the method which solved our classification problem. Namely, it's proved our procedure using functions of expenditures from (22.4) leads to a near result of the true classification, i.e., that errors have values in the order of $\varepsilon/\sqrt{R'}$.

This is only theoretical in practice we don't know either R' or ε. In addition there are complications concerned with different classification ratings of ξ_j, components of vector ξ. The ratings are necessary to sum information out of values $Y(j,r)$ and to form a suitable index of quality for each single expert.

The realized method embraces a set of procedures. There is little point in giving whole statement of our procedures within limits of the paper, but it is necessary to say more about some of them.

First, the analysis of $Y(j,r)$ as table data. In order to determine the quality of experts we worked with the variance analysis method using *weighted* averages. At that we consistently chose weights of components of ξ, calculated ratings of experts and set aside "the worst". After that we repeated the same actions without the removed expert and did it until we exhausted the experts. If *bad* experts differ from *good* experts "essentially" then the plan separates the set of *good* experts quite reliably.

22.4 Application

The project "Russia Anti-Corruption Diagnostics" finished in January 2002. It included two public opinion polls basing on representative samples of citizens (2017 respondents) and entrepreneurs (709 respondents).

Basing on identical questions, the two sample groups of individuals and entrepreneurs were referred to a number of typologies (classifications), such as "Understanding of Corruption", "Attitude towards Corruption" and others. The typology "Involvement in Corruption" was devised separately for individuals and entrepreneurs (based respectively on different sets of questions). And, finally, two additional typologies ("The Dependence of Business on Power" and "Business success") were specially devised for entrepreneurs.

Let's consider for example the typology "Attitude towards Corruption" represented by 5 classes. It had the following verbal description of classes:

Active rejection: Total rejection of corruption; readiness to make efforts to oppose corruption.

Passive rejection: Negative attitude to corruption and to oneself because of the need occasionally to resort to corruption without being ready to offer active resistance to it.

Resignation: Negative attitude to corruption combined with an idea of inevitability of this evil and hopelessness of attempts to combat it.

Self-justification: Dispassionate attitude to corruption and readiness to justify one's own behavior in case of personally taking part in corrupt deals.

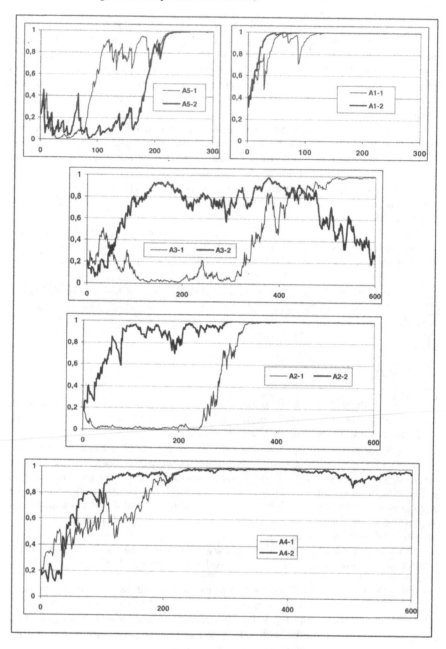

Fig. 22.1. Here each of the 5 classes have been represented by 2 trajectories simulated independently one from another

Active acceptance: Viewing corruption as a useful phenomenon in life and business.

This typology was constructed using seven questions including, for example, the following: *What is your general approach towards the fact that in order to resolve their problems citizens of this country have to quite often give bribes? Which of the statements below is the closest to your standpoint*:

– *Bribery is an indispensable part of our life. One can do nothing without it*
– *It can be avoided but it is easier to do things paying bribes*
– *It should be avoided since corruption demoralizes ourselves and our authorities*
– *Difficult to say*

or
Let's assume that you were told about a person who having learned that his boss had been accepting bribes, reported that to the prosecutor. Select one of the words below that from your standpoint best describes the person:

– *Hero*
– *Honest person*
– *Odd fellow*
– *Envious person*
– *Revengeful person*
– *Traitor*
– *Other*
– *Difficult to say*

The next table shows the result of the application of the method described above. In this case respondents were distributed into classes in accordance with the largest component of membership functions. The sizes of classes for citizens and entrepreneurs are both shown.

Name of the Class	The Size of the Class (Percent)	
	Citizens	Entrepreneurs
Active rejection	15.4	13.3
Passive rejection	29.5	33.3
Resignation	39.4	38.8
Self-justification	10.2	9.6
Active acceptance	5.5	5.1

On the one hand the dependable column nearness indicates a quality of our classification procedures and on the other hand it shows that in Russia citizens and entrepreneurs represent only one general assembly of statistical objects.

We should also notice that one of the most significant findings of this survey is the absence of correlation between involvement in corrupt business practices and success of business operations. Different techniques and indicators were used to check and double-check this factor, but the result remained the same.

22.5 Remark

We don't refer to scientific publications since the basic problems of statistical classification are well known while our workings present new procedures without history. Full information about this study is at http://www.indem.ru.